# Fmoc solid phase peptide synthesis

# The Practical Approach Series

SERIES EDITOR

**B. D. HAMES**
*Department of Biochemistry and Molecular Biology*
*University of Leeds, Leeds LS2 9JT, UK*

See also the Practical Approach web site at **http://www.oup.co.uk/PAS**
★ **indicates new and forthcoming titles**

# Fmoc solid phase peptide synthesis

## A Practical Approach

Edited by

### WENG C. CHAN

*School of Pharmaceutical Sciences,*
*University of Nottingham, University Park,*
*Nottingham NG7 2RD*

and

### PETER D. WHITE

*CN Biosciences, Padge Road, Beeston,*
*Nottingham NG9 2JR*

OXFORD
UNIVERSITY PRESS

*This book has been printed digitally and produced in a standard specification*
*in order to ensure its continuing availability*

# OXFORD
UNIVERSITY PRESS

Great Clarendon Street, Oxford OX2 6DP

Oxford University Press is a department of the University of Oxford.
It furthers the University's objective of excellence in research, scholarship,
and education by publishing world-wide in

Oxford New York

Auckland Bangkok Buenos Aires Cape Town Chennai
Dar es Salaam Delhi Hong Kong Istanbul Karachi Kolkata
Kuala Lumpur Madrid Melbourne Mexico City Mumbai Nairobi
São Paulo Shanghai Taipei Tokyo Toronto

Oxford is a registered trade mark of Oxford University Press
in the UK and in certain other countries

Published in the United States
by Oxford University Press Inc., New York

ISBN 978-0-19-963724-9

# Preface

This volume, like others in the Practical Approach Series, is intended to be primarily a practical guide to the subject, and as such, no attempt has been made by contributing authors to provide comprehensive reviews of the literature. As the title suggests, this work is devoted entirely to the Fmoc/tBu approach of solid phase peptide synthesis. This is not intended as any slight on the Merrifield technique; it was felt best to restrict the scope of this volume, in view of the limited space available, the number of similar works covering the Merrifield technique already in print, and the numerous innovations made in the Fmoc/tBu method over the last decade.

In the years since the publication of Atherton and Sheppard's seminal volume in this series on Fmoc/tBu solid phase peptide synthesis, the technique has matured considerably to become the standard approach for the routine production of peptides. The problems outstanding at the time of publication of this earlier work have now, for the most part, been solved. As a result, innovators in the field have been able to focus their efforts on developing new methodologies and chemistry for the synthesis of more complex structures. The focus of this new volume is therefore much broader, and covers not only the essential procedures for the production of linear peptides but also more advanced techniques for the preparation of cyclic, side-chain modified, phospho- and glycopeptides. Many other methods also deserving attention have been included: convergent peptide synthesis; peptide–protein conjugation; chemoselective ligation; and chemoselective purification. The difficult preparation of cysteine- and methionine-containing peptides is also covered, as well as methods for overcoming aggregation during peptide chain assembly.

Many of the techniques developed for the production of large arrays of peptides by parallel synthesis, such as T-bag, SPOT, and PIN synthesis, have naturally been included. Finally, a survey of available automated instrumentation has also been provided.

W.C.C.
P.D.W.

# Contents

## 4. Preparation and handling of peptides containing methionine and cysteine     77

*Fernando Albericio, Ioana Annis, Miriam Royo, and George Barany*

## 5. Difficult peptides     115

*Martin Quibell and Tony Johnson*

# 6. Synthesis of modified peptides     137

*Sarah L. Mellor, Donald A. Wellings, Jean-Alain Fehrentz,*
*Marielle Paris, Jean Martinez, Nicholas J. Ede,*
*Andrew M. Bray, David J. Evans, and G. B. Bloomberg*

# 7. Phosphopeptide synthesis     183

*Peter D. White*

# 8. Glycopeptide synthesis

*Jan Kihlberg*

**195** (8.)

# 9. Convergent peptide synthesis

*Kleomenis Barlos and Dimitrios Gatos*

**215** (9.)

**13.** Instrumentation for automated solid phase peptide synthesis     277

*Linda E. Cammish and Steven A. Kates*

**14.** Manual multiple synthesis methods     303

*B. Dörner, J. M. Ostresh, R. A. Houghten, Ronald Frank,
Andrea Tiepold, John E. Fox, Andrew M. Bray,
Nicholas J. Ede, Ian W. James, and Geoffrey Wickham*

# Contents

# Contributors

F. ALBERICIO
Department of Organic Chemistry, University of Barcelona, Marti i Franqués 1–11, Barcelona E-08028, Spain.

I. ANNIS
Union Carbide Corporation, PO Box 610, Bound Brook, NJ 08805, USA.

G. BARANY
Department of Chemistry, University of Minnesota, 207 Pleasant St SE, Minneapolis, MN 55455–0431, USA.

K. BARLOS
Department of Chemistry, University of Patras, 26 110 Patras, Greece.

G. BLOOMBERG
Department of Biochemistry, University of Bristol, Medical School, Bristol BS8 1ID, UK.

A. M. BRAY
Chiron Mimotopes Pty. Ltd., 11 Duerdin Street, Clayton, Victoria 3168, Australia.

L. E. CAMMISH
Mettler-Toledo Myriad Ltd., 2 Saxon Way, Melbourn, Royston, Hertfordshire SG8 6DN, UK.

W. C. CHAN
School of Pharmaceutical Sciences, University of Nottingham, University Park, Nottingham NG7 2RD, UK.

B. DÖRNER
Calbiochem-Novabiochem AG, Weidenmattweg 4, CH-4448 Läufelfingen, Switzerland.

J. W. DRIJFHOUT
Department of Immunohaematology and Blood Bank, Leiden University Medical Center, PO Box 9600, 2300 RC Leiden, The Netherlands.

N. J. EDE
Chiron Mimotopes Pty. Ltd., 11 Duerdin Street, Clayton, Victoria 3168, Australia.

D. J. EVANS
Avecia LSM, Gadbrook Park, Northwich, Cheshire CW9 7RA, UK.

# Contributors

J.-A. FEHRENTZ
Laboratoire de Chimie et Pharmacologie de Molécules d'Intérêt Biologique, Associé au CNRS, Faculté de Pharmacie, 15 av. C. Flahault, 34060 Montpellier, France.

J. E. FOX
Alta Bioscience, School of Biochemistry, University of Birmingham, PO Box 363, Edgbaston, Birmingham B15 2TT, UK.

R. FRANK
AG Molecular Recognition, Gesellschaft für Biotechnologische Forschung mbH, Mascheroder Weg 1, D-38124 Braunschweig, Germany.

D. GATOS
Department of Chemistry, University of Patras, 26 110 Patras, Greece.

P. HOOGERHOUT
RIVM—National Institute of Public Health and Environment, Laboratory for Vaccine Research, PO Box 1, 3720 BA Bilthoven, The Netherlands.

R. A. HOUGHTEN
Torrey Pines Institute for Molecular Studies, 3550 General Atomics Court, San Diego, CA 92121, USA.

I. W. JAMES
Chiron Mimotopes Pty. Ltd., 11 Duerdin Street, Clayton, Victoria 3168, Australia.

T. JOHNSON
Peptide Therapeutics Ltd., Peterhouse Technology Park, 100 Fulbourn Road, Cambridge CB1 9PT, UK.

S. A. KATES
PE Biosystems, 500 Old Connecticut Path, Framingham, MA 01701, USA.

J. KIHLBERG
Department of Organic Chemistry, Umeå University, SE-901 87 Umeå, Sweden.

Y.-A. LU
Department of Microbiology and Immunology, School of Medicine, Vanderbilt University, A-5119 Med. Ctr North, Nashville, TN 37232–2363, USA.

J.-A. MARTINEZ
Laboratoire de Chimie et Pharmacologie de Molécules d'Intérêt Biologique, Associé au CNRS, Faculté de Pharmacie, 15 av. C. Flahault, 34060 Montpellier, France.

P. MASCAGNI
Director of Research, Italfarmaco Research Centre, Via dei lavoratori 54, 20092 Cinisello Balsamo, Milano, Italy.

xviii

# Contributors

S. L. MELLOR
Department of Chemistry, Bedson Building, University of Newcastle upon Tyne, Newcastle upon Tyne NE1 7RU, UK.

J. M. OSTRESH
Torrey Pines Institute for Molecular Studies, 3550 General Atomics Court, San Diego, CA 92121, USA.

M. PARIS
Laboratoire de Chimie et Pharmacologie de Molécules d'Intérêt Biologique, Associé au CNRS, Faculté de Pharmacie, 15 av. C. Flahault, 34060 Montpellier, France.

M. QUIBELL
Peptide Therapeutics Ltd., Peterhouse Technology Park, 100 Fulbourn Road, Cambridge CB1 9PT, UK.

M. ROYO
Department of Organic Chemistry, University of Barcelona, Marti i Franqués 1–11, Barcelona E-08028, Spain.

R. C. SHEPPARD
15 Kinnaird Way, Cambridge CB1 4SN, UK.

J. P. TAM
Department of Microbiology and Immunology, School of Medicine, Vanderbilt University, A-5119 Med. Ctr North, Nashville, TN 37232–2363, USA.

A. TIEPOLD
AG Molecular Recognition, Gesellschaft für Biotechnologische Forschung mbH, Mascheroder Weg 1, D-38124 Braunschweig, Germany.

D. A. WELLINGS
Avecia LSM, Gadbrook Park, Northwich, Cheshire CW9 7RA, UK.

P. D. WHITE
CN Biosciences (UK) Ltd., Boulevard Industrial Park, Padge Road, Beeston, Nottingham NG9 2JR, UK.

G. WICKHAM
Chiron Mimotopes Pty. Ltd., 11 Duerdin Street, Clayton, Victoria 3168, Australia.

# Abbreviations

| | |
|---|---|
| Abu | 2-aminobutyric acid |
| Ac | acetyl |
| AcCl | acetyl chloride |
| AcOH | acetic acid |
| Acm | acetamidomethyl |
| AcN | acetonitrile |
| Ac$_2$O | acetic anhydride |
| ACP | acyl carrier protein |
| 1-Ada | 1-adamantyl |
| All | allyl |
| Alloc | allyloxycarbonyl |
| Boc | *tert*-butoxycarbonyl |
| Boc$_2$O | di-*tert*-butyl dicarbonate |
| Boc–N=N–Boc | bis(*tert*-butyl)azodicarboxylate |
| BOP | benzotriazol-1-yloxy tris(dimethylamino)phosphonium hexafluorophosphate |
| Bpoc | biphenylisopropoxycarbonyl |
| BrAc | bromoacetyl |
| BrAc-ONSu | *N*-succinimidyl bromoacetate |
| BSA | bovine serum albumin |
| Bum | *tert*-butoxymethyl |
| Bzl | benzyl |
| Cbz | carbobenzyloxy |
| Clt | 2-chlorotrityl |
| CLTR | 2-chlorotrityl resin |
| 2Cl-Z | 2-chlorobenzyloxycarbonyl |
| Cpn10 | chaperonin 10 kDa |
| Cpn60 | chaperonin 60 kDa |
| Cys | cysteine |
| DBU | 1,8-diazabicyclo[5.4.0]undec-7-ene |
| DCB | 2,6-dichlorobenzoyl chloride |
| DCC | 1,3-dicyclohexylcarbodiimide |
| DCE | dichloroethane |
| DCM | dichloromethane |
| Dde | 1-(4,4-dimethyl-2,6-dioxocyclohexylidene)ethyl |
| Ddiv | 1-(4,4-dimethyl-2,6-dioxocyclohexylidene)-3-methylbutyl (in some literature, this is also abbreviated to ivDde) |
| DhbtOH | 3,4-dihydro-3-hydroxy-4-oxo-1,2,3-benzotriazine |
| DIC | 1,3-diisopropylcarbodiimide |

| | |
|---|---|
| DIPEA | *N,N*-diisopropylethylamine |
| DMA | *N,N*-dimethylacetamide |
| Dmab | 4-[*N*-(1-(4,4-dimethyl-2,6-dioxocyclohexylidene)-3-methylbutyl)amino]benzyl |
| DMAP | 4-(*N,N*-dimethylamino)pyridine |
| Dmb | 2,4-dimethoxybenzyl |
| DMF | *N,N*-dimethylformamide |
| DMS | dimethyl sulphide |
| DMSO | dimethyl sulphoxide |
| DTNP | 2,2'-dithiobis(5-nitropyridine) |
| DTT | 1,4-dithiothreitol |
| EDT | 1,2-ethanedithiol |
| EDTA | ethylenediamine tetraacetic acid |
| EMS | ethylmethyl sulphide |
| EtOAc | ethyl acetate |
| Fmoc | 9-fluorenylmethoxycarbonyl |
| Fmoc-Cl | 9-fluorenylmethoxychloroformate |
| Gdm | guanidine |
| GSH | reduced glutathione |
| GSSG | oxidized glutathione |
| HAL | 5-(4-hydroxymethyl-3,5-dimethoxyphenoxy)valeryl |
| HATU | *N*-[(dimethylamino)-1*H*-1,2,3-triazolo[4,5-*b*]pyridin-1-ylmethylene]-*N*-methylmethanaminium hexafluorophosphate *N*-oxide |
| HBTU | *N*-[(1*H*-benzotriazol-1-yl)(dimethylamino)methylene]-*N*-methylmethanaminium hexafluorophosphate *N*-oxide |
| HEMA | poly(hydroxyethyl methacrylate) |
| HFIP | hexafluoroisopropanol |
| Hmb | 2-hydroxy-4-methoxybenzyl |
| HMBA | 4-hydroxymethylbenzoyl |
| HMPA | 4-hydroxymethylphenoxyacetyl |
| HMPB | 4-hydroxymethylphenoxybutyryl |
| HOAt | 1-hydroxy-7-azabenzotriazole |
| HOBt | 1-hydroxybenzotriazole |
| KLH | keyhole limpet haemocyanin |
| LCD | liquid crystal display |
| LC–MS | liquid chromatography–mass spectrometry |
| MAP | multiple antigen peptide |
| MBHA | 4-methylbenzhydrylamine |
| Meb | 4-methylbenzyl |
| MeIm | 1-methylimidazole |
| Met | methionine |
| MHPA | 4-hydroxy-3-methoxyphenoxyacetyl |
| Mmt | 4-methoxytrityl |

| | |
|---|---|
| Mob | 4-methoxybenzyl |
| MPS | multiple peptide synthesis |
| MSNT | 1-(mesitylene-2-sulphonyl)-3-nitro-1*H*-1,2,4-triazole |
| Mtr | 4-methoxy-2,3,6-trimethylbenzenesulphonyl |
| Mtt | 4-methyltrityl |
| NMA | *N*-methylmercaptoacetamide |
| NMM | *N*-methylmorpholine |
| NMP | *N*-methylpyrrolidone |
| NTN | *N*-terminal nucleophile |
| Npys | 3-nitro-2-pyridinesulphenyl |
| PAH | penicillin G acylase |
| PAL | 5-(4-aminomethyl-3,5-dimethoxyphenoxy)valeryl |
| PAM | 4-hydroxymethylphenylacetamidomethyl polystyrene |
| Pbf | 2,2,4,6,7-pentamethyldihydrobenzofuran-5-sulphonyl |
| PBS | phosphate-buffered saline |
| PDP | 3-(2-pyridyldithiol)propyl |
| Pfp | 2,3,4,5,6-pentafluorophenyl |
| PEGA | polyethylene glycol polyacrylamide copolymer |
| PEG–PS | polyethylene glycol–polystyrene |
| PEO–PS | polyethylene oxide–polystyrene |
| Phacm | phenylacetamidomethyl |
| Pic | 4-picolyl |
| Pip | phenylisopropyl |
| Pmc | 2,2,5,7,8-pentamethylchroman-6-sulphonyl |
| PS | polystyrene |
| PyAOP | 7-azabenzotriazol-1-yloxytris(pyrrolidino)phosphonium hexafluorophosphate |
| PyBOP | benzotriazol-1-yloxytris(pyrrolidino)phosphonium hexafluorophosphate |
| Pyr | 2-pyridinesulphenyl |
| RP–HPLC | reversed-phase high performance liquid chromatography |
| RT | room temperature |
| SAMA | *S*-acetylmercaptoacetyl |
| Scm | methoxycarbonylsulphenyl |
| SDS | sodium dodecylsulphate |
| Snm | *S*-(*N*-methyl-*N*-phenylcarbamoyl)sulphenyl |
| SPDP | *N*-succinimidyl 3-(2-pyridyldithiol)propionate |
| SPPS | solid-phase peptide synthesis |
| *t*Bu | *tert*-butyl |
| Tbdms | *tert*-butyldimethylsilyl |
| TBAF | tetrabutylammonium fluoride |
| TBDPS | *tert*-butyldiphenylsilyl |
| TBTU | *N*-[(1*H*-benzotriazol-1-yl)(dimethylamino)methylene]-*N*-methylmethanaminium tetrafluoroborate N-oxide |

| | |
|---|---|
| TCEP | tris(carboxyethyl)phosphine |
| TDO | 2,3-dihydro-2,5-diphenyl-4-hydroxy-3-oxothiophen-1,1-dioxide |
| TEA | triethylamine |
| TES | triethylsilane |
| TFA | trifluoroacetic acid |
| TFE | trifluoroethanol |
| TFFH | tetramethylfluoroformamidinium hexafluorophosphate |
| TFMSA | trifluoromethanesulphonic acid |
| THF | tetrahydrofuran |
| TIS | triisopropylsilane |
| TLC | thin-layer chromatography |
| Tmob | 2,4,6-trimethoxybenzyl |
| TMS | trimethylsilyl |
| TMSBr | trimethylsilyl bromide |
| Tmse | trimethylsilylethyl |
| TNBS | 2,4,6-trinitrobenzenesulphonic acid |
| Trt | trityl |
| TTd | tetanus toxoid |
| Xan | xanthen-9-yl |
| Z(2-Cl)-OSu | *N*-(2-chlorobenzyloxycarbonyl)succinimide |

# 1

# Introduction—a retrospective viewpoint

## R. C. SHEPPARD

The Chemical Society publication *Annual Reports on the Progress of Chemistry* for 1963 attempted to inform readers of all the highly significant advances in all the major fields of pure chemistry during that year. Fortunately, the section on peptide chemistry (1) drew attention to a paper by R. B. Merrifield which had just been published in the *Journal of the American Chemical Society* (2):

A novel approach to peptide synthesis has been the use of a chloromethylated poly-styrene polymer as an insoluble but porous solid phase on which the coupling reactions are carried out. Attachment to the polymer constitutes protection of the carboxyl group (as a modified benzyl ester), and the peptide is lengthened from its amino-end by successive carbodiimide couplings. The method has been applied to the synthesis of a tetrapeptide, but incomplete reactions lead to the accumulation of by products. Further development of this interesting method is awaited.

I remember thinking at the time that in this paper we had possibly seen both the beginning and the end of the interesting new technique of solid phase peptide synthesis. To many organic chemists, the described result was that anticipated—difficulty in bringing heterogeneous reactions to completion resulting in impure products. Both this and purification problems were expected to worsen as the chain length was increased beyond Merrifield's tetrapeptide limit. In fact, I probably had at the time an inadequate appreciation of the difference between truly heterogeneous or surface reactions and those in the solvated gel phase. The latter approaches much more closely the solution situation. However, the new technique also flouted many of the basic principles of contemporary organic synthesis which required rigorous isolation, purification, and characterization regimes following each synthetic step. In Merrifield's new technique, isolation consisted simply of washing the solid resin, there was no other purification of the products of each reaction, and little or no characterization of resin-bound intermediates was attempted. The first two of these are of course the important characteristics which give the method its speed and simplicity and contribute to its efficiency. Small wonder, though, that in many minds there was doubt about the future of the new technique.

Before the 1963 volume of *Annual Reports* was even published the situation had changed dramatically. A second paper from Merrifield appeared in the same journal (3) and was noted rather cryptically:

Added in proof: The method has now been developed further and used in a synthesis of bradykinin.

By any criterion this was an outstanding synthesis. Improvements in the chemistry were described which enabled the nonapeptide sequence of bradykinin to be assembled in four days and isolated, purified, and fully characterized in a further five. The overall yield of highly pure, fully active material was 68%. These results greatly exceeded any which could be achieved by contemporary solution methods and triggered an explosion of activity in solid phase peptide synthesis.

Merrifield's development was well timed, coming in a period when natural peptide hormones, antibiotics, and other biologically active peptides were being isolated apace and subjected to vigorous structural and functional study. The new technique was adopted with enthusiasm by biochemists, pharmacologists, and others who saw in solid phase peptide synthesis an answer to their pressing need for synthetic analogues of the new natural peptides. Some 10 years later, Meienhofer in his important 1973 review (4)* was already able to list more than 500 published solid phase syntheses. Such an output would have required commitment of prodigiously greater resources had it been achieved by contemporary solution methods. Application of solid phase peptide synthesis has continued to grow vigorously and expanded with equal success into the very important field of oligonucleotide synthesis.

Of course, not all of the early synthetic products were obtained without difficulty or in a satisfactory state of purity. Chemical problems, limitations in available purification procedures, and the absence of reliable analytical techniques applicable to the solid phase must have led to many failures, some going unrecognized. Peptide chemists in many laboratories had been active in solid phase synthesis during this time, but real understanding of the system was slow to emerge in a period when the emphasis was on applications. Of the many variants in detail suggested (4) during this early period, few achieved any widespread popularity. Indeed in some laboratories the 1964 Merrifield technique is still practised today with only minor changes.

The chemical problems associated with solid phase peptide synthesis are now better understood. Some, for example difficulties caused by vigorous reaction conditions and differential lability of temporary and permanent protecting groups, are associated with the particular chemical implementation of the method. Others, notably the need for near quantitative conversion at

---

* In this Introduction, references not given specifically are contained, prior to 1972, in Meienhofer's review (4), prior to 1989 in *Solid Phase Peptide Synthesis. A Practical Approach* (5), and prior to 1993 in the European Peptide Symposium *Josef Rudinger Lecture* (6). Reference 5 gives a general survey of the development of Fmoc-based solid phase synthesis up to 1989.

every stage and the effect of solvation and the nature of the polymer support on internal structure and reactivity, are intrinsic to the solid phase principle itself. Solvation effects within the solid support have proved to be highly significant. Initially, solid phase synthesis was widely thought to provide a complete solution to the solubility problems which had beset solution phase peptide synthesis for the previous two decades. In fact, solubility problems are not completely avoided, they simply reappear in different form (see later).

Merrifield's improved technique (3) (*Figure 1*) made use of protecting groups based predominantly on benzyl and *t*-butyl derivatives. Selective cleavage of the terminal *t*-butoxycarbonyl-amino derivative is required at every cycle of amino acid addition. Both this temporary protecting group and the more permanent benzyl derivatives used for side chain and carboxy terminal protection (peptide–resin linkage) are acid labile, and selectivity in cleavage cannot be absolute. Concomitant loss of some small proportion of side chain protecting groups and of the peptide–resin linkage must also occur at every cycle. The former results in liberation of reactive functional groups in the peptide sequence with potential for side reactions and impure final products; the latter in progressive loss of peptide from the polymer support with lowering of final yield and appearance of undesired functionality on the resin. It is a credit to the conditions devised for the Merrifield technique that both side reactions are usually held to acceptable levels.

A further inevitable consequence is that the more permanent benzyl derivatives will require substantially stronger acidic conditions for their eventual cleavage. Vigorous reagents such as liquid hydrogen fluoride are commonly used. Not all peptide sequences are entirely stable to such reagents, and as synthetic targets have become more and more ambitious, destructive side reactions have become potentially more serious. Liquid hydrogen fluoride is also a particularly unpleasant and hazardous reagent requiring special equipment for its safe handling.

Over the years, these considerations have concerned many chemists. Indeed Merrifield himself must have been early concerned about the need for mild reaction conditions. His original technique (2) employed the very stable benzyl–nitrobenzyl α-amino-protecting group–resin linkage combination, soon to be replaced in his bradykinin paper (3) by the more labile *t*-butyl/

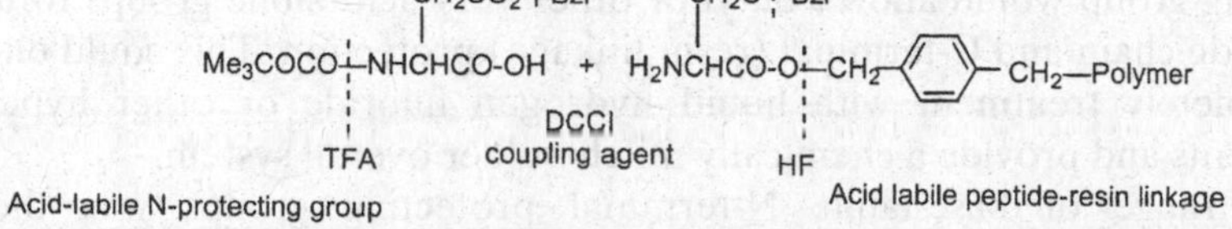

**Figure 1.** The Merrifield strategy illustrated for the synthesis of a dipeptide. Typical cleavage reagents for the various protecting groups are shown.

benzyl system. Other relevant suggestions during the early period (4) centred around development of more acid-labile α-amino-protecting groups. *N*-Trityl, *o*-nitrophenylsulphenyl, α,α-dimethyl-3,5-dimethoxybenzyloxycarbonyl, and biphenylisopropoxycarbonyl (Bpoc) were among those explored with perhaps the last being the most promising. In combination with a more acid-labile *p*-alkoxybenzyl ester peptide–resin linkage, Bpoc groups provided an alternative system (7) with milder reaction conditions, but it failed to displace or even seriously compete with the by now well-established Merrifield technique. Probably difficulties of compatible side chain protection and the lack of commercially available amino acid derivatives were significant factors at the time. The conservatism shown by many practitioners of solid phase peptide synthesis has been strongly evident throughout its history.

Development work on solid phase peptide synthesis began in our laboratory in Cambridge in 1972. Recognizing the need for accelerated methods of synthesis in a biology laboratory, we hoped to devise a new system which would avoid or at least ameliorate some of the difficulties mentioned above. As a first step, my colleague Eric Atherton successfully developed a new polar solid support compatible with polar reaction media which took into account the special solvation requirements of both solid phase peptide (8) and oligonucleotide (9) synthesis. For the test sequences studied it had advantages over the customary apolar polystyrene resin. In peptide synthesis it was used initially with a similar combination of acid-labile protecting groups to those of the standard Merrifield technique. As the new resin was applied to longer and more complex sequences, however, we felt that the substantial exposure to acidic regents involved in both the repeated cleavage of *t*-butoxycarbonyl (Boc) groups and in the final detachment reaction might now be a limiting factor. Furthermore, I could not fail to notice that the rigorous safety precautions in handling liquid hydrogen fluoride in the laboratory which were initially practised so assiduously seemed to become progressively less important. This was a matter of great concern, especially as I had recently encountered a peptide chemist friend at the European Peptide Symposium with a serious and long-lasting hydrogen fluoride burn on his wrist.

An obvious next step was to rearrange the overall protecting group strategy so that the exposure to acidic reagents was reduced. Eric Atherton had earlier shown in a trial assembly of bradykinin that basic reagents could be used to cleave peptides resin-bound through a base-labile *p*-carboxybenzyl alcohol linking agent. Alternative assignment of base-lability to the N-terminal protecting group would allow *t*-butyl or other very acid-labile groups to be used for side chain and C-terminal (resin linkage) protection. This could eliminate completely treatment with liquid hydrogen fluoride or other hyperacidic reagents and provide a chemically much milder overall system.

A range of base-labile N-terminal protecting groups was therefore investigated for application in solid phase peptide synthesis (5). Carpino's 9-fluorenylmethoxycarbonyl (Fmoc) group (10) soon proved to be quite exceptionally suitable. Carpino had proposed this amino-protecting group in 1972 but it had attracted little attention. The reason it had not found favour in solution chemistry is easy to see. Most of the established amino-protecting groups are cleaved with the formation of inert, often volatile or otherwise easily eliminated co-products. Fluorenylmethoxycarbonyl derivatives are cleaved by organic bases with the initial formation of non-volatile and reactive dibenzofulvene with the potential for addition or polymerization reactions. This problem is largely avoided in solid phase synthesis where soluble co-products are eliminated by simple washing. In continuous flow solid phase synthesis (see below), soluble co-products are removed from contact with the resin almost as soon as they are formed, reducing further the possibility of undesired side reactions.

We learned later that Hans Meienhofer and his colleagues at Hoffman-la Roche in Nutley, New Jersey, were following a similar line of development in polystyrene-based solid phase series. The results from the two laboratories were published nearly simultaneously (11, 12).

Fmoc-based chemistry (*Figure 2*) provided an efficient alternative to the Merrifield technique. The full range of Fmoc-amino acids were readily prepared using fluorenylmethyl chloroformate (5), although care was required in their purification. Some laboratories reported early difficulties now recognized as due to impure reagents. An early improvement was the use hydroxysuccinimide esters in the preparation of Fmoc-amino acids (13). Chemical companies were quick to recognize their potential and the full range of simple and side chain protected amino acid derivatives soon became commercially available. This was an important factor encouraging widespread adoption of the method. Pentafluorophenyl esters and, later, esters of dihydrooxobenzotriazine provided convenient storable reagents suitable for use in automatic synthesizers (5).

Our enthusiasm for the Fmoc-based solid phase procedure was finally confirmed in 1993 when an independent comparative evaluation of the Boc and Fmoc procedures was carried out simultaneously in a number of laboratories.

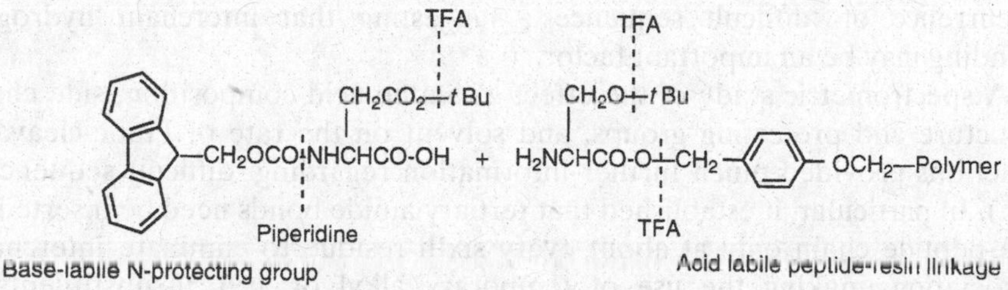

**Figure 2.** The Fmoc strategy illustrated for the synthesis of a dipeptide. A variety of coupling reagents (carbodiimides, pentafluorophenyl esters, aminium salts) may be used.

Compelling evidence was obtained for the benefit of the milder reaction conditions provided by the Fmoc method (14).

Introduction of the Fmoc procedure brought with it another advantage which has had far-reaching effects on the practise of the method. Fluorene derivatives have strong ultraviolet absorption in an accessible part of the spectrum. This property provided the basis for a simple method of analytical control in solid phase synthesis. Take-up of fluorene derivatives from solution onto the solid phase and vice versa could now be easily followed spectrometrically. It encouraged the development of the continuous flow method of synthesis in which reagents were pumped continuously through a stationary, physically supported resin bed (5). Inclusion of a UV cell within the flowing reagent stream provided a complete and immediate record of the acylation and deprotection steps of the synthesis as they were occurring. At its simplest it provided important reassurance to the chemist that all was progressing normally or enabled operator or synthesizer errors to be corrected in good time. At a higher level it enabled the construction of fully automatic synthesizers in which progression from one step to the next was controlled by computer interpretation of the analytical data. This is in contrast to earlier solid phase techniques which were commonly operated completely blind.

The ultraviolet absorption of fluorene derivatives also encouraged more detailed analytical studies of solid phase reactions. In particular, it allowed careful examination of the so-called 'difficult sequence' problem (6, 15). For some peptide sequences, sudden, sometimes catastrophic reduction in the rates of both deprotection and coupling reactions are observed as the resin-bound chains lengthen. This is believed to be a consequence of association of the peptide chains within the resin matrix, i.e. to inadequate solvation. Frequently synthesis cannot be completed satisfactorily because of these very low reaction rates. 'Difficult sequences' are encountered in using both Boc and Fmoc chemistry, although in the latter the generally lower solubility (solvation) of Fmoc derivatives may also contribute. They are minimized but not eliminated through the use of more powerfully solvating media of the dimethylformamide or dimethylsulphoxide types. Inclusion of tertiary amide bonds within the peptide sequence as with proline residues reduces the occurrence of 'difficult sequences', suggesting that interchain hydrogen bonding may be an important factor.

A spectrometric study of the effect of amino acid composition, side chain structure and protecting groups, and solvent on the rate of Fmoc-cleavage reactions provided much further information regarding 'difficult sequences' (15). In particular, it established that tertiary amide bonds need be inserted in the peptide chain only at about every sixth residue to eliminate interchain association, making the use of temporary alkyl or aryl N-substituents a practical possibility. Massive steric hindrance is usually encountered during acylation of, for example, *N*-benzyl peptide resins (as in **I**, $R_1 = R_2 = H$), but a substitution pattern in the aromatic ring (**I**, $R_1 = OH$; $R_2 = OMe$) was established which provided both an acid-labile, easily removed protecting group and a facile mechanistic pathway for the acylation reaction (16). Coupling of the incoming Fmoc-amino acid now occurs at the relatively unhindered *ortho*-hydroxy group with internal base catalysis provided by the nearby secondary amino function. An intramolecular O → N shift then completed the coupling reaction. This last migration step was not as rapid as we had anticipated, probably due to crowding in the cyclic intermediate, but it provided a very workable system which offers promise of a complete solution to the 'difficult sequences' problem. Equally importantly, it also solved the critical solubility problem which has for so long prevented the development of efficient and general solid phase fragment condensation procedures. Fmoc-based solid phase synthesis can now be used for the preparation of freely soluble, easily purified protected peptide fragments which can be efficiently assembled on a solid support into synthetic proteins (17).

$$\text{HN--CHR--CO--NH----resin}$$

I

Fmoc-based solid phased peptide synthesis is now firmly established alongside the Merrifield technique. It has proved both efficient and versatile, providing stepwise (single residue addition) and fragment assembly procedures, analytical and monitoring techniques, and solutions to some of the previously limiting chemical problems. It encouraged the development of continuous flow methods and the construction of sophisticated automatic peptide synthesizing equipment with analytical feedback control. It has been applied to the synthesis of modified peptides including phosphorylated and glycosylated derivatives, to the multiple synthesis of peptides, and in combinatorial techniques, and generally offers promise as the method of choice for the synthesis of peptides.

# References

1. Sheppard, R. C. (1963). *Ann. Rep.*, **60**, 448.
2. Merrifield, R. B. (1963). *J. Amer. Chem. Soc.*, **85**, 2149.
3. Merrifield, R. B. (1964). *J. Amer. Chem. Soc.*, **86**, 304.
4. Meienhofer, H. (1973). *Progr. Hormone Research*, **2**, 46.
5. Atherton, E. and Sheppard, R. C. (1989) *Solid phase peptide synthesis. A practical approach*. IRL Press, Oxford.
6. Sheppard, R. C. (1995). In *Peptides 1994. Proceedings of the 23rd European Peptide Symposium*, Braga, Portugal, p. 3. ESCOM, Leiden.
7. Wang, S. S. and Merrifield, R. B. (1969). *J. Amer. Chem. Soc.*, **91**, 6488.

8. Atherton, E., Clive, D. L. J., and Sheppard, R. C. (1975). *J. Amer. Chem. Soc.*, **97**, 6584.
9. Gait, M. J. and Sheppard, R. C. (1976). *J. Amer. Chem. Soc.*, **98**, 8514.
10. Carpino, L. A. and Han, G. Y. (1972). *J. Org. Chem.*, **37**, 3404.
11. Chang, C.-D. and Meienhofer, J. (1978). *Int. J. Peptide Prot. Res.*, **11**, 246.
12. Atherton, E., Fox, H., Harkiss, D., Logan, C. J., Sheppard, R. C., and Williams, B. J. (1978). *J. Chem. Soc., Chem. Comm.*, 537.
13. Sigler, G. F., Fuller, W. D., Chaturvedi, N. C., Goodman, M., and Verlander, M. (1983). *Biopolymers*, **22**, 2157.
14. Fields, G. B., Carr, S. A., Marshak, D. R., Smith, A. J., Stults, J. T., Williams, L. C., Williams, K. R., and Young, J. D. (1993). In *Techniques in protein chemistry IV* (ed. R. H. Angeletti), p. 229. Academic Press, New York.
15. Bedford, J., Hyde, C., Johnson, T., Wen, J. J., Owen, D., Quibell, M., and Sheppard, R. C. (1992). *Int. J. Peptide Prot. Res.*, **40**, 300.
16. Johnson, T., Quibell, M., and Sheppard, R. C. (1995). *J. Peptide Sci.*, **1**, 11.
17. Quibell, M., Packman, L. C., and Johnson, T. (1995). *J. Amer. Chem. Soc.*, **117**, 11656.

# Basic principles

PETER D. WHITE and WENG C. CHAN

## 1. The solid phase principle

Construction of a peptide chain on an insoluble solid support has obvious benefits: separation of the intermediate peptides from soluble reagents and solvents can be effected simply by filtration and washing with consequent savings in time and labour over the corresponding operations in solution synthesis; many of the operations are amenable to automation; excess reagents can be employed to help to drive reactions to completion; and physical losses can be minimized as the peptide remains attached to the support throughout the synthesis. This approach does, however, have its attendant limitations. By-products arising from either incomplete reactions, side reactions, or impure reagents will accumulate on the resin during chain assembly and contaminate the final product. The effects on product purity of achieving less than 100% chemical efficiency in every step are illustrated dramatically in *Table 1*. This has serious implications with regard to product purification as the impurities generated will, by their nature, be very similar to the desired peptide and therefore extremely difficult to remove. Furthermore, the analytical techniques employed for following the progress of reactions in solution are generally not applicable, and recourse must generally be made to simple qualitative colour tests to detect the presence of residual amines on the solid phase.

The principles of solid phase synthesis are illustrated in *Figure 1*. The C-terminal amino acid residue of the target peptide is attached to an insoluble support via its carboxyl group. Any functional groups in amino acid side chains must be masked with *permanent* protecting groups that are not affected by the reactions conditions employed during peptide chain assembly. The *temporary* protecting group masking the α-amino group during the initial resin loading is removed. An excess of the second amino acid is introduced, with the carboxy group of this amino acid being activated for amide bond formation through generation of an activated ester or by reaction with a coupling reagent. After coupling, excess reagents are removed by washing and the protecting group removed from the N-terminus of the dipeptide, prior to addition of the third amino acid residue. This process is repeated until the desired peptide

**Table 1.** Effects of accumulated errors on final product yields

| No. of reactions | | Yield of each reaction (%) | | | |
|---|---|---|---|---|---|
| | | 100 | 99 | 95 | 90 |
| 10 | Overall yields | 100 | 90 | 60 | 35 |
| 20 | | 100 | 81 | 36 | 12 |
| 30 | | 100 | 74 | 21 | 4 |
| 40 | | 100 | 67 | 13 | 1 |
| 50 | | 100 | 61 | 8 | < 1 |

sequence is assembled. In a final step, the peptide is released from the support and the side-chain protecting groups removed. Generally, side-chain protecting groups and resin linkage are chosen such that protecting groups are removed and the assembled peptide released under the same conditions.

For a general account of peptide synthesis and many of the techniques

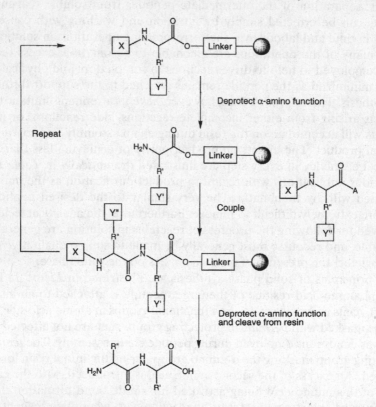

Deprotect α-amino function

Repeat

Couple

Deprotect α-amino function
and cleave from resin

X = Temporary amino protecting group
Y = Permanent side-chain protecting group
A = Carboxy activating group

**Figure 1.** The solid phase peptide synthesis (SPPS) principle.

described later in this volume, the reader's attention is drawn to the recently published book by Williams, Albericio and Giralt (1).

## 2. Merrifield SPPS

The principal features of the Merrifield technique (2–4), as it is now practised, are illustrated in *Figure 2*. The C-terminal amino acid is anchored to the support through formation of a benzyl ester with hydroxymethylphenylace-tamidomethyl polystyrene (PAM resin). (The original support, chloromethyl polystyrene, is now rarely utilized, as its use can give rise to side-reactions, especially during the synthesis of longer peptides, owing to the limited acid stability of peptide–benzyl ester linkage.) The *tert*-butoxycarbonyl (Boc) group is used for temporary protection of the α-amino group. Removal of this group is usually effected with neat trifluoroacetic (TFA) or TFA in dichloro-methane (DCM). The resulting trifluoroacetate is neutralized prior to coupling with diisopropylethylamine (DIPEA) in DCM, or neutralized *in situ* during the coupling reaction. Coupling was originally carried out by activation of the incoming amino acid with dicyclohexylcarbodiimide (DCC) in DCM, but nowadays the use of pre-formed amino acid symmetrical anhydrides or benzotriazolyl esters in DMF or *N*-methylpyrrolidone (NMP) is favoured. For masking of the side-chains of trifunctional amino acids, a range of benzyl-based protecting groups have been developed, chemically fine-tuned to suit the requirements of particular functional groups by substitution of the benzyl ring with appropriate electron donating or withdrawing groups. Release of the peptide from the resin and removal of the side-chain protecting groups is usually effected with anhydrous hydrogen fluoride.

With improvements in the quality of base-resin material and Boc-protected amino acids and the introduction of improved HF cleavage protocols, the Merrifield method has developed over the years into an extremely powerful tool, which in skilled hands has enabled the remarkably efficient synthesis of a number of large peptides and small proteins (2). However, the need to use highly toxic liquid HF in a special polytetrafluoroethylene (PTFE)-lined apparatus has generally deterred most newcomers to the field from taking up this method, with the unfortunate result that the number of practitioners of the Merrifield method is gradually dwindling.

## 3. Fmoc/*t*Bu SPPS

Unlike the Merrifield approach (*Figure 2*) which utilizes a regime of graduated acidolysis to achieve selectivity in the removal of temporary and permanent protection, the Fmoc/*t*Bu method (5) is based on an orthogonal protecting group strategy, using the base-labile *N*-Fmoc group for protection of the α-amino group and acid-labile side-chain protecting groups and resin-

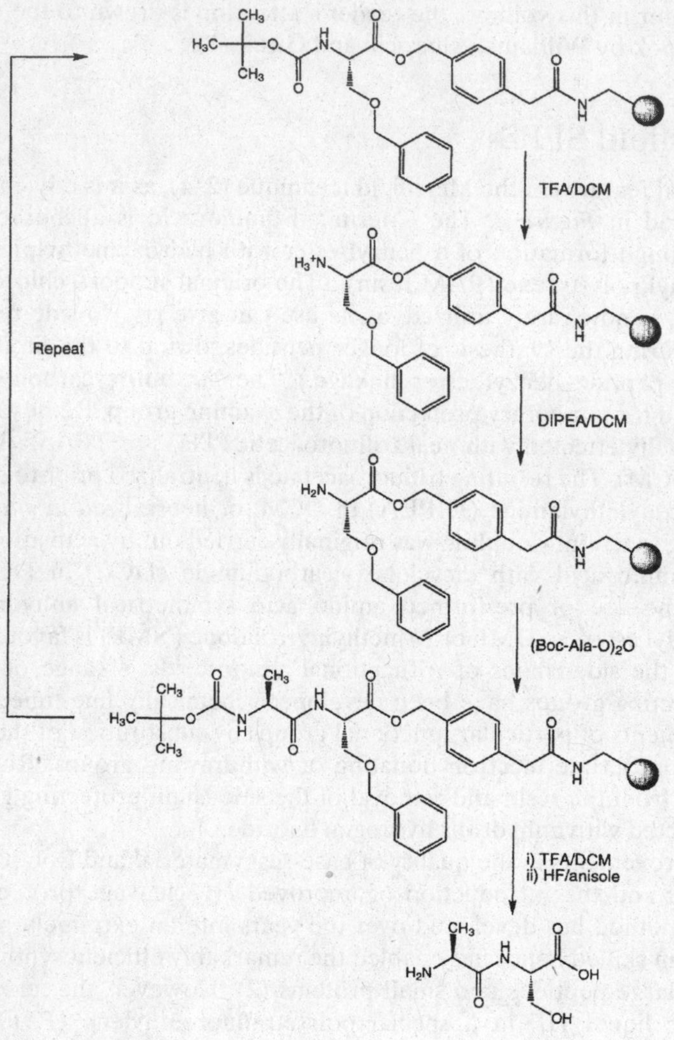

**Figure 2.** Merrifield SPPS.

linkage agents. Since removal of temporary and permanent protection is effected by completely different chemical mechanisms, side-chain protecting groups and linkage agents can be employed that are removed under considerably milder conditions than those used in the Merrifield method. In practice, *t*-butyl- and trityl-based side-chain protection and alkoxybenzyl-based linkers are used as they can be removed with TFA. This reagent is an excellent solvent for peptides, can be used in standard glass laboratory glassware, and being volatile is readily removed by evaporation. Indeed, it is this convenience of the cleavage reaction and the ease with which the method

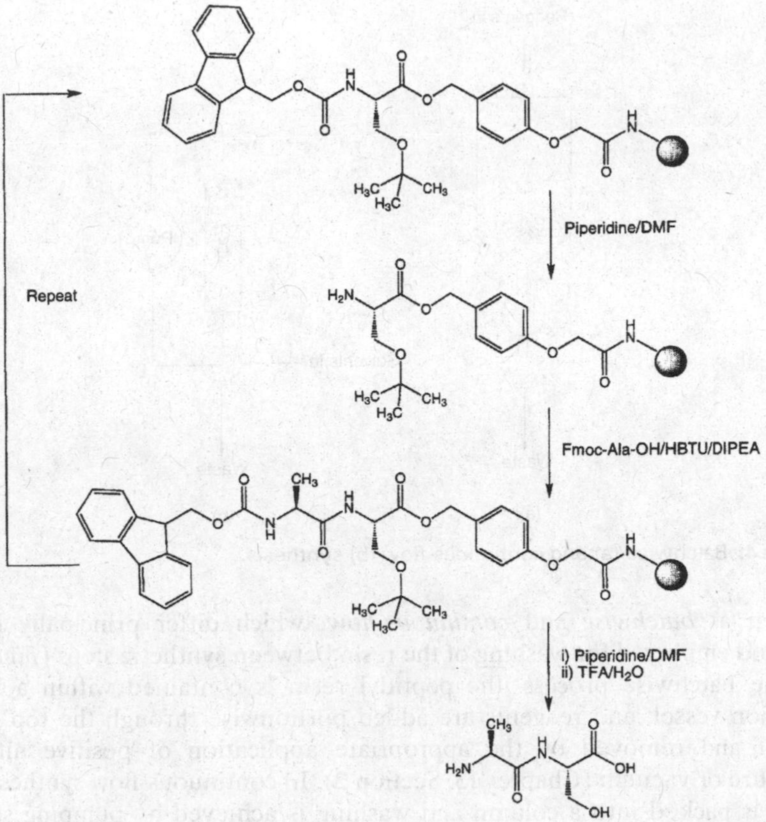

**Figure 3.** Fmoc SPPS.

can be adapted to multiple peptide synthesis that are undoubtedly the reasons for the popularity of the Fmoc/tBu approach.

The salient features of the Fmoc/tBu approach are summarized in *Figure 3*. The C-terminal residue is anchored to a TFA-labile linkage agent. The side-chain functionalities are protected with TFA-labile protecting groups. The temporary $N^\alpha$-Fmoc protecting group is removed with 20% piperidine in DMF. Coupling is typically carried out in DMF or NMP with pre-formed active esters or using activation reagents that generate *in situ* benzotriazolyl esters. Cleavage of the peptide from the resin and global side-chain deprotection is effected with 95% TFA. The background and development of the Fmoc approach to SPPS has been the subject of a number of excellent reviews (6, 7).

## 3.1 Resins

For the preparation of more than 50 μmol of peptide, synthesis is normally carried out on beaded resins. Two practical procedures are in common usage,

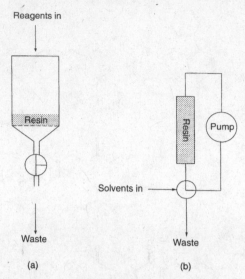

**Figure 4.** Batchwise (a) and continuous-flow (b) synthesis.

known as *batchwise* and *continuous-flow*, which differ principally in the method employed for washing of the resin between synthetic steps (*Figure 4*). In the batchwise process, the peptidyl resin is contained within a fritted reaction vessel, and reagents are added portionwise through the top of the vessel and removed by the appropriate application of positive nitrogen pressure or vacuum (Chapter 13, Section 2). In continuous-flow synthesis, the resin is packed into a column and washing is achieved by pumping solvent through the resin bed (see Chapter 13, Section 3).

For batchwise synthesis, the base matrix used is almost invariably 1% divinylbenzene cross-linked polystyrene (PS). It is relatively inexpensive to produce, swells in the solvents most commonly used in peptide synthesis, namely DCM, DMF, and NMP, and can be readily functionalized using the Friedel–Crafts reaction with chloromethyl, aminomethyl, and benzhydrylamino groups.

Polystyrene-based resins can also be used in continuous-flow synthesis, provided the beads are co-packed with glass beads and/or low flow rates are used. However, this arrangement is not entirely satisfactory, and the use of one of the supports especially manufactured for this purpose is preferred as they are designed to withstand the pressures generated in pumped-flow systems. The first commercially available continuous-flow supports consisted of a dimethylacrylamide carrier polymer contained within the pores of a rigid kieselguhr (Macrosorb®, NovaSyn K®) or polystyrene (Polyhipe®) matrix. These materials were somewhat friable and had a tendency to degrade over the course of long assemblies. Furthermore, being prepared from irregular-shaped particles, the beads did not always pack evenly, which led to problems

**14**

**Table 2.** Properties of base-matrices[a] used in Fmoc SPPS

| Resin | Particle size ($\mu$m) | Loading (mmol g$^{-1}$) | Swelling (ml g$^{-1}$) | |
|---|---|---|---|---|
| | | | DCM | DMF |
| Aminomethyl polystyrene | 75–150, 38–75 | 0.4–1.5 | 7 | 3 |
| Amino Tentagel | 90 | 0.2–0.3 | 5 | 5 |
| PEO–PS | 75–150 | 0.15–0.20 | | |
| PEGA | 150–300[b] | 0.2–0.4 | 13 | 11 |
| CLEAR | 90–250 | 0.2–0.4 | 7 | 6.5 |

[a] Before addition of linker.
[b] Swollen in water.

with channelling and uneven distribution of reagents through the resin bed. In recent years, these supports have been superseded by materials based on polyethylene glycol (PEG). The two most frequently utilized are Tentagel **1** and PEO–PS **2**, which are prepared by grafting of PEG to low cross-linked polystyrene. The beads of these materials are spherical, pack well into reaction columns, are able to withstand high back-pressures, and can be used in batchwise as well as continuous-flow synthesis. More recently, supports based on cross-linked PEG have also been developed (PEGA **3** and CLEAR **4**)—although not widely used, indications are that their performance compares favourably with the previously described composite PEG–PS supports. The properties of the resins discussed above are summarized in *Table 2*.

The requirement for small amounts of large numbers of peptides, particularly for antigen-mapping, has led to the development of various novel non-beaded supports designed especially for this purpose. The two most notable examples are PINS and SPOTS, the uses of which are described in Chapter 14.

## 3.2 Linkers

The purpose of the linker (or handle) is to provide a reversible linkage between the synthetic peptide chain and the solid support, and to protect the C-terminal $\alpha$-carboxyl group during the process of chain extension.

The choice of linker determines the C-terminal functional group in the final product. Most linkers are designed to release peptide acids or amides upon treatment with TFA, with the concentration of TFA required to effect cleavage dictating whether the product is released in a fully deprotected form or still retaining the acid-sensitive side-chain protecting groups. Linkers are also available in which the peptide–resin linkage is cleaved with nucleophiles, used for the preparation of C-terminally modified peptides such as peptide esters and secondary amides (see Chapter 6, Section 1.2), and by light, which are used for applications such as one-bead-one-compound libraries. The linkers most frequently used in Fmoc/tBu SPPS are listed in *Table 3*; those

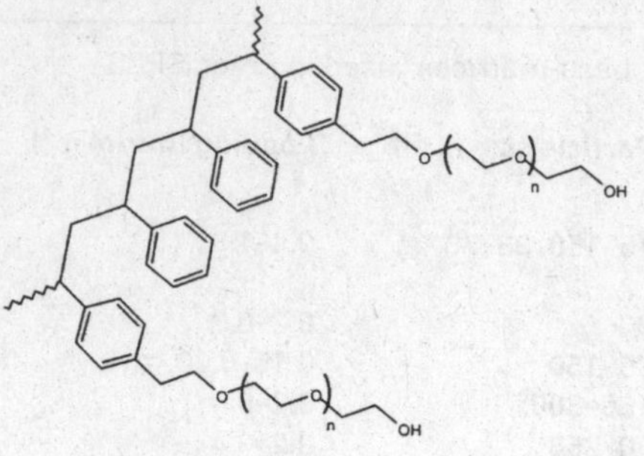

*Peter D. White and Weng C. Chan*

**1**

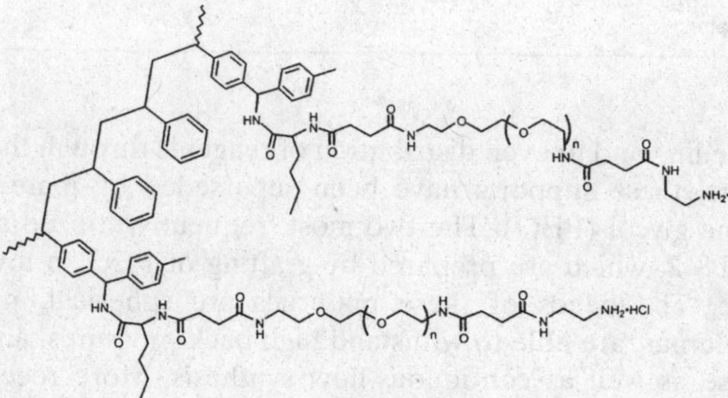

**2**

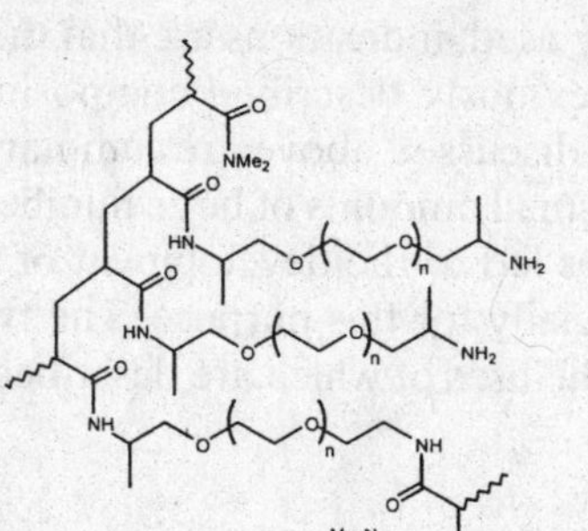

**3**

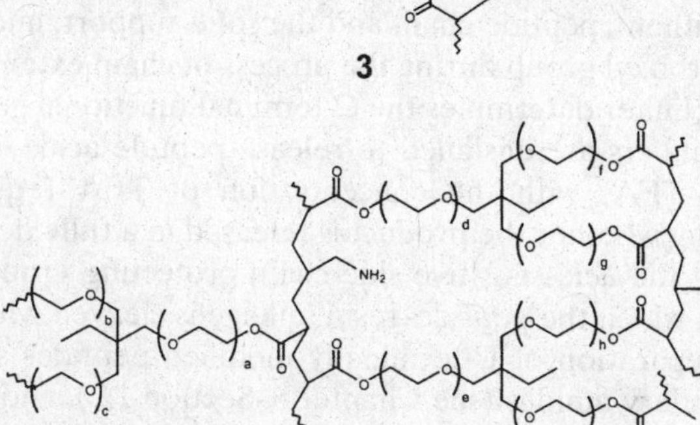

**4**

16

**Table 3.** Linker resins commonly used for the synthesis of peptide acids and peptide amides

| Linker resins | Cleavage conditions | Refs |
|---|---|---|

**Acidolysis to peptide acids**

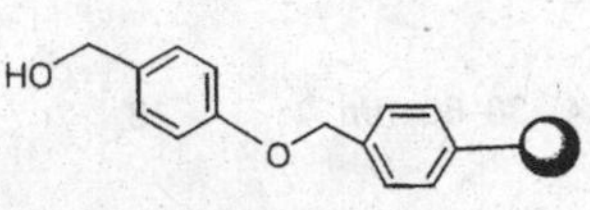

| 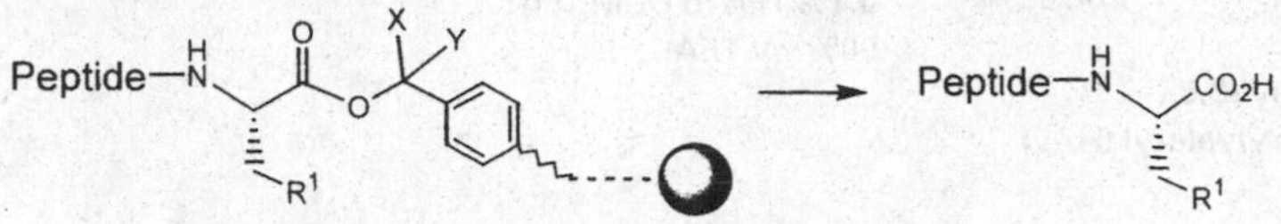 | 90–95% v/v TFA, 1–2 h [a, b] | 8 |

Wang resin

| | 90–95% v/v TFA, 1–2 h [a, b] | 9 |

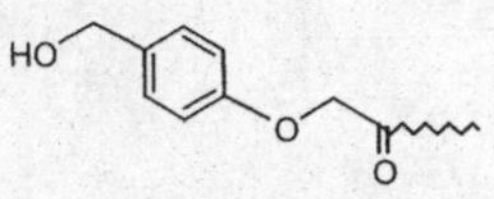

4-Hydroxymethylphenoxy acetyl
e.g. NovaSyn® TGA resin

**_Hyperacid labile linker_**

On treatment with a mild acidic solution, these typically yield protected peptide acids

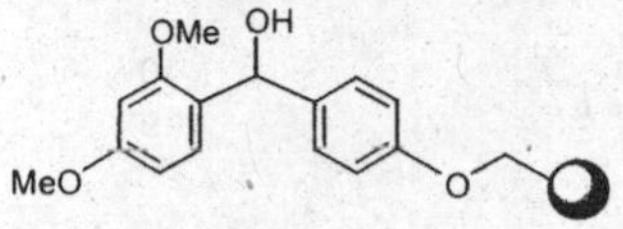

| | 10% AcOH in DCM, 1.5 h | 10 |
| | 0.2% TFA in DCM, 3 min | 10 |
| | 1–5% TFA in DCM, 5 min | |

Rink acid resin

| | 90% v/v TFA | 11 |
| | 1% TFA in DCM, 2–5 min | |

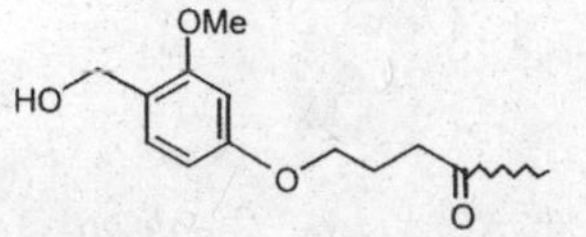

4-(4-Hydroxymethyl-3-
methoxyphenoxy)butyryl (HMPB)

| | 90% v/v TFA [a, b] | 12, 13 |
| | AcOH–TFE–DCM (1:1:8), 15–60 min | 14 |
| | 1–5% TFA in DCM, 1 min | |
| | 20% HFIP in DCM, 3–5 min | 15 |

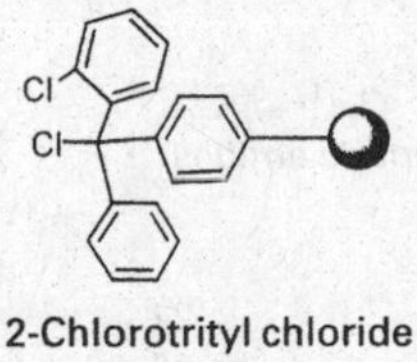

2-Chlorotrityl chloride resin

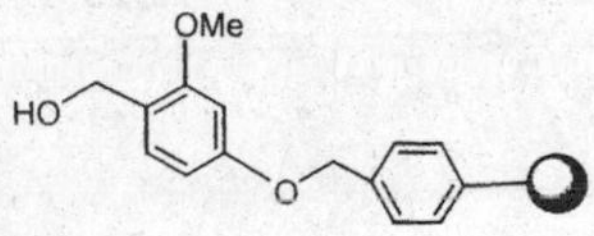

| | 1% TFA in DCM, 5–10 min | 16 |

SASRIN resin

**Table 3.** *Continued*

| Linker resins | Cleavage conditions | Refs |
|---|---|---|
| 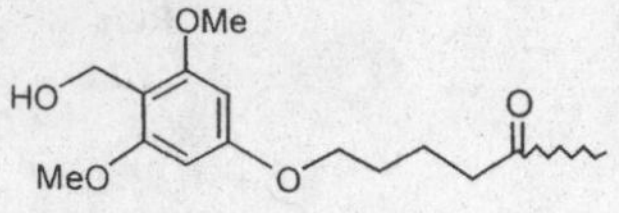 5-(4-Hydroxymethyl-3,5-dimethoxyphenoxy)valeryl (HAL) | 0.1% TFA in DCM, 1 h<br>90% v/v TFA | 17 |
| 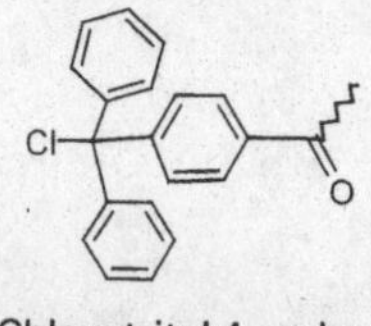 Chlorotrityl 4-carboxy | AcOH–TFE–DCM (5:1:4), 30–60 min<br>1–5% TFA in DCM<br>90% v/v TFA | 18 |

**Acidolysis to peptide amides**

| 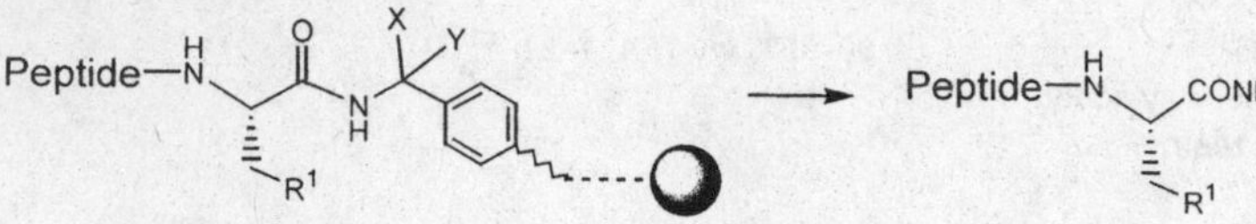 | | |
|---|---|---|
| 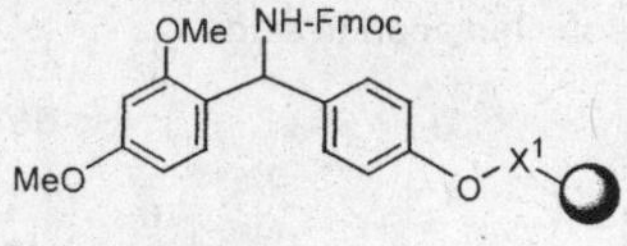 Rink amide resin 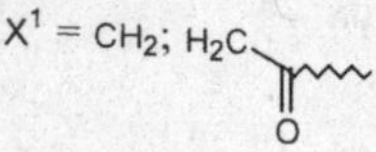<br><br>$X^1 = CH_2$; | 50% TFA in DCM, 1 h<br>90–95% v/v TFA, 1 h [a, b, c] | 10<br>19 |
| 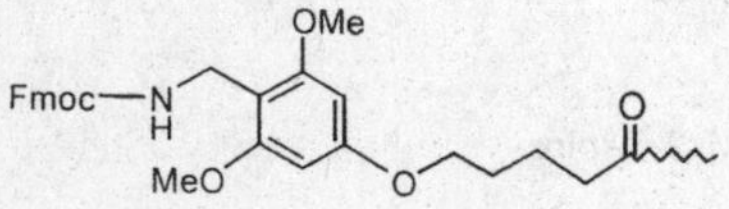 5-(4-*N*-Fmoc-aminomethyl-3,5-dimethoxyphenoxy)valeryl (PAL) | 90–95% v/v TFA, 1–2 h [a, b]<br>TFA–DCM–DMS (14:5:1), 2 h | 19, 20<br>20 |

**Hyperacid labile linker**

On treatment with a mild acidic solution, these typically yield protected peptide amides

| | | |
|---|---|---|
| 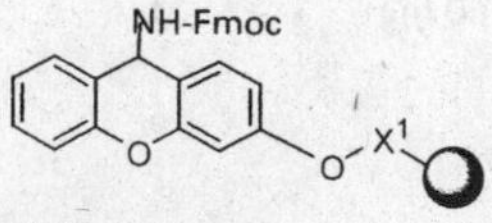 Sieber amide resin 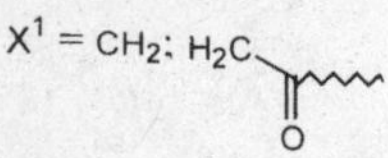<br><br>$X^1 = CH_2$; | 2% TFA in DCE, 5–10 min<br>1–5% TFA in DCM, 5–15 min<br>90% v/v TFA [c] | 21<br>22, 23<br>23 |

**Table 3.** *Continued*

| Linker resins | Cleavage conditions | Refs |
|---|---|---|

**Nucleophilic displacement to C-terminal modified peptides (Base-cleavable linkers)**

| | Cleavage conditions | Refs |
|---|---|---|
| 4-Hydroxymethyl benzoyl | Ammonia in MeOH, 24 h | 24, 25, 26 |
| 3-Nitro-4-hydroxymethyl benzoyl | $H_2O$–$Bu_4NF$, 40 min gives peptide acids; MeOH–KCN, 6 h gives peptidyl methyl esters; N-substituted amines in DCM, 15 h gives amides | 27 |
| 3-Carboxypropanesulphonamide (Ellman's 'safety-catch' linker) | (i) Iodoacetonitrile–DIPEA (20:5 eq.) in NMP, 24 h; followed by (ii) nucleophile (e.g. amines or alcohols, 2–5 eq.) in DCM, 12 h | 28 |

**Photolysis to peptide acids and amides (Photocleavable linkers)**

| | Cleavage conditions | Refs |
|---|---|---|
| $X^2$ = OH; NH-Fmoc 3-Nitro-4-hydroxymethyl benzoyl or 3-Nitro-4-*N*-Fmoc-aminomethyl benzoyl | h$\nu$ (350 nm), 12–24 h | 29 |
| $X^2$ = OH; NH-Fmoc $\alpha$-Methyl-6-nitro-veratrylamine (alcohol) based handles | Phosphate-buffered saline, pH 7.4 containing 5–50% DMSO, h$\nu$ (365 nm), 1–3 h (use of $NH_2NH_2$ as scavenger is recommended) | 30 |

[a] Recommended for routine synthesis of peptides.
[b] Typical recommended TFA-cocktails are: TFA–$H_2O$–EDT–triethylsilane/tri-isopropylsilane (90:6:3:1) or the TFA-reagent R (20) TFA–thioanisole–EDT–anisole (90:5:3:2). Under such conditions, global deprotection of acid-labile side-chain protecting groups is achieved.
[c] Prolonged treatment of benzylic resins ($X_1$ = $CH_2$) is not recommended.

**Table 4.** Common side-chain protecting groups used in Fmoc/tBu solid phase peptide synthesis

| Side-chain functionality of amino acids | Protecting groups | Abbreviations | Cleavage conditions | References |
|---|---|---|---|---|
| **Arg** | | Mtr | 90–95% v/v TFA, 4–6 h<br>TFA–anisole (9:1), 1 h | 31, 32<br>32 |
|  | | Pmc[a] | 50% v/v TFA in DCM, 1 h<br>TFA–anisole (9:1), 30 min<br>TFA–anisole–EDT–EMS (95:3:1:1), 1.5 h | 33, 34<br><br>34 |
|  | | Pbf | 95% v/v TFA, 30 min<br>(Deprotection rate 1.2–1.4 times faster than Pmc) | 35 |
| **Asp/Glu** | | O$^t$Bu[a] | 90% v/v TFA, 30 min | 36 |

Asn/Gln

| | | | |
|---|---|---|---|
| O1-Ada | TFA | 37, 38 |
| OAll [b] | Pd(Ph$_3$P)$_4$–AcOH–NMM Pd(Ph$_3$P)$_4$ (0.02 eq.)– PhSiH$_3$ (1.2–2 eq.) in DCM, 10–30 min | 39, 40 41 |
| ODmab [b] | 2% NH$_2$NH$_2$·H$_2$O in DMF, 5–10 min | 42 |
| OPip | 4% TFA in DCM, 15 min | 43 |
| Trt [a] | 90% v/v TFA, 30–60 min | 44 |
| Tmob | 90% v/v TFA, 1 h | 45, 46 |

**Table 4.** *Continued*

| Side-chain functionality of amino acids | Protecting groups | Abbreviations | Cleavage conditions | References |
|---|---|---|---|---|
| Cys ⌇SH | 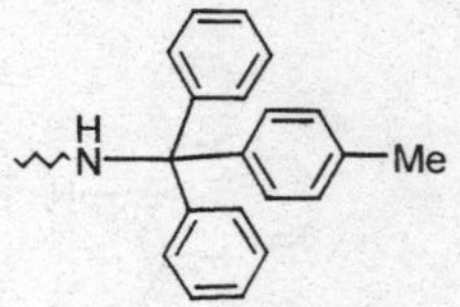 (trimethylphenylmethyl, N-linked) | Mtt[c] | 95% v/v TFA, 30 min | 47 |
| | 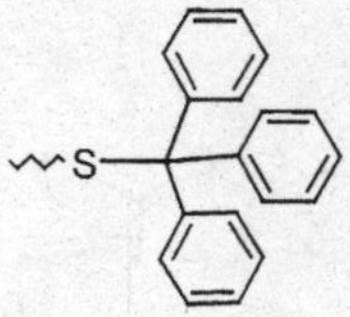 (triphenylmethyl, S-linked) | Trt[a] | 90% v/v TFA, 30 min | 48 |
| | 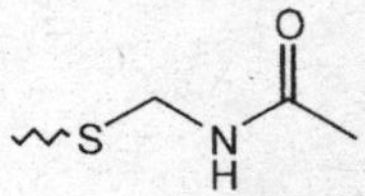 (acetamidomethyl) | Acm[b] | see Chapter 4 | 49, 50, 51 |
| | (S-tert-butyl) | S^tBu | see Chapter 4 | 52 |
| | 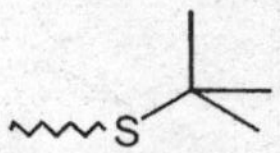 (tert-butyl) | ^tBu | see Chapter 4 | 53 |
| | 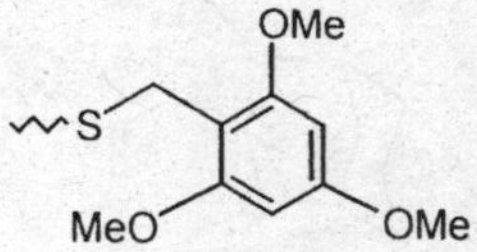 (trimethoxybenzyl) | Tmob | 5% TFA–3% TES in DCM | 54 |

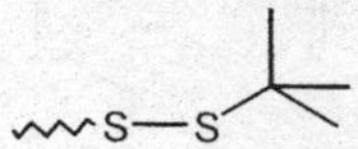

22

| | | | |
|---|---|---|---|
| His | Mmt | 0.5–1% TFA in DCM–TES (95:5), 30 min [d] 3% v/v TFA, 5–10 min [d] | 55 55 |
| | Trt [a] | 50% TFA in DCM, 30 min | 56 |
| | Boc | 90% v/v TFA, 30 min | 57 |
| | Bum [e] | 95% v/v TFA, 1–2 h | 58 |
| Lys/Orn | Boc [a] | 90% v/v TFA, 30 min | 59 |

23

**Table 4.** *Continued*

| Side-chain functionality of amino acids | Protecting groups | Abbreviations | Cleavage conditions | References |
|---|---|---|---|---|
| | 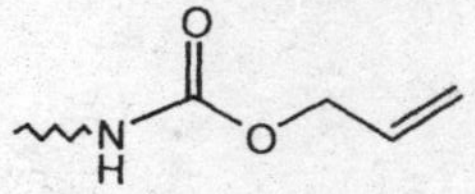 | Alloc[b] | Pd(Ph$_3$P)$_4$ (0.1 eq.)–PhSiH$_3$ (24 eq.) in DCM, 10 min | 41, 60 |
| | 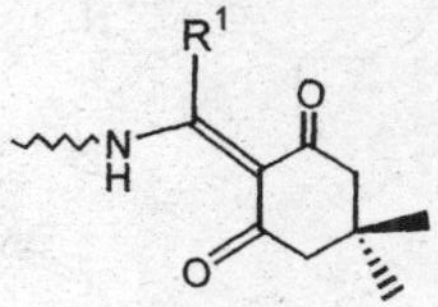 | Dde[b] (R$^1$ = Me) Ddiv[b,f] (R$^1$ = $^i$Bu) | 2% NH$_2$NH$_2$·H$_2$O in DMF, 5–10 min | 61, 62 |
| | | Mtt | 1% TFA in DCM, 30 min[d] AcOH–TFE–DCM (1:2:7), 1 h | 63 / 63 |
| Ser/Thr/Tyr | | $^t$Bu[a] | 90% v/v TFA, 30 min | 64, 65 |
| | 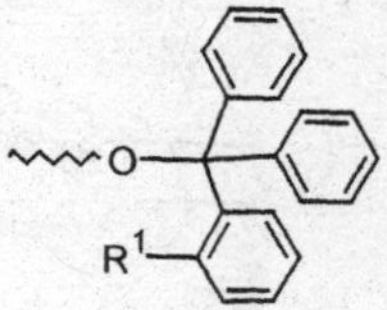 | Trt (R$^1$ = H) Clt (R$^1$ = Cl) for Tyr | 1–5% TFA in DCM, 2–5 min[d] | 66, 67 |

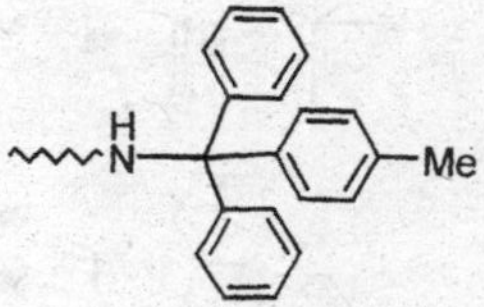

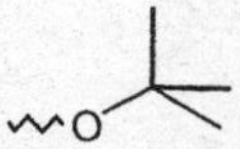

24

| | | |
|---|---|---|
| Tbdms | 68 | 0.1 M TBAF–DMF, 15 min (Tyr) |
| | 68 | TFA, 15 min (Ser/Thr) |
| Boc[a] | 69 | (i) 90% v/v TFA, 1 h; followed by (ii) 1% aq. TFA, 1–2 h |

Trp

[a] Recommended for routine synthesis of peptides.
[b] Stable to TFA.
[c] Strongly recommended when Asn or Gln resides at the N-terminus.
[d] Use a sufficient volume of the deprotection solution, in order to account for the buffering capacity of the peptide chain and solid support. It is generally recommended that up to 5% v/v TES or TIS should be included as a scavenger in the deprotection mixture.
[e] Guarantee no enantiomerization during carboxy-activation.
[f] 'Completely' stable to repeated treatment with 20% v/v piperidine in DMF.

recommended for the routine synthesis of peptide acids and carboxamides are highlighted.

## 3.3 Side-chain protecting groups

More than half of the amino acids commonly encountered in proteins have side-chains that contain reactive functional groups. In solid phase synthesis, it is usual for all of these potentially reactive groups to be masked because of the rather harsh conditions employed and the need to achieve the highest level of efficiency in all chemical reactions. For routine synthesis, protecting groups that are removed with TFA are usually employed as this allows the peptide to be globally deprotected at the same time as it is cleaved or released from the support. Furthermore, a wide range of groups is also available which can be selectively removed on the solid phase, thus enabling the selective modification of side-chains of individual residues within the peptide chain. These find application in the synthesis of cyclic peptides, phosphopeptides, and biotinylated peptides (see Chapter 6). *Table 4* lists the most commonly used side-chain protecting groups, together with the conditions required for their removal; those recommended for routine use are highlighted.

## 3.4 First residue attachment

The first step in the process of solid phase peptide synthesis is the attaching, or loading, of the resin linker with the C-terminal amino acid. The satisfactory execution of this process is particular important because, first, the extent of this reaction will determine the yield of the final product and, second, sites on the resin not reacted in this initial process can potentially be acylated in subsequent cycles, leading to the generation of related C-terminally truncated by-products. In the case of resins in which the anchorage point is a hydroxyl group, this process is often accompanied by enantiomerization, owing to the harshness of the conditions applied to effect this esterification. The problem is most serious with histidine and cysteine, and for these residues the use of trityl-based resins is recommended as these are loaded under conditions which do not cause loss of chiral integrity (Chapter 3, Section 5.2 and Chapter 4, Section 3.2.1).

Peptides containing proline or N-alkylated amino acids in the C-terminal dipeptide sequence present special problems because of the ease with which these dipeptides cyclize to give the corresponding diketopiperazine (*Figure 5*). This not only results in a reduction in yield of the desired product, but may also lead to the generation of truncated sequences through subsequent acylation of the regenerated starting resin. This side-reaction is most problematic with supports in which the dipeptide in anchored by a benzyl ester, and for this reason resins in which attachment is via a more hindered trityl ester should be used (Chapter 3, Section 5.2).

26

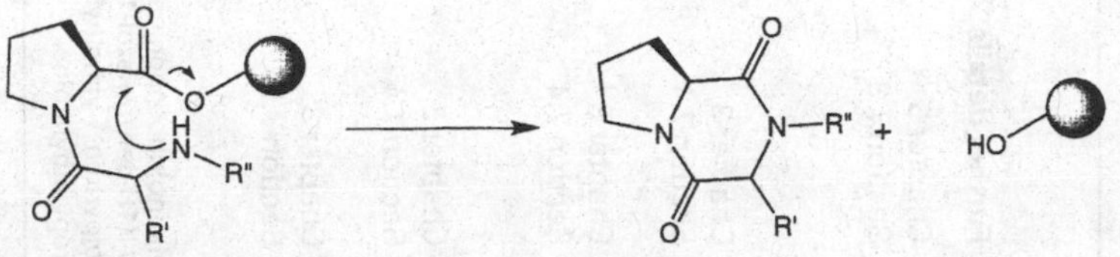

**Figure 5.** Diketopiperazine formation.

## 3.5 Coupling step

Most methods of amide bond formation involve chemical activation of the carboxy component. Those commonly employed in organic synthesis are generally regarded as too harsh to be used in peptide synthesis, leading to the formation of over-activated intermediates, which are unselective in their reactions and consequently prone to side-reactions. Peptide chemists have therefore sought milder activating methods, mostly based on the formation of active esters, pre-formed or generated *in situ*.

Those in most frequent use are listed in *Table 5*, together with cross-reference to the relevant protocols in later chapters.

## 3.6 *N*-Fmoc deprotection reaction

In solid phase synthesis, removal of the *N*-Fmoc group is usually achieved by treatment with 20–50% v/v piperidine in DMF. The mechanism of the Fmoc-deprotection reaction is shown in *Figure 6*. The key step is initial deprotonation of the fluorene ring to generate the aromatic cyclopentadiene-type intermediate. This rapidly eliminates to form dibenzofulvene, which is scavenged by piperidine to afford the adduct **5**. The products of the deprotection reaction absorb UV strongly, offering potential for monitoring of this reaction. In continuous-flow peptide synthesizers this is achieved by following the change in optical density of the column effluent with time. The curve shown in *Figure 7a* was obtained at 304 nm using a 0.1 mm path-length quartz flow-cell and is typical of that obtained during a deprotection reaction that follows normal reaction kinetics. For such reactions (providing flow rate and column volume are kept constant), the progress of peptide chain assembly can be assessed by comparing the area under consecutive curves. With sluggish deprotection reactions (*Figure 7b*), such as those encountered with aggregated sequences, the area under the deprotection is reduced and can no longer be meaningfully compared with that of the previous cycle.

In synthesizers operating in the batchwise mode of synthesis, monitoring of the deprotection reaction is carried out by taking aliquots and measuring the change over time of the optical density or conductivity of the sample.

Whilst deprotection with piperidine is effective in most cases, it has been shown that for long peptides incomplete Fmoc deprotection can occur even in

**Table 5.** Coupling methods used in Fmoc/tBu SPPS

| Coupling reagent | Additive | Active species | Conditions | Comments | Further details |
|---|---|---|---|---|---|
| DIC (or DCC) | | Symmetrical anhydride | Fmoc-amino acid/DIC (2:1) in DCM | Anhydride, generated in DCM, but used in DMF. Wastes 1 eq. of amino acid derivative. | Chapter 3, Section 7.3 |
| DIC (or DCC) | HOBt | Benzotriazolyl ester | Fmoc-amino acid/DIC/ HOBt (1:1:1) in DMF | Activation in DMF is slow | Chapter 3, Section 7.1 |
| PyBOP (or TBTU, HBTU) | HOBt | Benzotriazolyl ester | Fmoc-amino acid/PyBOP/ HOBt/DIPEA (1:1:1:2) in DMF | Most popular coupling method. Activation process extremely fast | Chapter 3 Section 7.4 |
| HATU | | 9-Azabenzotriazolyl ester | Fmoc-amino acid/HATU/ DIPEA (1:1:2) in DMF | Excellent method for difficult couplings | Chapter 3, Section 7.4 |
| TFFH | | Acid fluoride | Fmoc-amino acid/TFFH/ DIPEA (1:1:2) in DMF | Particularly useful for coupling N-alkyl and α-substituted residues | Chapter 3: Section 7.5 |

DIC, diisopropylcarbodiimide; DCC, dicyclohexylcarbodiimide; HOBt, 1-hydroxybenzotriazole; PyBOP, benzotriazol-1-yloxytris(pyrrolidino)phosphonium hexafluorophosphate; TBTU, N-[(1H-benzotriazol-1-yl)(dimethylamino)methylene]-N-methylmethanaminium tetrafluoroborate; HBTU, N-[(1H-benzotriazol-1-yl)(dimethylamino)methylene]-N-methylmethanaminium hexafluorophosphate N-oxide; HATU, N-[(dimethylamino)-1H-1,2,3-triazolo[4,5-b]pyridin-1-ylmethylene]-N-methylmethanaminium hexafluorophosphate N-oxide; TFFH, tetramethylfluoroformamidinium hexafluorophosphate; DIPEA, diisopropylethylamine.

28

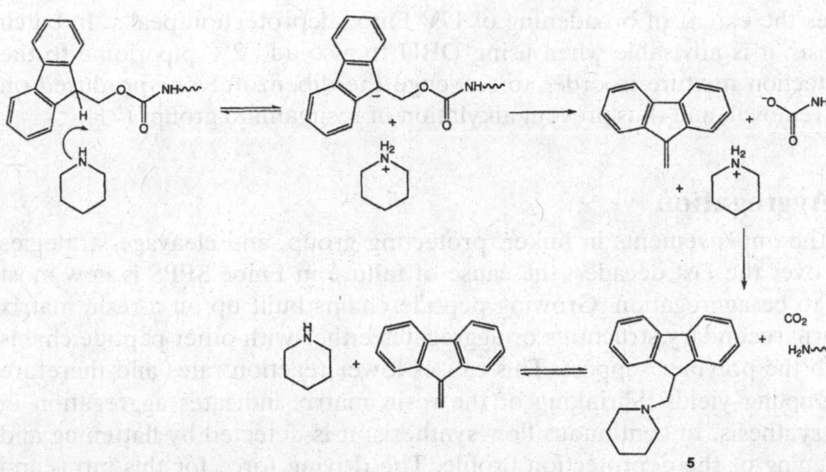

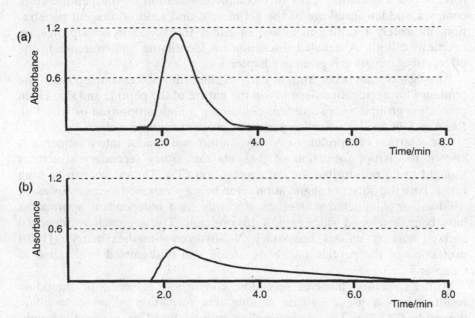

**Figure 6.** *N*-Fmoc removal reaction.

**Figure 7.** Fmoc deprotection profiles obtained from continuous-flow synthesis. (a) Typical Fmoc deprotection profile; (b) extended deprotection indicative of aggregation.

the presence of high concentrations of piperidine. In these cases, it is advisable to increase the time required for deprotection or to use a stronger base such as 1,8-diazabicyclo[5.4.0]undec-7-ene (DBU) (70). This tertiary base appears to be a very good alternative to piperidine since it causes rapid deprotection, less enantiomerization of resin-bound C-terminal Cys(Trt), and reduces the extent of broadening of UV Fmoc-deprotection peaks. In batch synthesis, it is advisable when using DBU to also add 2% piperidine to the deprotection mixture in order to scavenge the dibenzofulvene produced on Fmoc removal, and thus prevent alkylation of resin amino groups (71).

## 3.7 Aggregation

With the improvements in linker, protecting group, and cleavage strategies made over the last decades, the cause of failure in Fmoc SPPS is now most likely to be aggregation. Growing peptide chains built up on a resin matrix can form secondary structures or aggregates either with other peptide chains or with the polymer support. This causes lower reaction rates and therefore low coupling yields. Shrinking of the resin matrix indicates aggregation in batch synthesis. In continuous flow synthesis, it is detected by flattening and broadening of the deprotection profile. The driving force for this intra- and interchain association is thought to be hydrogen bonding and hydrophobic forces. This association leads to incomplete solvation of the peptide–resin complex, sudden shrinkage of the gel matrix, and reduced reagent penetration, ultimately resulting in failure of either the acylation or deprotection reaction or both. A detailed discussion on identifying and overcoming the effects of aggregation is given in Chapter 5.

Aggregation can occur from as early as the fifth residue coupled (72). The tendency for aggregation depends on the nature of the peptide and side-chain protecting groups, with sequences containing a high proportion of Ala, Val, Ile, Asn, Gln residues showing particular propensity for this effect.

The insertion of a proline or N-alkyl amino acid residue into a sequence is known to disrupt formation of β-sheets and other secondary structures thought to be responsible for the aggregation (73). The effects can be long range, with the outset of aggregation often being postponed for as many as six residues, or eliminated altogether. Recently, two independent approaches have been developed which exploit this principle. The approach by Sheppard and co-workers utilizes temporary N-2-hydroxy-4-methoxybenzyl (Hmb) protection of the peptide backbone amide and is discussed in detailed in Chapter 5.

Mutter's method involves reversible conversion of serine or threonine residues into a pseudoproline residue, via formation of an oxazolidine dipeptide (74, 75). The pseudoproline is introduced as a dipeptide unit, overcoming the problems normally associated with acylation of secondary amino acids. This has the added advantage of extending the peptide chain by two residues in one step. As the pseudoproline unit is stable to AcOH/trifluoroethanol/DCM, peptides prepared on extremely acid-labile resins, such as 2-chlorotrityl chloride, can be isolated with the pseudoproline moiety still in place. This can be advantageous when preparing peptides for use in fragment condensation reactions or when dealing with intractable

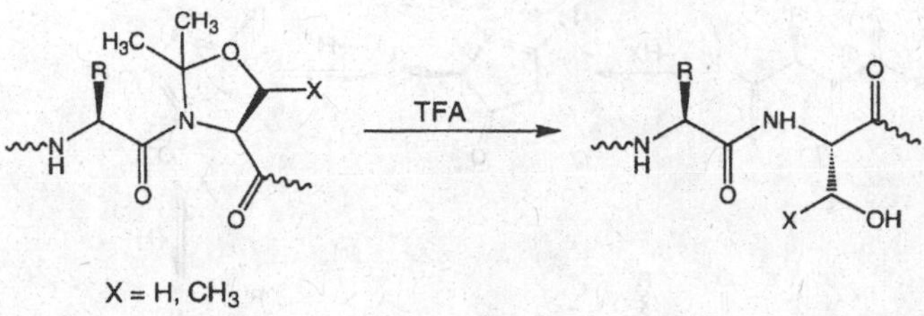

X = H, CH₃

**Figure 8.** Ring-opening of oxazolidine dipeptides.

sequences, since peptides containing pseudoproline residues often exhibit markedly improved solubility properties. Regeneration of the Thr or Ser residue from the oxazolidine can be effected by treatment with TFA in the normal manner (*Figure 8*).

## 3.8 Enantiomerization

Of the 20 amino acids that are commonly found in proteins, with the exception of glycine, all have a chiral centre of L-configuration at their α-carbon atoms, and two, isoleucine and threonine, also have a chiral centre in their side-chains. The biological properties of proteins and peptides are critically dependent on the configuration of the backbone chiral centres, so maintaining the integrity of these centres is of paramount importance in peptide synthesis.

In all 19 optically active amino acids, one of the substituents on the α-carbon is a potentially acidic hydrogen atom. Removal and subsequent reattachment of this proton represents a potential mechanism for enantiomerization of this chiral centre. Chiral integrity is particularly at risk during coupling as the acidity of this proton is greatly enhanced when the carboxy group is activated. In practice, direct enolization does not appears to be an important mechanism for enantiomerization except for amino acids such as phenylglycine, which offer an additional mode of enol stabilization.

The other mechanism by which enantiomerization occurs involves deprotonation and ring opening of an oxazolone intermediate, which is generated by attack on the activated carboxy group of the adjacent amide bond (*Figure 9*). Oxazolone formation occurs easily with carboxy-activated peptides, but can also occur with *N*-urethane-protected amino acids if the carboxy group is strongly activated. Oxazolones derived from urethane-protected amino acids are quite resistant to deprotonation and, in general, enantiomerization is rarely encountered in stepwise synthesis, except with the special cases of histidine and cysteine, and during attachment of the C-terminal residue to hydroxy-functionalized resins (Chapter 3, Section 5). In practice enantiomerization via oxazolone formation is mainly associated with the fragment condensation approach as this inevitably involves carboxy activation of peptide fragments. For this reason, fragments are normally

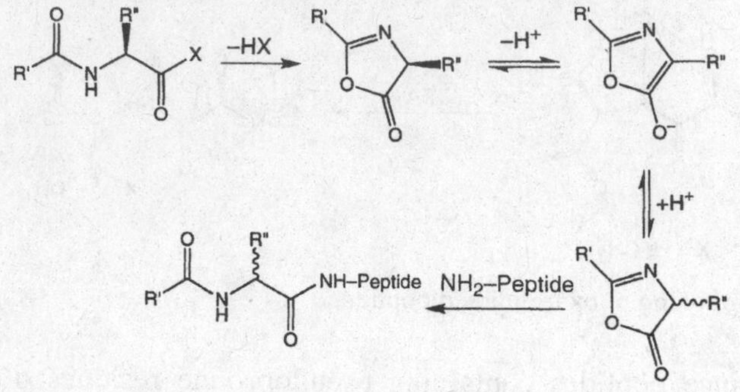

**Figure 9.** Enantiomerization via oxazolone formation.

joined at glycine, which is achiral, or proline, which is resistant to oxazolone formation.

Enantiomerization of cysteine can occur during coupling and during piperidine-mediated removal of $N^\alpha$-Fmoc, when the cysteine residue is anchored to the resin via an ester bond, and may involve a β-elimination-type mechanism; these problems are discussed in Chapter 4. With histidine, it is the proximity of the basic π-nitrogen of the imidazole side-chain to the α-hydrogen is thought to account for the facile enantiomerization of carboxy-activated histidine derivatives. This is not normally a problem during routine chain extension using Fmoc-His(Trt) derivatives, but may become an issue with aggregated sequences where slow reactions are encountered. In such cases, the use of derivatives in which the π-nitrogen is blocked, such as Fmoc-His(Bum)-OH (58), is recommended.

## 3.9 Side reactions

Newcomers to solid phase peptide synthesis reading any review on the subject may be forgiven for believing that the technique is fraught with difficulties and that it is virtually impossible to prepare any peptide without encountering major side-reactions. Side-reactions do certainly occur, but most are well documented and can be generally avoided by careful planning of the synthesis and by the appropriate selection of protecting groups and resin linker.

The side-reactions that can occur during chain assembly are listed in *Table 6*; those associated with the cleavage reaction are given in Chapter 3, *Table 5*. Aspartimide formation requires special mention as this is the side-reaction most likely to be encountered in routine synthesis, the others being normally only observed if the recommendations given in this and subsequent chapters are not followed. The reaction involves attack of the nitrogen attached to the α-carboxy group of aspartic acid or asparagine on the side-chain ester or amide group respectively, resulting in formation of a five-membered imide. This intermediate can suffer a number of fates: it can undergo ring opening

**Table 6.** Side-reactions that can occur during chain assembly

| Residue affected | Occurrence | Structure/s formed | Comments | Possible action |
|---|---|---|---|---|
| Arg | A | | Ornithine formation. Occurs when protonation or acyl-based groups are used for protection of the guanidine side-chain | Prevented through the use of sulphonyl-based protection, such as Pbf and Pmc |
| | A | | δ-Lactam formation. Results from attack of the Nᵟ-atom on the activated carboxy group. Most problematic with Arg PSAs | Addition of HOBt inhibits lactam formation. Slow reactions may need repeating with fresh reagents |
| Asn | A | | Cyanoalanine formation. Occurs during the activation of Asn when the side-chain carboxamide is not protected | Use pre-formed active esters, such as Fmoc-Asn-OPfp, or use a side-chain protected derivative, such as Fmoc-Asn(Trt)-OH |
| Asn | P | | Deamidation. Can occur in aqueous media, particularly under basic conditions, leading to formation of α- and β-aspartyl peptides. Most problematic with peptides containing Asn-Gly and Asn-Ser | Avoid prolonged exposure |

33

**Table 6.** *Continued*.

| Residue affected | Occurrence | Structure/s formed | Comments | Possible action |
|---|---|---|---|---|
| Asp | S | | Aspartimide formation. Can result in formation of α- and β-piperidides β-aspartyl peptides. Most problematic with Asp(OtBu)–X (X = Gly, Asn(Trt), Ser, Thr, Arg(Pmc) sequences | Incorporate a N-Hmb-protected amino acid prior to addition of Asp |
| Asp-Pro | P | | The Asp–Pro amide bond is slowly cleaved in acidic aqueous media | Avoid prolonged exposure |
| Cys | A | | Enantiomerization | Use a pre-formed OPfp or DIC/HOBt activation for addition of Cys (Chapter 4, Section 3.2.2) |
| Cys | C | | Enantiomerization. Occurs during attachment of Cys to hydroxy-functionalized resins and subsequent Fmoc removal | Use a trityl-based resin (Chapter 4, Section 3.2.1) |
| Cys | C | | Piperidinylalanine formation. Can occur with Cys attached to hydroxy-functionalized resins | Does not occur with Cys attached to amino-functionalized resins (Chapter 4, Section 3.2.3) |
| Glu | N | | Pyroglutamate formation. May occur during N-Fmoc removal with peptides containing N-terminal Glu, side-chain protected with benzyl-based groups | Keep deprotection time to a minimum |

34

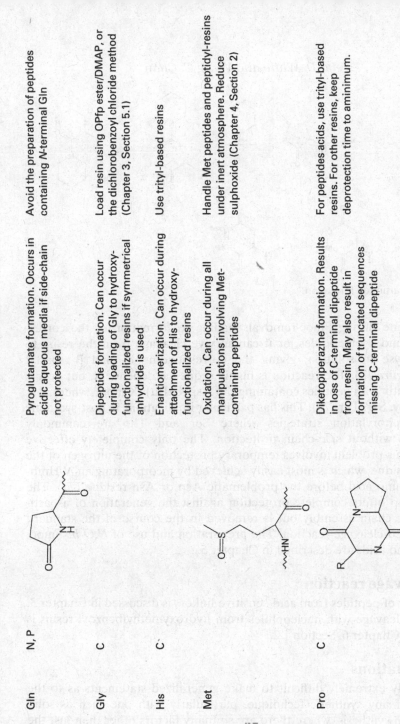

| | | | |
|---|---|---|---|
| Gln | N, P | Pyroglutamate formation. Occurs in acidic aqueous media if side-chain not protected | Avoid the preparation of peptides containing N-terminal Gln |
| Gly | C | Dipeptide formation. Can occur during loading of Gly to hydroxy-functionalized resins if symmetrical anhydride is used | Load resin using OPfp ester/DMAP, or the dichlorobenzoyl chloride method (Chapter 3, Section 5.1) |
| His | C | Enantiomerization. Can occur during attachment of His to hydroxy-functionalized resins | Use trityl-based resins |
| Met | | Oxidation. Can occur during all manipulations involving Met-containing peptides | Handle Met peptides and peptidyl-resins under inert atmosphere. Reduce sulphoxide (Chapter 4, Section 2) |
| Pro | C | Diketopiperazine formation. Results in loss of C-terminal dipeptide from resin. May also result in formation of truncated sequences missing C-terminal dipeptide | For peptides acids, use trityl-based resins. For other resins, keep deprotection time to a minimum. |

C, when located at C-terminus; N, when located at N-terminus; A, occurs during coupling; S, may occur during chain-assembly; P, post-synthetic modification. PSA, pre-formed symmetrical anhydride.

35

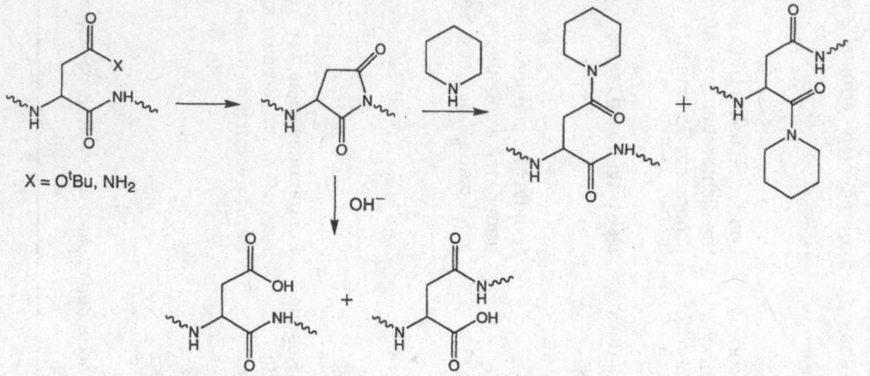

X = O$^t$Bu, NH$_2$

**Figure 10.** Aspartimide formation.

with piperidine during Fmoc-removal, leading to formation of the corresponding α- and β-piperidides, or it can survive cleavage from the resin, to later hydrolyse in solution, giving the corresponding α- and β-aspartyl peptides (*Figure 10*). The reaction is highly sequence dependent, but occurs most frequently with peptides containing the Asp(OtBu)–X motif, where X = Asn(Trt), Gly, Ser, Thr (76). This has particular implications in post-synthetic global phosphorylation strategies where Ser and Thr are commonly incorporated without side-chain protection. The only completely effective solution to this problem involves temporary protection of the nitrogen of the preceding residue, which is most easily achieved by incorporating an *N*-Hmb-protected amino acid before the problematic Asp or Asn residue (77). The *N*-Hmb group offers complete protection against the generation of aspartimides during chain assembly but is removed in the course of the standard TFA-mediated cleavage reaction. The preparation and use of *N,O-bis*Fmoc-*N*-Hmb-amino acids are described in Chapter 5.

## 3.10 Cleavage reaction

The cleavage of peptides from acid-sensitive linkers is discussed in Chapter 3, Section 10; cleavage with nucleophiles from hydroxymethylbenzoyl resins is dealt with in Chapter 6, Section 1.2.

## 3.11 Limitations

It is obviously extremely difficult to make generalized statements as to the limitations of any synthetic technique, particularly with one such as solid phase peptide synthesis, where there are so many factors other than just the success of chain assembly that can affect the overall overcome of the procedure. Nevertheless, it is not unreasonable to expect peptide sequences of up to 50 residues in length to be prepared in moderate to good purity by the standard methods, subject to the important caveat that aggregation during chain assembly did not occur.

However, as the chain length increases, separation of the target peptide from the accumulated by-products becomes increasingly difficult. With particularly long peptides, the small and almost undetectable levels of impurities in protected amino acid derivatives and solvents, together with obvious by-products arising from incomplete Fmoc-removal and coupling reactions, begin to have an ever-increasing impact on the outcome of the synthesis. For these reasons, methods of chemoselective ligation, a technique whereby smaller, more manageable peptides are coupled together in aqueous media, and chemoselective purification, a method whereby the target peptide can be chemically tagged to assist purification, have been increasingly exploited for the production of large peptides and small proteins by chemical synthesis. These powerful techniques are dealt with in detail in Chapters 11 and 12.

# References

1. Lloyd-Williams, P., Albericio, F., and Giralt, E. (1997). *Chemical approaches to the synthesis of peptide and proteins* (Series ed. C. W. Rees). CRC Press, New York.
2. Kent, S. B. H. (1988). *Ann. Rev. Biochem.*, **57**, 957.
3. Barany, G., Kneib-Cordnier, N., and Mullen, D. G. (1987). *Int. J. Peptide Protein Res.*, **1987**, 705.
4. Stewart, J. M. and Young, J. D. (ed.) (1984). *Solid phase peptide synthesis* (2nd edn). Pierce Chemical Company, Rockford.
5. Atherton, E., Fox, H., Harkiss, D., and Sheppard, R. C. (1978). *J. Chem. Soc., Chem. Commun.*, 539.
6. Fields, G. B. and Noble, R. L. (1990). *Int. J. Peptide Protein Res.*, **35**, 161.
7. Grant, G. (ed.) (1992). *Synthetic peptides*. W. H. Freeman & Co., New York.
8. Wang, S.-S. (1973). *J. Am. Chem. Soc.*, **95**, 1328.
9. Sheppard, R. C. and Williams, B. J. (1982). *Int. J. Peptide Protein Res.*, **20**, 451.
10. Rink, H. (1987). *Tetrahedron Lett.*, **28**, 3787.
11. Flörsheimer, A. and Riniker, B. (1991). *Peptides 1990: Proceedings of the 21st European Peptide Symposium* (ed. E. Giralt and D. Andreu), p. 131. ESCOM, Lieden.
12. Barlos, K., Gatos, D., Kallitsis, J., Papaphotiu, G., Sotiriu, P., Wenqing, Y., and Schafer, W. (1989). *Tetrahedron Lett.*, **30**, 3943.
13. Barlos, K., Gatos, D., Kapolos, S., Papaphotiu, G., Schafer, W., and Wenqing, Y. (1989). *Tetrahedron Lett.*, **30**, 3947.
14. Barlos, K., Chatzi, O., Gatos, D., and Stavropoulos, G. (1991). *Int. J. Peptide Protein Res.*, **37**, 513.
15. Bollhagen, R., Schmiedberger, M., Barlos, K., and Grell, E. (1994). *J. Chem. Soc., Chem. Commun.*, 2559.
16. Mergler, M., Nyfeler, R., Tanner, R., Gosteli, J., and Grogg, P. (1988). *Tetrahedron Lett.*, **29**, 4009.

17. Albericio, F. and Barany, G. (1991). *Tetrahedron Lett.*, **32**, 1015.
18. Bayer, E., Clausen, N., Goldammer, C., Henkel, B., Rapp, W., and Zhang, L. (1994). In *Peptides: chemistry, structure and biology* (ed. R. S. Hodges and J. A. Smith), p. 156. ESCOM, Leiden.
19. Bernatowicz, M. S., Daniels, S. B., and Köster, H. (1989). *Tetrahedron Lett.*, **30**, 4645.
20. Albericio, F., Kneib-Cordonier, N., Biancalana, S., Gera, L., Masada, R. I., Hudson, D., and Barany, G. (1990). *J. Org. Chem.*, **55**, 3730.
21. Sieber, P. (1987). *Tetrahedron Lett.*, **28**, 2107.
22. Chan, W. C., White, P. D., Beythien, J., and Steinauer, R. (1995). *J. Chem. Soc., Chem. Commun.*, 589.
23. Han, Y., Bontems, S. L., Hegges, P., Munson, M. C., Minor, C. A., Kates, S. A., Albericio, F., and Barany, G. (1996). *J. Org. Chem.*, **61**, 6326.
24. Bray, A. M., Valerio, R. M., and Maeji, N. J. (1993). *Tetrahedron Lett.*, **34**, 4411.
25. Bray, A. M., Jhingran, A.G., Valerio, R. M., and Maeji, N. J. (1994). *J. Org. Chem.*, **59**, 2197.
26. Atherton, E., Logan, C. J., and Sheppard, R. C. (1981). *J. Chem. Soc., Perkin Trans. 1*, 538.
27. Nicolás, E., Clemente, J., Perelló, M., Albericio, F., Pedroso, E., and Giralt, E. (1992). *Tetrahedron Lett.*, **33**, 2183.
28. Backes, B. J., Virgilio, A. A., and Ellman, J. A. (1996). *J. Am. Chem. Soc.*, **118**, 3055.
29. Hammer, R. P., Albericio, F., Gera, I., and Barany, G. (1990). *Int. J. Peptide Protein Res.*, **35**, 31.
30. Holmes, C. P. and Jones, D. G. (1995). *J. Org. Chem.*, **60**, 2318.
31. Atherton, E., Sheppard, R. C., and Wade, J. D. (1983). *J. Chem. Soc., Chem. Commun.*, 1060.
32. Fujino, M., Wakimsu, M., and Kitadu, C. (1981). *Chem. Pharm. Bull.*, **29**, 2825.
33. Ramage, R. and Green, J. (1987). *Tetrahedron Lett.*, **28**, 2287.
34. Ramage, R., Green, J., and Blake, A. J. (1991). *Tetrahedron*, **47**, 6353.
35. Carpino, L. A., Shroff, H., Triolo, S. A., Mansour, M. E., Wenschuch, H., and Albericio, F. (1993). *Tetrahedron Lett.*, **34**, 7829.
36. Schwyzer, R. and Dietrich, H. (1961). *Helv. Chim. Acta*, **44**, 2003.
37. Okada, Y., Iguchi, S., and Kawasaki, K. (1987). *J. Chem. Soc., Chem. Commun.*, 1532.
38. Okada, Y. and Iguchi, S. (1988). *J. Chem. Soc., Perkin Trans. 1*, 2129.
39. Trzeciak, A. and Bannwarth, W. (1992). *Tetrahedron Lett.*, **33**, 4557.
40. Kates, S. A., Solé, N. A., Johnson, C. R., Hudson, D., Barany, G., and Albericio, F. (1993). *Tetrahedron Lett.*, **34**, 1549.
41. Dessolin, M., Guillerez, M.-G., Thieriet, N., Guibé, F., and Loffet, A. (1995). *Tetrahedron Lett.*, **36**, 5741.
42. Chan, W. C., Bycroft, B. W., Evans, D. J., and White, P. D. (1995). *J. Chem. Soc., Chem. Commun.*, 2209.
43. Yue, C. W., Thierry, J., and Potier, P. (1993). *Tetrahedron Lett.*, **34**, 323.
44. Sieber, P. and Riniker, B. (1991). *Tetrahedron Lett.*, **32**, 739.
45. Wegand, F., Steglich, W., Bjarnason, J., Ahktar, R., and Chytil, N. (1968). *Chem. Ber.*, **101**, 3623.
46. Wegand, F., Steglich, W., and Bjarnason, J. (1968). *Chem. Ber.*, **101**, 3642.

47. Friede, M., Denery, S., Neimark, J., Kieffer, S., Gausepohl, H., and Briand, J. P. (1992). *Peptide Res.*, **5**, 145.
48. Hiskey, R. G., Mizoguchi, T., and Igeta, H. (1966). *J. Org. Chem.*, **31**, 1188.
49. Veber, D. F., Milkowski, J. D., Varga, S. L., Denkewalter, R. G., and Hirschmann, R. (1972). *J. Am. Chem. Soc.*, **94**, 5456.
50. Kamber, B. (1971). *Helv. Chim. Acta*, **54**, 927.
51. Fujii, N., Okata, A., Funakoshi, S., Bessho, K., and Yajima, H. (1987). *J. Chem. Soc., Chem. Commun.*, 163.
52. Weber, U. and Hartter, P. (1970). *Hoppe-Seyler's Zeitschrift für Physiologische Chemie*, **351**, 1384.
53. McCurdy, S. N. (1989). *Peptide Res.*, **2**, 147.
54. Munson, M. C., Garcia-Echeverria, C., Albericio, F., and Barany, G. (1992). *J. Org. Chem.*, **57**, 3013.
55. Barlos, K., Gatos, D., Hatzi, O., Koch, N., and Koutsogianni, S. (1996). *Int. J. Peptide Protein Res.*, **47**, 148.
56. Sieber, P. and Riniker, B. (1987). *Tetrahedron Lett.*, **28**, 6031.
57. Yamashiro, D., Blake, J., and Li, C. H. (1972). *J. Am. Chem. Soc.*, **94**, 2855.
58. Colombo, R., Colombo, F., and Jones, J. H. (1984). *J. Chem. Soc., Chem. Commun.*, 292.
59. Chang, C. D., Wakai, M., Ahmed, M., Meienhofer, J., Lundell, E. D., and Huang, J. D. (1980). *Int. J. Peptide Protein Res.*, **15**, 59.
60. Thieriet, N., Alsina, J., Giralt, E., Guibé, F., and Albericio, F. (1997). *Tetrahedron Lett.*, **38**, 7275.
61. Bycroft, B. W., Chan, W. C., Chhabra, S. R., and Hone, N. D. (1993). *J. Chem. Soc., Chem. Commun.*, 778.
62. Chhabra, S. R., Hothi, B., Evans, D. J., White, P. D., Bycroft, B. W., and Chan, W. C. (1998). *Tetrahedron Lett.*, **39**, 1603.
63. Aletras, A., Barlos, K., Gatos, D., Koutsogianni, S., and Mamos, P. (1995). *Int. J. Peptide Protein Res.*, **45**, 488.
64. Beyerman, H. C. and Bontekoe, J. S. (1961). *Proc. Chem. Soc.*, 249.
65. Callahan, F. M., Anderson, G. W., Paul, R., and Zimmermann, J. E. (1963). *J. Am. Chem. Soc.*, **85**, 201.
66. Barlos, K., Gatos, D., Koutsogianni, S., Schafer, W., Stavropoulos, G. and Yao, W. Q. (1991). *Tetrahedron Lett.*, **32**, 471.
67. Barlos, K., Gatos, D., and Koutsogianni, S. (1998). *J. Peptide Res.*, **51**, 194.
68. Fischer, P. M. (1992). *Tetrahedron Lett.*, **33**, 7605.
69. White, P. (1992). *Peptides: chemistry and biology* (ed. J. A. Smith and J. E. Rivier), p. 537. ESCOM, Leiden.
70. Wade, J. D., Bedford, J., Sheppard, R. C., and Tregear, G. W. (1991). *Peptide Res.*, **4**, 194.
71. Kates, S. A., Nuria, A., Solé, M., Beyermann, M., Barany, G., and Albericio, F. (1996). *Peptide Res.*, **9**, 106.
72. Bedford, J., Hyde, C., Johnson, T., Jun, W., Owen, D., Quibell, M., and Sheppard, R. C. (1992). *Int. J. Peptide Protein Res.*, **40**, 300.
73. Toniolo, C., Bonora, G. M., Mutter, M., and Pillai, V. N. R. (1981). *Makromol. Chem.*, **182**, 2007.
74. Mutter, M., Nefzi, A., Sato, T., Sun, X., Wahl, F., and Wöhr, T. (1995). *Peptide Res.*, **8**, 145.

75. Wöhr, T., Wahl, F., Nefzi, A., Rohwedder, B., Sato, T., Sun, X., and Mutter, M. (1996). *J. Am. Chem. Soc.*, **118**, 9218.

76. Lauer, J. L., Fields, C. G., and Fields, G. B. (1994). *Lett. Pept. Sci.*, **1**, 197.

77. Quibell, M., Owen, D., Packman, L. C., and Johnson, T. (1994). *J. Chem. Soc., Chem. Commun.*, 2343.

# 3

# Basic procedures

WENG C. CHAN and PETER D. WHITE

## 1. Introduction

A number of excellent descriptions of the techniques related to peptide chain assembly have already been published (1–3). These processes are also described in the operator manuals supplied by the peptide synthesis instrument manufacturers. Accordingly, the treatment of the subject presented here has been kept brief in order to provide more space in this volume for those topics not covered in detail in other publications of this type.

The protocols have been written as they would be carried out using a manual peptide synthesis vessel. Whilst it is appreciated that most scientists preparing peptides will be using automated peptide synthesizers, it is not possible, given the wide variation in operating procedures, to describe how such methods may be applied to individual instruments. Particular emphasis has been given here to those operations which are typically carried out off-instrument, such as first residue attachment and peptide-resin cleavage.

## 2. Manual synthesis

The operations described in this chapter can be carried out in a purpose-built peptide synthesis vessel or in a sintered glass funnel fitted with a three-way stopcock (*Figure 1*). The operation of the system is extremely simple: solvents are added from the top of the vessel, ensuring any resin adhering to the sides is rinsed down into the resin bed; the resin bed is agitated by setting the tap to position 1 to allow flow of nitrogen to the reaction vessel; solvents and reagents are removed by setting the tap to position 2 to connect the vessel to the vacuum. The use of such vessels has previously been described in detail (4).

## 3. Resin handling

Peptide synthesis resins are extremely fragile and the beads, if wrongly handled, can easily fracture, leading to the generation of fines which can block reaction vessel filter-frits and solvent lines.

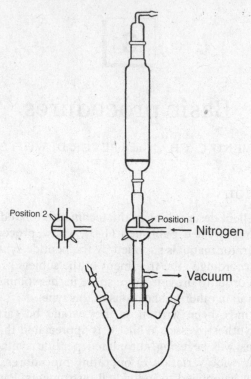

**Figure 1.** Manual peptide synthesis bubbler vessel.

It is particularly important that the correct method is used for mixing the resin and soluble reactants. Polystyrene-based supports are best agitated by bubbling an inert gas through the resin bed, or by shaking or vortexing the reaction vessel. Whilst all of these approaches are employed in commercial synthesizers, gas-bubbling and shaking are the most appropriate for use in manual synthesis. PEG–PS and PEGA resins once swollen are sufficiently rigid to be packed in columns and washed in a continuous-flow fashion by pumping liquid through the resin bed. They can also be agitated by gas-bubbling and shaking. Agitation of any resin with a magnetic stirrer is not recommended as the action results in attrition of the resin particles.

All resins must be swollen with an appropriate solvent before use, with the exception of PEGA resins which are supplied as a slurry in methanol or water. Underivatized polystyrene resins do not swell particularly well in dimethyl-formamide (DMF), hence the resin must be first swollen in dichloromethane (DCM). In contrast, PEG–PS resins can be packed dry into the reaction vessel and be swollen directly with DMF. In continuous-flow instruments, pumping the solvent up the reaction column is an effective way to remove air bubbles trapped in the resin bed.

Resins are normally washed between reaction steps with DMF as it is an excellent solvent both for protected amino acid derivatives and for most reagents used in peptide synthesis.

At the end of the synthesis or before long-term storage, resins must be shrunk and dried to remove all traces of solvent from within the resin particles as described in *Protocol 1C*. Many resins, when dry, adhere strongly to glass and plastic surfaces, making it difficult to remove the last traces of the material from the reaction vessel or filter funnel. Any resin remaining after removal of the bulk of the material can be washed with DMF down into the bottom corner of the filter-frit, washed with DCM, being careful not to spread the resin back up the sides of the vessel, excess DCM removed, and the damp resin removed with a spatula.

PEGA resins must always be handled in the swollen state as the resin beads have no structure when dry. Following chain assembly the resin can be washed successively with DMF and DCM, excess DCM removed, and the resin transferred to the cleavage reactor as a damp cake.

---

**Protocol 1.**   Basic resin-handling procedures

A. *Swelling (Polystyrene-based resins)*
1. Place the dry resin in a peptide synthesis reactor (*Figure 1*).
2. Add sufficient DCM to cover resin with three-times bed volume.
3. Shake reaction vessel gently to remove air bubbles and form a suspension of resin.
4. Leave for 30 min.
5. Apply vacuum to remove DCM.

B. *Washing*
1. Add three bed volumes of solvent.
2. Agitate for 1 min.
3. Apply vacuum to drain resin.
4. Repeat step 1 four times.

C. *Drying-down resin*
1. Wash resin successively (as described in B) with either: DMF, DCM, MeOH, and/or hexane (polystyrene-based resins); or DMF, isopropanol, hexane (loaded 2-chlorotrityl resins); or DMF, DCM, ether (PEG–PS resins).
2. Air-dry the resin by application of the vacuum for 10 min.
3. Transfer resin to a sample tube.
4. Dry *in vacuo* for 18 h.

---

# 4. Resin functionalization

Attachment of linkage agents bearing carboxyl functional groups to amino-functionalized resins can be effected using any of the standard peptide coupling methods described in *Section 7*. However, for linkers containing hydroxyl groups, such as hydroxymethylphenoxyacetyl (HMPA) and hydro-methylbenzoyl (HMBA), the use of coupling methods which do not involve the addition of a tertiary base are recommended if self-acylation is to be avoided. An unusual amino acid such as norleucine inserted between the resin and the linker can serve as an internal standard, enabling attachment and cleavage yields to be easily determined by amino acid analysis following acid hydrolysis. A protocol for loading amino-functionalized resin with HMBA linker is provided in Chapter 6, Section 1.2.2.

Polystyrene, PEG–PS, and PEGA-based resins pre-functionalized with all the standard linkers required for the production of peptide acids, peptides amides, and protected peptide fragments are commercially available.

# 5. Attachment of the first residue

Anchoring of the first residue to the synthesis support is one of the most critical steps in peptide synthesis; an injudicious combination of linker, loading method, or amino acid derivative can have a highly deleterious effects on the purity of the final product, and the use of poorly loaded resins is simply just a waste of expensive reagents.

The properties of the various linker-derivatized resins used in Fmoc SPPS have already been discussed and were summarized in Chapter 2, *Table 2*. For the purpose of selecting the appropriate procedure for attachment of the first residue, these can be conveniently separated into three categories on the basis of their reactive functional groups (hydroxymethyl-, trityl chloride-, and aminomethyl-based resins); the mostly commonly used are listed in *Table 1*.

## 5.1 Hydroxymethyl-based resins

The protocols described in this section are applicable to all linker-derivatized resins in which loading involves esterification of the first amino acid to a linker hydroxyl group. The ease of esterification with Fmoc-amino acids increases in the order $5 < 3 \approx 4 < 6 < 1 \approx 2$.

Formation of ester bonds is considerably more difficult than that of amide bonds; harsher conditions than those normally used for chain elongation must be employed which, unless carried out under controlled conditions, can lead to low substitution, enantiomerization, and dipeptide formation. To ensure satisfactory results when using these protocols it is important that the following precautions are observed:

(a) All reagents and glassware must be dried thoroughly before use as the presence of moisture can severely affect loading efficiencies.

**Table 1.** Commonly used resins for SPPS

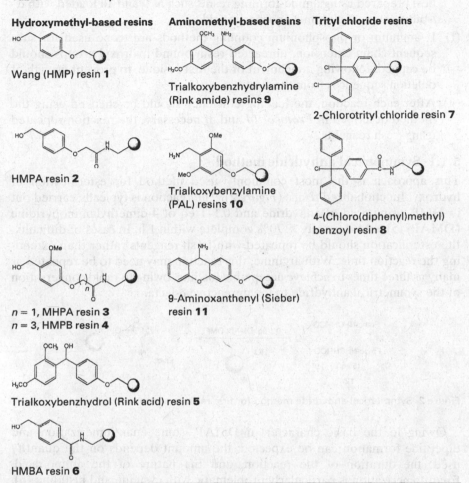

**Hydroxymethyl-based resins**

Wang (HMP) resin **1**

HMPA resin **2**

$n = 1$, MHPA resin **3**
$n = 3$, HMPB resin **4**

Trialkoxybenzhydrol (Rink acid) resin **5**

HMBA resin **6**

**Aminomethyl-based resins**

Trialkoxybenzhydrylamine (Rink amide) resins **9**

Trialkoxybenzylamine (PAL) resins **10**

9-Aminoxanthenyl (Sieber) resin **11**

**Trityl chloride resins**

2-Chlorotrityl chloride resin **7**

4-(Chloro(diphenyl)methyl) benzoyl resin **8**

(b) Amino acids containing water of recrystallization should be dried either by dissolving in ethyl acetate, drying the solution over anhydrous magnesium sulphate, and isolating by evaporation; or by repeated evaporation from dioxane.

(c) Reactions should be conducted using the minimum volumes of solvent, to ensure maximum reagent concentration.

(d) For the synthesis of peptide acids containing C-terminal cysteine, histidine, proline, methionine and tryptophan residues, it is better to use trityl-based resins, such as 2-chlorotrityl resin; for details see Section 5.2. Alternative protocols for attachment of Cys are also provided in Chapter 4, Section 3.2.1.

(e) Peptides containing C-terminal glutamine and asparagine residues are best prepared using amide-forming resins such as **9** and **10** loaded with α-*t*-butyl esters of glutamic or aspartic acid.

(f) If aminium or phosphonium coupling methods are to be used for subsequent chain extension, unreacted resin-bound hydroxyl groups should be capped, following attachment of the first residue, to prevent growth of deletion sequences from these sites.

(g) After each reaction, the loading efficiency should be checked using the method described in *Protocol 14* and, if necessary, the reaction repeated using fresh reagents.

### 5.1.1 Symmetrical anhydride method

This approach is the most commonly used method for esterification of hydroxyl-functionalized resins (*Figure 2*). The reaction is typically carried out in DMF, using 5 eq of anhydride and 0.1–1 eq of 4-dimethylaminopyridine (DMAP) (5), and is usually > 70% complete within 1 h. In cases of difficulty, the esterification should be repeated with fresh reagents rather than extending the reaction time. With arginine, this reaction may need to be repeated as many as three times to achieve acceptable loading, owing to rapid conversation of the symmetrical anhydride to the unreactive δ-lactam.

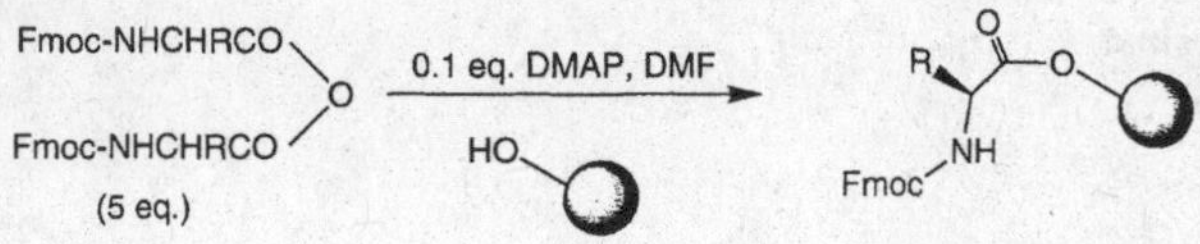

**Figure 2.** Symmetrical anhydride method for first residue attachment.

Owing to the basic character of DMAP, some enantiomerization and dipeptide formation can be expected; the amount depends on the quantity used, the duration of the reaction, and the nature of the amino acid. Enantiomerization is particularly problematic with cysteine and histidine (6). For most other amino acids, however, the levels of enantiomerization can be generally controlled to within acceptable limits by keeping the amount of DMAP used to a minimum. Dipeptide formation can be a problem with this method, particularly during the loading of glycine (7).

---

**Protocol 2.** Attachment to alcohol-based resins using symmetrical anhydrides

*Reagents*
- Fmoc-amino acid derivatives
- DIC
- Hydroxyl-functionalized peptide synthesis resin
- Dry, acid-free DCM
- DMF
- DMAP

---

*Method*

1. Place the resin in a dry reaction vessel.

2. Swell and wash with DMF as described in *Protocol 1.*

3. Prepare the appropriate Fmoc-amino acid anhydride (5 eq)[a] according to *Protocol 9.* Dissolve in minimum volume of DMF and add to resin. If necessary add extra DMF to ensure complete coverage of the resin bed.

4. Dissolve DMAP (0.1 eq)[a] in DMF, and add this solution to the resin/ amino acid mixture. Stopper the vessel and allow the mixture to agitate at room temperature for 1 h.

5. Wash the resin with DMF (5 times).

6. Transfer 10 mg of resin with a wide-mouth Pasteur pipette to a small sintered glass funnel. Wash and dry-down as described in *Protocol 1C.* Determine the extent of loading using *Protocol 14.* If the loading is below 70%, repeat steps 2–5.

7. Add benzoic anhydride (5 eq)[a] and pyridine (1 eq)[a] in DMF, and agitate for 30 min.

8. Wash and shrink-down the resin as described in *Protocol 1C.*

[a] Relative to resin substitution level (loading).

## 5.1.2 Dichlorobenzoyl chloride method

This method, although requiring longer reaction times than the symmetrical method, results in products virtually free of impurities arising from enantiomerization and dipeptide formation (6). It involves formation of a mixed anhydride between an Fmoc-amino acid derivative and 2,6-dichlorobenzoyl chloride (DCB) in DMF/pyridine (*Figure 3*). This anhydride effects esterification of resin-bound hydroxyl groups in yields of typically > 70% in 18 h. Levels of D-isomer and dipeptide formation for most amino acids are typically less than 1%, with *S*-acetamidomethyl cysteine giving the most enantiomerization (6.5% D-isomer) of the amino acids tested (8).

Repeating this reaction with fresh reagents appears to result in little improvement in yield. Sites not acylated by Fmoc-amino acid are apparently capped, presumably with 2,6-benzoyl groups. For this reason, it is not essential to cap following this reaction.

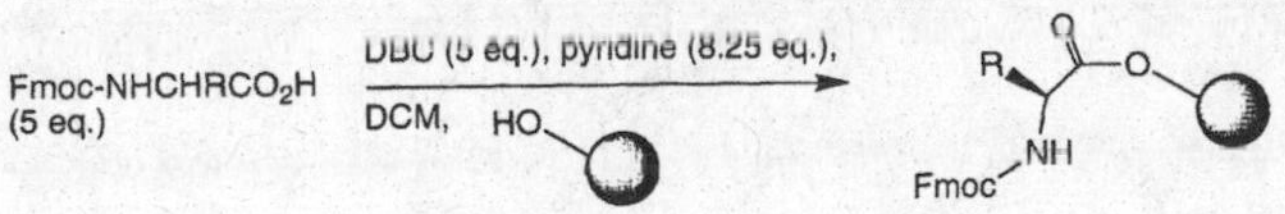

Fmoc-NHCHRCO₂H (5 eq.) — DBU (5 eq.), pyridine (8.25 eq.), DCM, HO—

**Figure 3.** DCB method for first residue attachment.

---

**Protocol 3.** DCB method

*Reagents*
- Fmoc-amino acid derivatives
- DCB
- Hydroxyl-functionalized peptide synthesis resin
- DMF
- Pyridine

*Method*

1. Place the resin in a dry reaction vessel.

2. Swell and wash with DMF, as described in *Protocol 1*. Add extra DMF if necessary to ensure complete coverage of the resin.

3. Add the appropriate Fmoc-amino acid (5 eq)[a], followed by pyridine (8.25 eq)[a]. Gently agitate until the amino acid has dissolved completely.

4. Add DCB (5 eq)[a] and agitate the mixture gently for 18 h.

5. Wash and shrink-down the resin as described in *Protocol 1C*.

[a] Relative to resin loading.

---

### 5.1.3 MSNT/MeIm method

Originally developed for DNA synthesis, this method has in recent years been increasingly used in peptide synthesis for the esterification of protected amino acids to solid supports (9). It has proved particularly effective in loading Fmoc-amino acids to the relatively unreactive hydroxyl group of HMBA-derivatized resins (10) and in situations where enantiomerization is a problem (11). The reaction involves treating the protected amino acid in DCM with 1-(mesitylene-2-sulphonyl)-3-nitro-*1H*-1,2,4-triazole (MSNT) **13** and 1-methyl-imidazole (MeIm) (*Figure 4*). The process is extremely moisture sensitive and so it is essential that all glassware and reagents are dried thoroughly before use.

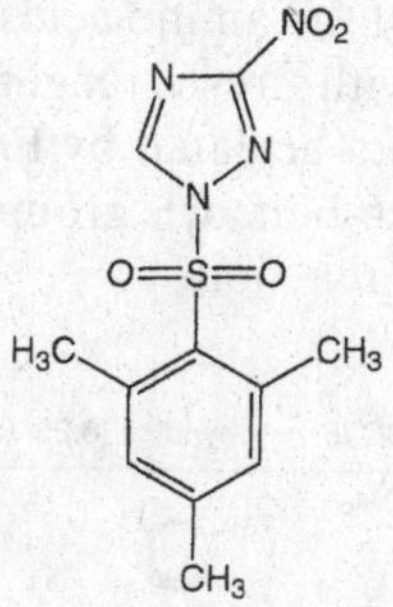

**13**

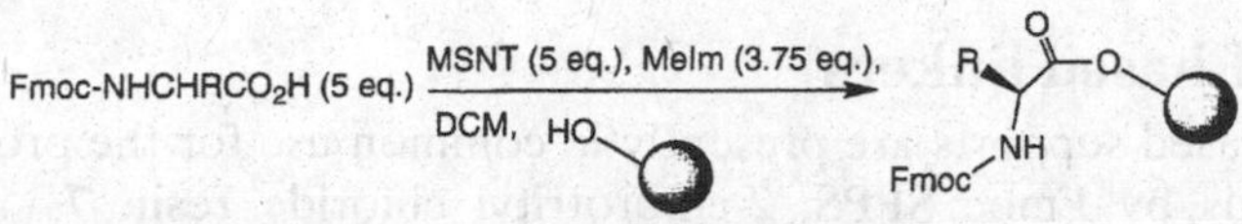

Fmoc-NHCHRCO$_2$H (5 eq.)   —MSNT (5 eq.), MeIm (3.75 eq.), DCM, HO→

**Figure 4.** MSNT/MeIm method for first residue attachment.

---

## Protocol 4.   MSNT/MeIm method

### Reagents

- Fmoc-amino acid derivatives
- MSNT
- Dry THF
- Hydroxyl-functionalized peptide synthesis resin
- Dry, acid-free DCM
- MeIm
- DMF

### Method

1. Place the resin in a dry reaction vessel.

2. Swell and wash with DCM as described in *Protocol 1*. Add sufficient DCM to cover resin and flush vessel with nitrogen. Seal with a septum.

3. Weigh the appropriate Fmoc-amino acid (5 eq) into a dry, round-bottomed flask, equipped with a magnetic stirrer. Add dry DCM to dissolve the amino acid derivative (approximately 3 ml/mmol). One or two drops of THF can be added to aid complete dissolution.

4. Add MeIm (3.75 eq) followed by MSNT (5 eq). Flush flask with nitrogen and seal with a septum, then stir the mixture until the MSNT has dissolved.

5. Using a syringe, transfer the amino acid solution to the vessel containing the resin.

6. Allow the mixture to stand at room temperature for 1 h, with gentle agitation.

7. Wash with DCM (5 times) and DMF (5 times).

8. Transfer 10 mg of resin with a wide-mouthed Pasteur pipette to a small, sintered glass funnel. Wash and dry-down as described in *Protocol 1C*. Determine the extent of loading using *Protocol 14*. If the loading is below 70%, repeat steps 2–7.

9. Add benzoic anhydride (5 eq) and pyridine (1 eq) in DMF and agitate gently for 30 min.

10. Wash and dry-down the resin as described in *Protocol 1C*.

## 5.2 Trityl-based linkers

Two trityl-based supports are presently in common use for the production of peptide acids by Fmoc SPPS: 2-chlorotrityl chloride resin **7** (12) and 4-(chloro(diphenyl)methyl)benzoyl resin **8** (13).

These resins are extremely moisture sensitive, and so it is essential that all reagents and glassware should be dried thoroughly before use. 2-Chlorotrityl chloride resin can be stored desiccated at room temperature; 4-(chloro-(diphenyl)methyl)benzoyl resins should be generated from the precursor trityl alcohol immediately before use as described in *Protocol 5*. The chemical characteristics of **7** and **8** are identical.

2-Chlorotrityl chloride resin is normally supplied with a displaceable chlorine content of 1.0–1.6 mmol/g. For the purposes of peptide synthesis, this substitution can sometimes be too high and can be reduced by treating the resin with a sub-stoichiometric amount of amino acid derivative and then capping unreacted sites with MeOH.

Since coupling to trityl chloride-type resins does not involve activation of the incoming amino acid derivative the process is free from enantiomerization and dipeptide formation (14), making these supports ideal tools for the preparation of peptides containing C-terminal histidine or cysteine residues. Their use is also recommended for the synthesis of peptides containing C-terminal proline (15), methionine, or tryptophan residues (14). Proline-containing dipeptide attached to standard benzyl alcohol-based supports, such as **1** and **6**, can undergo diketopiperazine formation with concomitant loss of peptide chains during Fmoc group removal. Methionine and typtophan residues can be alkylated by the cations generated from a benzyl-based linker during the TFA-mediated cleavage reaction, leading to reattachment of the peptide to the solid support with a corresponding reduction in yield. The extent of both of these side-reactions appears to be significantly reduced when trityl-based resins are used, a fact which has been attributed to the bulk of the linker group.

---

**Protocol 5.** Chlorination of trityl alcohol resins

*Reagents*
- Dry toluene
- 4-(Hydroxy(diphenyl)methyl)benzoyl aminomethyl resin
- Freshly distilled acetyl chloride (AcCl)
- Dry DCM

*Method*
1. Place the resin in a sintered glass funnel.
2. Swell and wash with DCM as described in *Protocol 1*.
3. Wash with toluene (5 times).

---

4. Transfer resin as a damp cake to a round-bottomed flask. Add sufficient toluene to cover resin bed.
5. Add AcCl (1 ml/g resin). Fit flask with reflux condenser and CaCl₂ guard tube.
6. Heat mixture at 60°C for 3 h.
7. Allow to cool to room temperature.
8. Slurry resin with toluene into a dry reaction vessel and wash with DCM.
9. Load immediately as described in Chapter 9, *Protocol 1*.

## 5.3 Aminomethyl-based linkers

For the addition of the first amino acid to supports derivatized with peptide amide-forming linkers, such as trialkoxybenzhydrylamine **9**, trialkoxybenzyl-amine **10**, and aminoxanthenyl **11** resins, any of the standard peptide coupling methods described in *Section 7* can be used. Since basic reagents such as 4-dimethylaminopyridine are not required to effect this transformation, there is no risk of enantiomerization or dipeptide formation. Many resins of this type are supplied Fmoc-protected and these should be treated with piperidine prior to the addition of the first amino acid. With automated peptide synthesizers, Fmoc removal and loading with the C-terminal residue can be programmed to occur as part of the normal coupling cycle.

The addition of the first residue to dimethoxybenzhydrylamine-based resins can be sluggish, particularly the β-branched amino acids, such as valine and isoleucine. Repeating the coupling after 1 h with fresh reagents generally leads to complete resin derivatization.

Coupling through the side-chains of the α-*t*-butyl esters of glutamatic acid or aspartic acid to amide-forming linkages provides a facile route to C-terminal glutamine- and asparagine-containing peptides (16, 17).

## 6. Fmoc removal

Removal of the Fmoc group from the N-terminus of the resin-bound peptide chain is normally achieved by treating the peptidyl resin with 20–50% piperidine in DMF. The reaction is typically complete within 4–10 min, depending on the nature of the peptide being synthesized. With peptides containing aspartic acid and asparagine, inclusion of 0.1 M HOBt in the deprotection mixture has been found to be partially effective in suppressing aspartimide formation (18).

In the case of sequences which have become aggregated the standard treatment with piperidine in DMF may not always be effective. Even extending the time of deprotection reaction considerably does not always result in complete removal of Fmoc groups; in such cases, the use of the non-nucleophilic base 1,8-diazabicyclo[5.4.0]undec-7-ene (DBU) is recommended (19). It is normally employed in combination with piperidine as a solution in DMF (20)—piperidine is included to scavenge the dibenzofulvene that is formed on Fmoc removal, thus preventing alkylating of the newly liberated amino group. The efficacy of this reagent has been demonstrated in the synthesis of the notoriously difficult polyalanine sequence, H-(Ala)$_{10}$-Val-OH (20). However, care should be exercised with peptides containing aspartic acid and asparagine residues as it has been found to promote aspartimide formation (18, 21).

---

**Protocol 6.**  Manual removal of Fmoc groups

*Reagents*
• 20% Piperidine in DMF (v/v) or DBU/piperidine/DMF (1:1:48 v/v)

*Method*
1. Place the peptidyl resin in a sintered glass funnel or manual SPPS reaction vessel.
2. If the resin is dry, swell the resin as described in *Protocol 1A*.
3. Wash resin with DMF (5 times).
4. Add deprotection reagent to cover resin.
5. Agitate gently for 2 min. Drain off reagent.
6. Repeat steps 4–5 twice.
7. Wash resin with DMF (5 times).

---

# 7. Coupling methods

The stepwise introduction of N$^\alpha$-protected amino acids in solid phase synthesis normally involves *in situ* carboxy activation of the incoming amino acid or the use of pre-formed activated amino acid derivatives. In order to drive the acylation to completion an excess of activated amino acid derivative is utilized, typically 2–10 times the resin functionality. The excess used depends on the nature of the activated species and on the resin loading and void volume of the reaction vessel employed. The most important consideration is to maintain as high an effective concentration of reagents as possible. In small-scale synthesis and with low-functionality resins, large excesses may need to be used as the dead volume of such systems tends to be quite high; conversely with resins of high functionality lower excesses can be utilized.

The coupling time required to effect complete acylation depends on the nature of the active species and the peptide sequence being assembled and on the concentration of the reagents and resin functionality. For highly reactive species, such as those generated *in situ* using activating reagents like TFFH **14** (22), PyBOP **15** (23), TBTU **16** (24), and HATU **17** (25), 30 min is usually sufficient. However, when coupling together hindered residues, such as β-branched amino acids or N-substituted amino acids, or when preparing peptides that have become aggregated, reaction time may need to be increased considerably. In manual synthesis, it is standard practice to check the completeness of the coupling reaction before proceeding to the next step in the synthesis cycle. For this purpose a number of sensitive tests have been developed to detect the presence of residual amine on the solid support (*Protocol 13*).

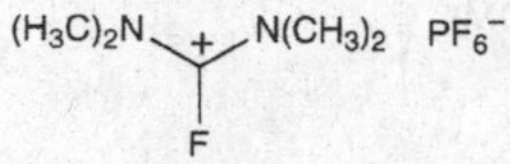

**14**

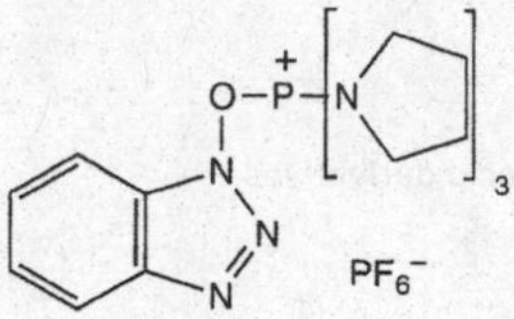

**15**

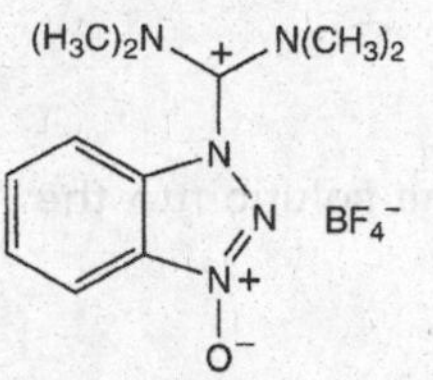

**16**

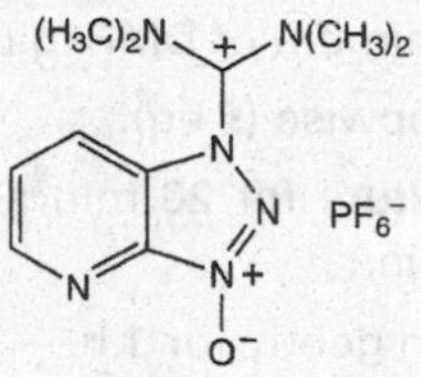

**17**

With peptides that have become aggregated it is frequently not possible to obtain complete reaction, even after carrying out multiple acylation reactions. In such cases it is often necessary to repeat the synthesis, incorporating an Hmb-protected residue (Chapter 5) or an oxazolidine-dipeptide (26–28) two to four residues before the region prone to aggregation.

The order of efficiency of the methods described below is: OPfp esters/HOBt < DIC/HOBt < HBTU ≈ PyBOP < HATU. The reader is directed to a recent discussion on coupling reagents by Albericio and Carpino (29).

## 7.1 DIC/HOBt

The reaction of an Fmoc-amino acid with an equivalent amount of DIC and HOBt results in formation of the corresponding OBt ester **18**. The reaction is fastest in non-polar solvents like DCM (30) but also works well in DMF, provided the mixture is allowed to pre-activate before being added to the resin. The use of DIC is preferred to that of dicyclohexylcarbodiimide (DCC) because the urea generated by the former is soluble in DMF. This procedure is also effective for the generation of OAt esters **19** (25).

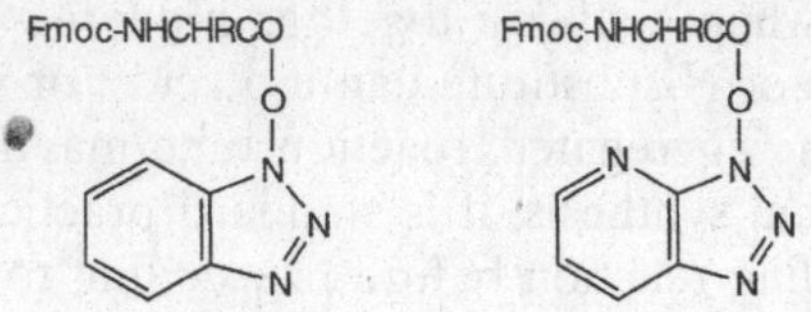

Fmoc-NHCHRCO

**18**

Fmoc-NHCHRCO

**19**

---

**Protocol 7.** Synthesis of OXt esters

*Reagents*
- Fmoc-amino acid derivatives
- DIC
- HOXt (HOBt or HOAt)
- DMF

*Method*

1. Place the appropriate Fmoc-amino acid (5 eq) and HOXt (5 eq)[a] in a dry round-bottomed flask or sample vial, equipped with a magnetic stirrer.
2. Add the minimum of DMF to dissolve.
3. Add DIC dropwise (5 eq).
4. Stir the mixture for 20 min and add the solution to the N-deblocked peptidyl resin.
5. Agitate resin gently for 1 h.
6. Transfer 10 mg of resin with a wide-mouthed Pasteur pipette to a small, sintered glass funnel. Wash and dry-down as described in *Protocol 1C*. Perform Kaiser or TNBS test (*Protocol 13*). If positive, repeat steps 5 and 6.
7. If resin still gives a positive colour test after 4 h, wash resin with DMF (5 times) and repeat coupling reaction with fresh reagents.
8. Collect the derivatized resin as outlined previously.

[a] Relative to resin functionality.

---

## 7.2 Active esters

Of the various active esters that have been exploited in solution phase peptide synthesis, only the pentafluorophenyl esters (OPfp) **20** have found wide application in Fmoc/tBu solid phase synthesis. They are efficient acylation agents, particularly when used in conjunction with an equivalent of HOBt, and their use is generally regarded to be free from side-reactions (31). Derivatives are commercially available for most proteinous amino acids.

---

**Protocol 8.** OPfp esters

*Reagents*
- Fmoc-amino acid pentafluorophenyl ester
- DMF
- HOBt

*Method*

1. Weigh the appropriate Fmoc-amino acid pentafluorophenyl ester (5 eq)[a] and HOBt (5 eq) in a dried sample vial.

2. Add the minimum volume of DMF to dissolve.

3. Add the solution to the N-protected peptidyl resin.

4. Agitate resin gently for 1 h.

5. Transfer 10 mg of resin with a wide-mouthed Pasteur pipette to a small, sintered glass funnel. Wash and dry-down as described in *Protocol 1C*. Perform Kaiser or TNBS test (*Protocol 13*). If positive, repeat steps 4 and 5.

6. If resin still gives a positive colour test after 4 h, wash resin with DMF (5 times) and repeat coupling reaction with fresh reagents.

7. Collect the derivatized resin as outlined previously.

[a] Relative to resin functionality.

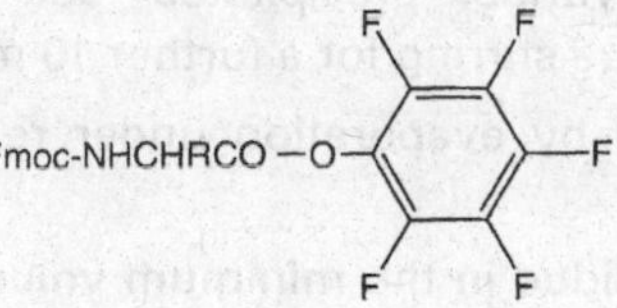

---

Fmoc-NHCHRCO—O— (pentafluorophenyl ring with F, F, F, F, F)

**20**

# 7.3 Symmetrical anhydride

Reaction of an Fmoc-amino acid with 0.5 eq of a carbodiimide in DCM generates the corresponding symmetrical anhydride **12** (*Figure 5*). If the symmetrical anhydride is to be used directly in solid phase synthesis, DIC should be used as the dehydrating reagent.

Some Fmoc-amino acids are not readily soluble in DCM and will require the addition of DMF to assist complete dissolution. Many Fmoc-amino anhydrides are also poorly soluble in DCM and will precipitate from the reaction mixture as they are formed. However, this is not a problem, since they re-dissolve in

$$\text{Fmoc-NHCHRCO}_2\text{H (2 eq.)} \xrightarrow{\text{DIC (1 eq.), DCM}} \begin{array}{c} \text{Fmoc-NHCHRCO} \\ \text{O} \quad \text{(1 eq.)} \\ \text{Fmoc-NHCHRCO} \end{array}$$

**12**

**Figure 5.** Formation of symmetrical anhydride.

DMF following removal of DCM by evaporation. This procedure is not appropriate for activation of asparagine residues with unprotected amide side-chains as this can lead to cyanoalanine formation (32).

---

### Protocol 9. Symmetrical anhydrides

*Reagents*
- Fmoc-amino acid derivatives
- DIC
- Dry, acid-free DCM
- DMF

*Method*

1. Place the appropriate Fmoc-amino acid (10 eq)[a] in a dry, round-bottomed flask, equipped with a magnetic stirrer.

2. Add dry DCM to dissolve the amino acid derivative, using approximately 3 ml/mmol. One or two drops of DMF may be needed to aid complete dissolution.

3. Add a solution of DIC (5 eq)[a] in dry DCM to the amino acid solution.

4. Stir the mixture for 10 min at 0°C, keeping the reaction mixture free of moisture with a calcium chloride drying tube. If a gelatinous solid (symmetrical anhydride) precipitates, add dry DMF dropwise to redissolve. Continue stirring for a further 10 min.

5. Remove the DCM by evaporation under reduced pressure using a rotary evaporator.

6. Redissolve the residue in the minimum volume of DMF, and add the solution to the N-deprotected peptidyl resin.

7. Agitate resin gently for 1 h.

8. Transfer 10 mg of resin with a wide-mouthed Pasteur pipette to a small, sintered-glass funnel. Wash and dry-down as described in *Protocol 1C*. Perform Kaiser or TNBS test (*Protocol 13*). If positive, repeat steps 7–8.

9. If resin still gives a positive colour test after 4 h, wash resin with DMF (5 times) and repeat coupling reaction with fresh reagents.

[a] Relative to resin functionality.

---

## 7.4 Aminium/phosphonium activation methods

In recent years, aminium- (until recently referred to as uronium) and phosphonium-based derivatives have become the preferred tools for *in situ* carboxyl activation. The most popular reagents PyBOP **15** (23), TBTU **16** (24), and HBTU **21** (24, 33) smoothly convert Fmoc-amino acids, in the presence of a tertiary base, to their corresponding OBt esters (*Figure 6*). Recently, analogous derivatives HATU **17** (25) and PyAOP **22** (34) which generate OAt esters have become commercially available. These are used in an identical manner but have been shown to give superior results in terms of both coupling efficiency and suppression of enantiomerization (34–37).

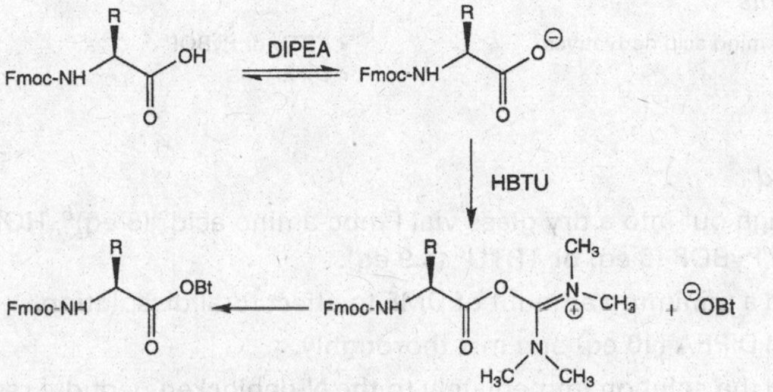

**Figure 6.** Generation of an OBt ester using HBTU.

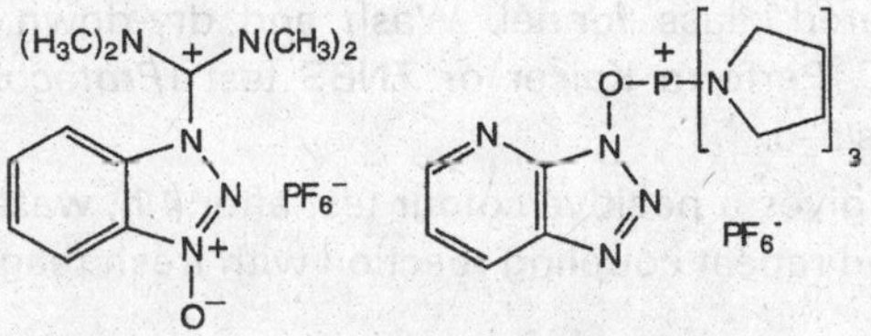

The reactivities of analogous aminium and phosphonium derivatives are essentially equivalent. However, aminium-based reagents can, if used inappropriately, cause capping of resin-bound amino groups through formation of N-terminally guanidinated peptides **23** (38). This side-reaction occurs most

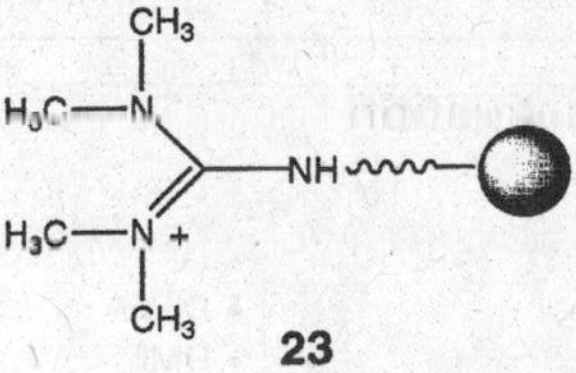

57

frequently during on-resin cyclization reactions where, by the very nature of the reaction, exposure of the amino component to the activator cannot be avoided. Hence for this application, phosphonium derivatives such as PyBOP or PyAOP should be used.

On automated peptide synthesizers, the activation reaction is generally carried out by dispensing solutions of reagents to vials containing the appropriate amino acid derivatives. TBTU is preferred to PyBOP for this application as solutions of TBTU in DMF have greater stability.

---

**Protocol 10.** TBTU/PyBOP activation

*Reagents*
- Fmoc-amino acid derivatives
- HOBt
- DMF
- TBTU or PyBOP
- DIPEA

*Method*

1. Weigh out into a dry glass vial Fmoc-amino acid[a] (5 eq)[b], HOBt[c] (5 eq), and PyBOP (5 eq) or TBTU[d] (4.9 eq).

2. Add a minimum amount of DMF to effect total dissolution.

3. Add DIPEA (10 eq) and mix thoroughly.

4. Add the solution immediately to the N-deblocked peptidyl resin.

5. Agitate resin gently for 1 h.

6. Transfer 10 mg of resin with a wide-mouthed Pasteur pipette to a small, sintered glass funnel. Wash and dry-down as described in *Protocol 1C*. Perform Kaiser or TNBS test (*Protocol 13*). If positive, repeat steps 5–6.

7. If resin still gives a positive colour test after 4 h, wash resin with DMF (5 times) and repeat coupling reaction with fresh reagents.

[a] Considerable enantiomerization occurs during aminium- or phosphonium-mediated coupling of Fmoc-protected cysteine derivatives (39, 40). Cysteine is best introduced using the appropriate symmetrical anhydride or OPfp ester.
[b] Relative to resin functionality.
[c] HOBt may be omitted, although this may increase the risk of enantiomerization of susceptible amino acids.
[d] It is advisable to use a slight deficiency of aminium reagent to ensure that no unreacted material remains in the reaction mixture.

---

**Protocol 11.** HATU activation

*Reagents*
- Fmoc-amino acid derivatives
- HATU
- DIPEA
- DMF

*Method*

1. Weigh out into a dry glass vial Fmoc-amino acid[a] (5 eq)[b] and HATU (4.9 eq)[c].
2. Add a minimum amount of DMF to effect total dissolution.
3. Add DIPEA (10 eq) and mix thoroughly.
4. Add the solution immediately to the N-deprotected peptidyl resin.
5. Agitate resin gently for 30 min.
6. Transfer 10 mg of resin with a wide-mouthed Pasteur pipette to a small, sintered glass funnel. Wash and dry-down as described in *Protocol 1C*. Perform Kaiser or TNBS test (*Protocol 13*). If positive, repeat steps 5–6.
7. If resin still gives a positive colour test after 4 h, wash resin with DMF (5 times) and repeat coupling reaction with fresh reagents.

[a] Considerable enantiomerization occurs during aminium- or phosphonium-mediated coupling of Fmoc-protected cysteine derivatives (39, 40). Cysteine is best introduced using the appropriate symmetrical anhydride or OPfp ester.
[b] Relative to resin functionality.
[c] It is advisable to use a slight deficiency of aminium reagent to ensure that no unreacted material remains in the reaction mixture.

## 7.5  Acid fluorides

Acid fluorides are less reactive than acid chlorides (41) but, nevertheless, are extremely powerful acylating agents. Fmoc-amino fluorides are stable in the presence of tertiary amines, unlike the corresponding acid chlorides which are rapidly converted to less reactive oxazolones (42), making them ideal reagents for peptide synthesis (43). They can be generated and isolated using reagents such as TFFH (44), diethylaminosulphur trifluoride (DAST) (45), or cyanuric fluoride (46). However, the most convenient strategy is to generate the acid fluoride *in situ* using TFFH **14** as a coupling reagent (47). The reaction is carried out in DMF under conditions compatible with normal protocols used for solid phase synthesis, by reacting the protected amino acid derivative with TFFH in the presence of 2 eq of DIPEA.

Fmoc-amino acid fluorides have proved to be extremely effective tools for the preparation of peptides containing contiguous *N*-alkyl and α,α-disubstituted amino acid residues (43, 47, 48).

**Protocol 12.**  TFFH activation

*Reagents*

- Fmoc-amino acid
- TFFH
- DIPEA
- DMF

**Protocol 12.** *Continued*

*Method*

1. Weigh out into a dry glass vial Fmoc-amino acid (5 eq) and TFFH (4.9 eq).

2. Add a minimum amount of DMF to effect total dissolution.

3. Add DIPEA (10 eq) and mix thoroughly.

4. Add the solution immediately to the N-deblocked peptidyl resin.

5. Agitate resin gently for 1 h.

6. Transfer 10 mg of resin with a wide-mouthed Pasteur pipette to a small, sintered glass funnel. Wash and dry-down as described in *Protocol 1C*. Perform Kaiser or TNBS test (*Protocol 13*). If positive, repeat steps 5–6.

7. If resin still gives a positive colour test after 4 h, wash resin with DMF (5 times) and repeat coupling reaction with fresh reagents.

# 8. Assembly of the peptide chain

The addition of each successive amino acid in the peptide chain involves carrying out a combination of the protocols described above. A typical procedure for effecting addition of one amino acid is given in *Table 2*. DMF is usually employed as the wash solvent, although NMP or DMA can also be used. If the synthesis has to be interrupted, it is best to halt the synthesis after the DMF wash step following removal of the N-terminal Fmoc group. Fmoc-protected peptidyl resins have a tendency to lose Fmoc on storage, generating dibenzofulvene, which is able to react with and effectively block the newly exposed N-terminal amino group.

**Table 2.** Typical reaction cycle for addition of one amino acid residue, starting from washed, Fmoc-protected resin

|  | Procedure | Time (min) | Repeat | Protocol |
|---|---|---|---|---|
| Deprotect | 1. Add 20% piperidine, agitate for 2 min and drain | 6 |  | 6, steps 4–6 |
| Wash | 2. Wash resin with DMF | 1 | 5 |  |
| Acylation | 3. Add activated amino acid derivative dissolved in DMF to resin | | | 7–12 |
| | 4. Agitate gently | 30 | | |
| Test | 5. Remove resin sample for colour test | | | |
| | 6. Continue agitation if necessary | | | |
| Wash | 7. Wash resin with DMF | 1 | 5 | |

The exceptions are peptides containing an N-terminal glutamine residue without side-chain protection and polymer-bound dipeptides—these should be stored Fmoc-protected, in order to prevent intramolecular cyclization. For a period of a few days, the peptidyl resin can be stored swollen in DMF at 5°C. For longer periods, the resin should be stored in the dried state; such resins should be re-swollen in DMF for 18 h before use.

# 9. Analytical procedures

## 9.1 Resin tests

The ninhydrin test, devised by Kaiser (49), is the most widely used qualitative test for the presence or absence of free amino groups (*Protocol 13A*). The test is simple and rapid, although it should be noted that some deprotected amino acids do not show the expected dark blue colour typical of free primary amino groups (50) (e.g. serine, asparagine, aspartic acid) and that proline, being a secondary amino acid, does not yield a positive reaction. Furthermore, occasionally false negative results are observed, particularly with strongly aggregated sequences. Other methods such as picric acid monitoring (51) or the TNBS test (52) (*Protocol 13B*) are also available. For proline, the chloranil test (53) is recommended (*Protocol 13C*).

---

**Protocol 13.** Qualitative amine tests

A. *Kaiser test*

*Reagents*

- 5% Ninhydrin in ethanol (w/v)
- 80% Phenol in ethanol (w/v)
- KCN in pyridine (2 ml 0.001 M KCN in 98 ml pyridine)

*Method*

1. Sample a few resin beads and wash several times with ethanol.
2. Transfer to a small glass tube and add 2 drops of each of the solutions above.
3. Mix well and heat to 120°C for 4–6 min.
4. The presence of resin-bound free amine is indicated by blue resin beads.

B. *TNBS test*

*Reagents*

- 5% DIPEA in DMF (v/v)
- 1% TNBS in water (w/v)

*Method*

1. Sample a few resin beads and wash several times with ethanol.

---

**Protocol 13.** *Continued*

**2.** Put the sample on a microscope slide, add 1 drop of each solution.

**3.** Watch the sample under the microscope and look for changes in colour.[a]

**4.** The test is positive when the resin beads turn yellow or red within 10 min and negative when the beads remain colourless.

C. *Chloranil test*

*Reagents*

- 2% p-Chloranil in DMF (w/v)
- 2% Acetaldehyde in DMF (v/v)

*Method*

**1.** Add 1 drop of acetaldehyde solution and 1 drop of *p*-chloranil solution to a few mg of resin beads placed on a microscope slide.

**2.** Allow to stand at room temperature for 10 min.

**3.** Blue stained resin beads indicate the presence of amines.

[a] It may be more convenient to perform this procedure by placing the resin beads in a 2 ml ignition tube, adding 3 drops of each reagents, and leaving the suspension for 10 min. Decant off the supernatant, re-suspend the resin beads in HPLC-grade acetone or DMF, and look for coloration of the beads.

## 9.2 Bromophenol blue monitoring

Bromophenol blue monitoring is a non-destructive technique for following the progress of coupling reactions (54). Bromophenol is an indicator which changes from blue to yellow on protonation. It binds to unacylated resin-bound amino groups, staining the resin beads blue. The test is carried out by simply adding 3 drops of 1% bromophenol in DMA to the coupling reaction. As the reaction proceeds the blue coloration fades to yellow, indicating complete acylation.

## 9.3 Fmoc determination

The method described below provides a quick and easy means of obtaining the substitution of any Fmoc-functionalized resin. The theoretical substitution of a peptidyl resin can be calculated from the substitution of the base resin using *Equation 1*.

$$A = B \times 1000/[1000 + (B \times (M - X))] \qquad [1]$$

$A$ = theoretical substitution (mmol/g);
$B$ = substitution of starting resin (mmol/g);
$M$ = molecular weight of target peptide, plus all protecting groups;
$X$ = 18 for hydroxymethyl; 36 for trityl chloride; 17 for aminomethyl-based resins, but 239 if resin is initially protected with Fmoc.

---

**Protocol 14.** Estimation of level of first residue attachment

*Equipment and reagents*
- Absorption spectrophotometer
- Fmoc-amino acid resin
- 20% Piperidine in DMF

*Method*
1. Take 3 × 10 mm matched silica UV cells.
2. Weigh dry Fmoc-amino acid resin (approx. 1 μmol with respect to Fmoc) into two of the cells.
3. Dispense freshly prepared 20% piperidine in DMF (3.00 ml) into each of the three UV cells.
4. Agitate the resin suspension with the aid of a Pasteur pipette for 5–10 min.[a]
5. Place the reference cell containing only the 20% piperidine solution into a spectrophotometer and zero at 290 nm.
6. Place, in turn, the cells containing the settled resin into the spectrophotometer and read the absorbance at 290 nm. Take an average of these two values.
7. Calculate the loading using *Equation 2*.

$$\text{Loading (mmol/g)} = (\text{Abs}_{sample}) / (\text{mg of sample} \times 1.75)^b \qquad [2]$$

[a] For polystyrene-based resins that have high substitution levels (> 0.4 mmol/g) and have been extensively dried *in vacuo*, leave the suspension (with occasional agitation) for 2–3 h.
[b] Based on $\varepsilon = 5253$ M$^{-1}$ cm$^{-1}$.

---

## 9.4 HPLC analysis

Reverse-phase HPLC is the standard analytical tool used for the control of the purity of synthetic peptides. Analysis is most commonly performed on reversed-phase HPLC columns, consisting of 3–5 μm, 100–300 Å pore size C8 or C18 silica; the larger pore material is generally favoured for the analysis of larger peptides (> 30 residues). Samples are eluted with a gradient formed between water and acetonitrile. An ion-pairing reagent is added to both solvents to improve resolution and selectivity. The most popular buffer systems are listed in *Table 3*.

For all new peptides, it is good practice to use an exploratory gradient starting at 5% acetonitrile, rising to 90% over a period of 20–40 min, depending on the particle size of silica used and the length and diameter of the column. Once the chromatographic behaviour of the peptide has been established, shorter or shallower gradients can be used for more detailed analysis. Monitoring is normally performed at 210–230 nm.

When preparing samples for analysis, it is essential that the peptide is

**Table 3.** HPLC systems

| Starting buffer | Developing buffer |
|---|---|
| 0.1% TFA in water | 0.1% TFA in 90/10 acetonitrile/water (v/v) |
| 0.01 M Ammonium acetate in water | 0.01 M Ammonium acetate in 90/10 acetonitrile/water (v/v) |
| 0.05 M Triethylammonium phosphate (pH 2.25) in water | 0.05 M Triethylammonium phosphate (pH 2.25)/acetonitrile; 2:3 (v/v) |

completely dissolved, otherwise an unrepresentative result may be obtained. Basic peptides will generally dissolve in dilute aqueous acetic acid. For acidic peptides, addition of a drop of ammonia will generally effect dissolution. Mixtures of acetonitrile and water are excellent solvents for synthetic peptides, although the acetonitrile concentration must be kept to a minimum or peak splitting can occur. DMF can also be used in small amounts, provided that the elution gradient is maintained at 5–10% acetonitrile until the DMF has been fully eluted. Some peptides dissolve in denaturants such as guanidine hydrochloride or urea. For extremely intractable sequences, trifluoroacetic acid can be employed but, if used at high concentrations, it can dissolve the silica at the top of the column, leading to peak broadening and splitting.

Liquid chromatography–mass spectrometry (LC–MS) is an extremely power tool for the analysis of peptides, providing not only information on the purity of the product but also conformation of structures. The system typically consists of a microbore HPLC system coupled to an electrospray mass spectrometer. Using such a system the composition of a crude peptide mixture can be quickly determined and by-products identified, enabling synthetic protocols to be rapidly optimized.

Further, detailed information on HPLC and other techniques that are applied for the analysis of synthetic peptides has been reviewed extensively(55, 56).

# 10. TFA-mediated cleavage

This section describes the procedures used for cleavage of peptides from acid-sensitive linkages. The protocols for the release of peptides from base-sensitive linkages are given in Chapter 6.

## 10.1 Preparing the resin for cleavage

Before the cleavage reaction can be performed, it is essential that the peptidyl resin is thoroughly washed to remove all traces of DMF, as described in *Protocol 1C*.

## 10.2 Cleavage reactions releasing fully deprotected peptides

Owing to the variability in the behaviour of different peptide resins, it is recommended that a preliminary small-scale cleavage of peptide–resin using 20–30 mg sample be carried out to determine the optimum cleavage conditions, such as the choice of scavenger(s) and length of reaction. This will enable the extent of cleavage (e.g. by quantitative analysis of the reference amino acid attached to the linker, where appropriate) and the quality of the crude cleaved peptide (by LC–MS) to be determined.

### 10.2.1 TFA cleavage

Release of peptides anchored to alkoxybenzyl alcohol and trialkoxybenzyl-amine linkages is generally achieved by treating the peptidyl-resin with 95% TFA for 1–3 h. Under these conditions, *t*-butyl-based protecting groups, Pmc and Pbf from arginine, and Trt groups from asparagine, glutamine, histidine, and cysteine are also cleaved, resulting in the formation of the fully side-deprotected peptide. During this process highly reactive cationic species are generated from the protecting groups and resin-linkers which can, unless trapped, react with and modify those amino acids containing electron-rich functional groups: tyrosine, tryptophan, methionine, and cysteine. For this reason various nucleophilic reagents, known as scavengers, are added to the TFA to quench these ions.

The most frequently used scavenger is water, which is moderately effective at scavenging *t*-butyl cations and the products of the cleavage of arylsulphonyl-based protecting groups. 1,2-Ethanedithiol (EDT) is the best scavenger for *t*-butyl cations and, like water, also offers some protection to unprotected tryptophan against sulphonation. It also assists in the removal of the trityl protecting group from cysteine. Suppression of methionine oxidation can be effected by addition of ethylmethyl sulphide, ethanedithiol, or thioanisole. The last reagent is also known to accelerate removal of Mtr, Pmc, and Pbf from arginine residues; however, care should be exercised in its use as there is some evidence to suggest that it can cause partial removal of Acm, *t*-butylthio, or *t*-butyl protecting groups from cysteine residues (57).

Phenol is thought to offer some protection to Tyr and Trp residues (58). Trialkylsilanes, such as triisopropylsilane (TIS) and triethylsilane (TES), have been shown to be effective, non-odorous substitutes for EDT (59), particularly for peptides containing Arg(Pmc)/Arg(Pbf) and Trp(Boc) (60, 61). These reagents are also very efficient at quenching highly stabilized cations liberated on cleavage of Trt (59), Tmob (61, 62), and the Rink amide linker, and hence their use is strongly recommended when these moieties are present.

A number of universal cleavage mixtures have been advocated, the most popular of which are Reagent K (TFA/thioanisole/water/phenol/EDT 82.5:5:5:5:2.5 v/v) and Reagent R (TFA/thioanisole/anisole/EDT 90:5:3:2 v/v) (58). However, with recent developments in protecting group chemistry, in

**Table 4.** Cleavage cocktails

| | Cocktail | Time |
|---|---|---|
| Peptides containing all amino acids except Arg(Mtr) Cys(Trt), Met, and unprotected Trp | TFA/TIS/water 95:2.5:2.5 (v/v) | 1.5–3 h |
| Peptides containing all amino acids except Arg(Mtr) and/or unprotected Trp | TFA/TIS/water/EDT 94:1:2.5:2.5 (v/v) | 1.5–3 |
| All peptides | TFA/thioanisole/water/ phenol/EDT 82.5:5:5:5:2.5 (v/v) | 1.5–18 h[a] |

[a] Depending on number of Arg(Mtr) residues present.

particular the introduction of Trp(Boc) and Arg(Pbf/Pmc), the use of such complex mixtures containing toxic and odorous reagents are no longer necessary, except in certain circumstances. In general, the cleavage cocktails listed in *Table 4* will give excellent results in most cases. In those instances where difficulties are encountered, *Table 5* can help in the identification of the problem and in the location of a possible solution.

---

**Protocol 15.** General TFA cleavage

*Method*

1. Place dry resin in a flask and add cleavage reagent (*Table 4*).

2. Flush flask with N$_2$, stopper and leave to stand at room temperature with occasional swirling for 1.5–18 h, depending on sequence (*Table 4*).

3. Remove the resin by filtration under reduced pressure through a sintered glass funnel. Wash the resin twice with clean TFA.[a]

4. Combine filtrates and isolate the peptide as described in *Protocol 17*.

[a] The expended resin should not be discarded but retained, in case it should prove necessary to repeat the cleavage reaction.

---

### 10.2.2 TMSBr cleavage

Trimethylsilyl bromide (TMSBr) in TFA (63) cleanly removes all acid-labile protecting groups used in Fmoc/*t*Bu SPPS. The reaction is extremely rapid, with the complete deprotection of a peptide containing four Mtr protected arginine residues having been achieved within 15 min (64). The use of this reagent also suppresses the formation of sulphonated by-products arising from inter- and intra-molecular transfer of sulphonyl-based protecting groups from arginine to the indole ring of tryptophan.

**Table 5.** Side reactions encountered during cleavage reactions

| Mass difference | Residue affected | Possible explanation | Possible remedies |
|---|---|---|---|
| −71 | Cys | Loss of Acm | Repeat cleavage omitting thioanisole from cleavage cocktail |
| +2 | Trp | Reduction of the indole ring of Trp | Omit silane scavengers from the cleavage cocktail and substitute for EDT |
| +16 | Met | Sulphoxide formation. A small amount of sulphoxide formation is not uncommon (~5%) | Use peroxide-free ether for peptide precipitation and include EDT or EMS in cleavage cocktail. Ensure Met used is free of sulphoxides. Reduce sulphoxides using method in Chapter 4, *Protocol 1* |
| +44 | Trp(Boc) | Incomplete decarboxylation of Trp(CO$_2$H) intermediate | Lyophilize peptide from 0.1% AcOH aq. Repeat as necessary |
| +56 | | Incomplete removal of $t$-Bu groups | Re-treat peptide with cleavage cocktail, monitoring reaction by HPLC |
| +56 | Cys, Met, Tyr, Trp | $t$-Butylation | Increase EDT content of cleavage cocktail. Resynthesize peptide using Trt-protected derivatives for introduction of Ser, Thr, and Tyr, and Fmoc-Trp(Boc) for addition of Trp |

**Table 5.** *Continued*

| Mass difference | Residue affected | Possible explanation | Possible remedies |
|---|---|---|---|
| +172 | Trp 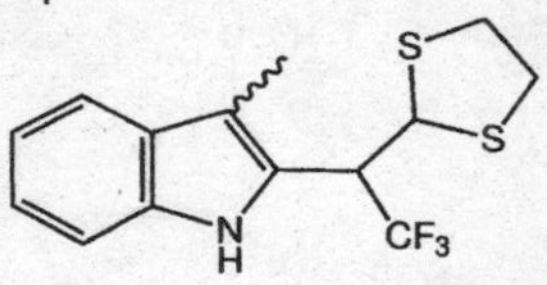 | Dithioketal formation. Associated with the prolonged exposure of Trp or EDT-containing cleavage cocktails, usually observed during the removal of Mtr protection from Arg | Repeat cleavage, monitoring reaction by HPLC to ensure minimum reaction time. Repeat cleavage using TMSBr in TFA. Resynthesize peptide using Fmoc-Trp(Boc) |
| +212 | Arg | Incomplete removal of Mtr. | Re-treat peptide with cleavage cocktail, monitoring reaction by HPLC. Repeat cleavage using TMSBr in TFA. Resynthesize peptide using Fmoc-Arg(Pbf). |
| +212 | Arg(Mtr) and Trp 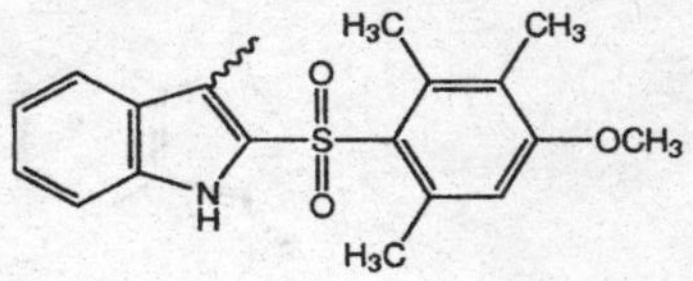 | Modification of Trp by Mtr. Identified by characteristic UV spectrum (*Figure 7*). Extending the cleavage reaction has no effect on product composition | Repeat cleavage using TMSBr in TFA. Resynthesize peptide using Fmoc-Arg(Pbf) and Fmoc- Trp(Boc). |
| +222 | n/a | Omitted to remove Fmoc group prior to cleavage reaction | Treat peptide with 20% piperidine in DMF for 20 min. Evaporate and triturate with ether |
| +242 | Cys | Incomplete removal of Trt | Re-treat peptide with cleavage cocktail containing EDT and TIS. Precipitate product directly with ether, without evaporation of TFA. Repeat process until no Trt remains |

68

| | | | |
|---|---|---|---|
| +242 | N-terminal Asn(Trt) | Incomplete removal of Trt | Re-treat peptide with cleavage cocktail, monitoring reaction by HPLC. Resynthesize peptide using Fmoc-Asn(Mtt) |
| +254/266 | Arg(Pbf)/(Pmc) | Incomplete removal of Pbf/Pmc | Re-treat peptide with cleavage cocktail, monitoring reaction by HPLC |
| +254/266 | Arg(Pbf)/(Pmc) and Trp | Modification of Trp by Pbf/Pmc. Identified by characteristic UV spectrum/ Extending the cleavage reaction has no effect on product composition | Repeat cleavage using TMSBr in TFA. Resynthesize peptide using Fmoc-Trp(Boc). |

Pbf, n=1
Pmc, n=2

---

**Protocol 16.**   TMSBr cleavage

*Reagents*
- TFA
- EDT
- TMSBr

- *m*-Cresol
- Thioanisole
- Peptidyl resin

*Method*

1. Place dry resin in a flask and add *m*-cresol (0.25 ml/g resin), EDT (1.25 ml/g resin), thioanisole (2.35 ml), and TFA (18 ml/g resin).

2. Add peptide resin and cool to 0°C.

3. Flush flask with $N_2$ and quickly add TMSBr (3.3 ml/g resin).

4. Flush flask again with $N_2$ and stopper.

5. Leave stand for 15 min with occasional agitation.

6. Remove the resin by filtration under reduced pressure through a sintered glass funnel. Wash the resin twice with clean TFA.

7. Combine filtrates and isolate the peptide as described in *Protocol 17*.

---

## 10.3  Peptide isolation

Following the cleavage reaction, the peptide is usually precipitated by the addition of cold diethyl or *t*-butyl methyl ether. The precipitation can be done directly from the TFA solution, or following evaporation of the TFA and volatile scavengers. The latter approach generally gives better yields, particularly of small peptides, but can lead to incomplete deprotection of *S*-trityl cysteine residues due to the reversibility of the reaction.

Since it is beyond the scope of this book to give a detailed descriptions of the techniques employed for the purification of synthetic peptides, readers are recommended to refer to the following reviews (55, 65).

---

**Protocol 17.**   Peptide isolation

*Equipment and reagents*
- Rotary evaporator equipped with $CO_2$/acetone cold finger and oil pump protected by a soda-lime trap

- Peroxide-free diethyl ether or *t*-butyl methyl ether

A. *Quick method*

1. Transfer the cleavage mixture to an appropriate sized round-bottomed flask.

2. Evaporate the TFA and scavenger mixture to a glassy film.[a]

---

3. Add cold ether slowly (equal to volume of original solution), being careful not to disturb the film of peptide adhering to the walls of the flask.

4. Wash the peptide film by gently swirling the ether around the flask.

5. Decant the ether[b] and repeat the washing process four times.

6. Allow the peptide to air-dry.

7. Dissolve in a suitable aqueous buffer and lyophilize.

## B. *Filtration*

1. Carry out steps 1–3 of *Protocol 17A*, or add the TFA solution (*i.e.* cleavage mixture) directly to a 10-fold volume of cold ether.

2. Suspend the peptide in the ether solution by triturating the sides of the flask with a spatula.

3. Filter the precipitated peptide through a hardened filter paper in a Hirsch funnel under light vacuum.

4. Wash the precipitate further with cold ether.[b]

5. Dissolve the peptide in a suitable aqueous buffer and lyophilize.

## C. *Centrifugation*

1. Carry out steps 1–3 of *Protocol 17A*, or add the TFA solution directly to a 10-fold volume of cold ether.

2. Suspend the peptide in the ether solution by triturating the sides of the flask with a spatula.

3. Transfer the suspension to a centrifuge tube and seal.

4. Place the tube inside a larger tube and seal.

5. Centrifuge for 10 min using a bench-top instrument. **It is essential that a spark-proof centrifuge is used for this purpose**.

6. Carefully decant the ether from the tube.[b]

7. Add fresh ether, seal and shake the tube to re-suspend the peptide.

8. Centrifuge.

9. Repeat steps 6–8 four times.

10. Dissolve the pellet in a suitable aqueous buffer and lyophilize.

---

[a] For small-scale multiple synthesis the evaporation step is best carried out in a multi-sample concentrator such as a Genevac®.
[b] The ether washings should be retained until the yield of product has been established. If a poor yield is obtained, the washings should be evaporated *in vacuo* to dryness and the precipitation procedure repeated.

## 10.4 Monitoring the cleavage reaction

It is often helpful to follow the progress of the cleavage reaction, particularly if the peptide of interest contains residues that are difficult to deprotect, such as Arg(Mtr), or it is suspected that side-reactions are occurring during this process. This is done by removing a small portion of the solution, evaporating the TFA and isolating the peptide using one of the methods given in *Protocol 17*. The peptide can then be analysed by LC–MS or by HPLC equipped with diode-array detection. The latter method is a particularly valuable tool for identifying modifications to tryptophan residues that cannot be unequivocally assigned by MS (*Figure 7*). If such post-column detection methods are unavailable, the following rules of thumb may prove helpful in the interpretation of HPLC elution profiles:

1. Partially protected and modified peptides generally elute later than the corresponding unmodified peptide; the greater the number of protecting groups remaining, the more strongly it is retained on the reversed-phase HPLC column.

2. Partially protected and modified peptides usually can be distinguished by performing a time course study; the proportion of the former will decrease during the course of the reaction, whereas the converse is true for the latter.

3. Methionine sulphoxide-containing peptides invariably elute earlier than the corresponding unoxidized material.

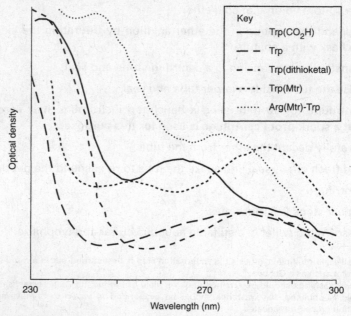

**Figure 7.** UV spectra of some commonly encountered Trp modifications.

## 10.5  Release of fully protected peptides from hyper-acid labile supports with 1% TFA

Peptides attached to resins **3**, **4**, **5**, **7**, **8**, and **11** can be released under conditions that leave most standard side-chain protecting groups intact. These supports are therefore valuable tools for the preparation of fully protected peptides for use in fragment condensation or cyclization strategies. Peptide release is normally effected by repetitive treatment with 1% TFA in DCM; for resins **5**, **7**, and **8** this can also be achieved under even milder conditions using TFE in DCM, as described in Chapter 9, *Protocol 3*.

Careful experimentation is essential if premature loss of side-chain protecting groups is to be avoided. Ideally, the cleavage should be carried out in a sealed manual peptide synthesis reaction vessel, and the filtration effected by applying nitrogen pressure rather than by the use of a vacuum, to prevent evaporation of the highly volatile DCM. For peptides containing methionine or tryptophan, 1% EDT should be added to the cleavage mixture to prevent reattachment of the peptide; this is not essential with trityl-based resins. If the peptide contains a C-terminal tryptophan residue, the use of Trp(Boc) is strongly recommended (66).

The purification of protected peptide fragments is described in Chapter 9, *Protocol 4*.

---

**Protocol 18.  Cleavage with dilute TFA solution**

*Reagents*
- 1% TFA in DCM
- 10% Pyridine in MeOH

*Method*

1. Pre-swell the peptidyl-resin (1 mmol) with DCM in a manual peptide synthesis vessel, and then wash with DCM (3 times).

2. Drain off excess DCM.

3. Add the TFA solution (10 ml), stopper the funnel and agitate gently for 2 min.

4. Remove stopper and replace with quick-fit adaptor connected to a low-pressure nitrogen supply.

5. Filter solution by applying nitrogen pressure into a flask containing pyridine solution (2 ml).

6. Repeat steps 3–5 up to 10 times.

7. Wash the residual protected peptide from the resin with DCM and MeOH.

8. Check filtrates by TLC[a] or HPLC.

---

**Protocol 18.** *Continued*

9. Combine product-containing filtrates and evaporate under reduced pressure to 5% of the volume.

10. Add water (40 ml) to the residue and cool mixture with ice to aid precipitation of the product.

11. Filter the precipitated peptide through a hardened filter paper in a Hirsch funnel under light vacuum.

12. Wash product consecutively with water (3 times), 5% NaHCO₃ aq. (2 times), water (3 times), 0.05 M KHSO₄ (2 times) and water (6 times).

13. Dry the peptide sample in a desiccator *in vacuo* over KOH pellets.

[a] TLC Analysis of protected peptide fragments is described in Chapter 9, *Protocol 2*. Typically, the maximal concentration of peptide may be contained in the first wash or in one of the later washes, depending on the buffering capacity of the amide bonds and other functional groups present.

# References

1. Atherton, E. and Sheppard, R. C. (1989). In *Solid phase peptide synthesis: a practical approach* (Series ed. D. Rickwood and B. D. Hames). IRL Press, Oxford.
2. Stewart, J. M. and Young, J. D. (ed.) (1984). *Solid phase peptide synthesis*, 2nd edn. Pierce Chemical Company, Rockford.
3. Wellings, D. A. and Atherton, E. (1997). In *Methods in enzymology* (ed. G. B. Fields), Vol. 289, p. 44. Academic Press, New York.
4. Atherton, E. and Sheppard, R. C. (1989). In *Solid phase peptide synthesis: a practical approach* (Series ed. D. Rickwood and B. D. Hames), pp. 137–139. IRL Press, Oxford.
5. Atherton, E. and Sheppard, R. C. (1989). In *Solid phase peptide synthesis: a practical approach* (Series ed. D. Rickwood and B. D. Hames), p. 134. IRL Press, Oxford.
6. Sieber, P. (1987). *Tetrahedron Lett.*, **28**, 6147.
7. Pedroso, E., Grandas, A., Saralalegui, M. A., Giralt, E., Granier, C., and van Rietschoten, J. (1982). *Int. J. Peptide Protein Res.*, **15**, 301.
8. Pharmacia LKB Biochrom, Biological Chemistry Report 89:3.
9. Blankemeyer, B., Nimtz, M., and Frank, R. (1990). *Tetrahedron Lett.*, **31**, 1701.
10. Meldal, M., Svendsen, L., Juliano, M. A., Del Nery, E., and Scharfstein, J. (1998). *J. Peptide Sci.*, **4**, 83.
11. Harth-Fritschy, E. and Cantacuzène, D. (1997). *J. Pept. Res.*, **50**, 415.
12. Barlos, K., Gatos, D., Kallitsis, K., Papaphotiu, G., Sotiriu, P., Wenqing, Y., and Schäfer, W. (1989). *Tetrahedron Lett.*, **30**, 3943.
13. Bayer, E., Clausen, N., Goldammer, C., Henkel, B., Rapp, W., and Zhang, L. (1994). In *Peptides: chemistry, structure and biology* (ed. R. S. Hodges and J. A. Smith), p. 156. ESCOM, Lieden.
14. Barlos, K., Chatzi, O., Gatos, D., and Stavropoulos, G. (1991). *Int. J. Peptide Protein Res.*, **37**, 513.

15. Steinauer, R., and White, P. (1994). In *Innovations and perspectives in solid phase synthesis, 3rd International Symposium* (ed. R. Epton), p. 689. Mayflower Worldwide Ltd., Birmingham.
16. Breipohl, G., Knoll, J., and Stüber, W. (1990). *Int. J. Peptide Protein Res.*, **35**, 281.
17. Albericio, F., van Abel, R., and Barany, G. (1990). *Int. J. Peptide Protein Res.*, **35**, 284.
18. Lauer, J. L., Fields, C. G., and Fields, G. B. (1994). *Lett. Pept. Sci.*, **1**, 197.
19. Wade, J. D., Bedford, J., Sheppard, R. C., and Tregear, G. W. (1991). *Pept. Res.*, **4**, 194.
20. Kates, S. A., Nuria, A., Solé, M., Beyermann, M., Barany, G., and Albericio, F. (1996). *Pept. Res.*, **9**, 106.
21. Kitas, E., Knorr, R., Trzeciak, A., and Bannwarth, W. (1991) *Helv. Chim. Acta*, **74**, 1314.
22. Carpino, L. A. and El-Faham, A. (1995). *J. Am. Chem. Soc.*, **117**, 5401.
23. Coste, J., Le-Nguyen, D., and Castro, B. (1990). *Tetrahedron Lett.*, **31**, 205.
24. Knorr, R., Trzeciak, A., Bannwarth, W., and Gillessen, D. (1989). *Tetrahedron Lett.*, **30**, 1927.
25. Carpino, L. A. (1993). *J. Am. Chem. Soc.*, **115**, 4397.
26. Mutter, M., Nefzi, A., Sato, T., Sun, X., Wahl, F., and Wöhr, T. (1995). *Pept. Res.*, **8**, 145.
27. Wöhr, T., Wahl, F., Nefzi, A., Rohwedder, B., Sato, T., Sun, X., and Mutter, M. (1996). *J. Am. Chem. Soc.*, **118**, 9218.
28. White, P., Bloomberg, G., and Munns, M. (1998). Poster 19 presented at the 25th European Peptide Symposium, Budapest.
29. Albericio, F. and Carpino, L. A. (1997). In *Methods in enzymology* (ed. G. B. Fields), Vol. 289, p. 104. Academic Press, London.
30. Rich, D. H. and Singh, J. (1979). In *The peptides* (ed. E. Gross and J. Meinenhofer), Vol. 1, p. 241. Academic Press, New York.
31. Atherton, E., Holder, J., Meldal, M., Sheppard, R. C., and Valerio, R. M. (1988). *J. Chem. Soc., Perkin Trans. 1*, 2887.
32. Mosjov, S., Mitchell, A. R., and Merrifield, R. B. (1980). *J. Org. Chem.*, **45**, 555.
33. Fields, C. G., Lloyd, D. H., Macdonald, R. L., Otteson, K. M., and Noble, R. L. (1991). *Pept. Res.*, **4**, 95.
34. Albericio, F., Cases, M., Alsina, J., Triolo, S. A., Carpino, L. A., and Kates, S. A. (1997) *Tetrahedron Lett.*, **38**, 4853.
35. Carpino, L. A., El-Faham., A., Minor, C., and Albericio, F. (1994). *J. Chem. Soc., Chem. Commun.*, 201.
36. Carpino, L. A. and El-Faham, A. (1994). *J. Org. Chem.*, **59**, 695.
37. Carpino, L. A., El-Faham, A., and Albericio, F. (1994). *Tetrahedron Lett.*, **35**, 2279.
38. Gausepolh, H., Pieles, H., and Frank, R. W. (1992). In *Peptides: chemistry, structure, and biology* (ed. J. A. Smith and J. E. Rivier), p. 523. ESCOM, Lieden.
39. Kaiser, T., Nicholson, G. J., Kohlbau, H. J., and Voelter, W. (1996). *Tetrahedron Lett.*, **37**, 1187.
40. Han, Y. X., Albericio, F., and Barany, G. (1997). *J. Org. Chem.*, **62**, 4307.
41. Carpino, L. A., Ionescu, D., and El-Faham, A. (1998). *Tetrahedron Lett.*, **39**, 241.
42. Carpino, L. A., Chao, H. G., Beyermann, M., and Bienert, M. (1991). *J. Org. Chem.*, **56**, 2635.

43. Carpino, L. A., Beyermann, M., Wenschuh, H., and Bienert, M. (1996). *Acc. Chem. Res.*, **29**, 268.

44. Carpino, L. A. and El-Faham, A. (1995). *J. Am. Chem. Soc.*, **117**, 5401.

45. Kaduk, C., Wenschuh, H., Beyermann, M., Forner, K., and Carpino, L. A. (1996). *Lett. Pept. Sci.*, **2**, 285.

46. Bertho, J. N., Loffet, A., Pinel, C., Reuther, F., and Sennyey, G. (1991). *Tetrahedron Lett* , **32**, 1303.

47. Triolo, S. A., Ionescu, D., Wenschuh, H., Solé, N., El-Faham, A., Carpino, L. A., and Kates, S. A. (1998). In *Peptides 1996* (ed. R. Epton and R. Ramage), p. 839. Mayflower Scientific Ltd., Birmingham.

48. Wenschuh, H., Beyermann, M., Winter, R., Bienert, M., Ionescu, D., and Carpino, L. A. (1996). *Tetrahedron Lett.*, **37**, 5483.

49. Kaiser, E., Colescott, R. L., Bossinger, C. D., and Cook, P. I. (1970). *Anal. Biochem.*, **34**, 595.

50. Fontenot, J. D. (1991). *Pept. Res.*, **4**, 19.

51. Oded, A. and Houghten, R. A. (1990). *Pept. Res.*, **3**, 42.

52. Hancock, W. S. and Battersby, J. E. (1976). *Anal. Biochem.*, **71**, 260.

53. Vojkovsky, T. (1995). *Pept. Res.*, **8**, 236.

54. Krchñák, V., Vágner, J., and Lebl., M. (1988). *Int. J. Peptide Protein Res.*, **32**, 415.

55. Dunn, B. M. and Pennington, M. W. (1994). In *Peptide analysis protocols* (ed. J. Walker). Humana Press, Totowa.

56. G. B. Fields (ed.) (1997). *Methods in enzymology* (Series ed. J. N. Abelson and M. I. Simon), Vol. 289. Academic Press, New York.

57. Atherton, E., Sheppard, R. C., and Ward, P. (1985). *J. Chem. Soc., Perkin Trans. 1*, 2065.

58. King, D. S., Fields, C. G., and Fields, G. B. (1990). *Int. J. Peptide Protein Res.*, **36**, 255.

59. Pearson, D., Blanchette, M., Baker, M. L., and Guidon, C. A. (1989). *Tetrahedron Lett.*, **30**, 2739.

60. White, P. (1993). Novabiochem Innovations, 2/93.

61. Fields, C. G. and Fields, G. B. (1993). *Tetrahedron Lett.*, **34**, 6661.

62. Solé, N. A. and Barañy, G. (1992). *J. Org. Chem.*, **57**, 5399.

63. Funakoshi, S., Murayama, E., Guo, L., Fujii, N., and Yajima, H. (1988). *J. Chem. Soc., Chem. Commun.*, 382.

64. Penrose, A. and White, P. (1989). Poster 487 presented at the 19th meeting of the Federation of European Biochemical Societies, Rome.

65. Atherton, E. and Sheppard, R. C. (1989). In *Solid phase peptide synthesis: a practical approach* (Series ed. D. Rickwood and B. D. Hames), p. 156. IRL Press, Oxford.

66. Kamber, B. and Riniker, B. (1992). In *Peptides: chemistry, structure, and biology* (Series ed. J. A. Smith and J. E. Rivier), p. 525. ESCOM, Lieden.

# Preparation and handling of peptides containing methionine and cysteine

FERNANDO ALBERICIO, IOANA ANNIS, MIRIAM ROYO, and
GEORGE BARANY

## 1. Introduction

Among the genetically encoded amino acid residues, methionine (Met) and cysteine (Cys) are special because they each contain an atom of sulphur. The present chapter describes how these residues are incorporated into peptides in the context of an Fmoc/tBu solid-phase synthesis strategy, as well as further considerations once the synthetic peptide is released from the support. Of added interest, some manipulations of Cys are advantageously performed at the level of the assembled peptide-resin, prior to cleavage. Many of the aspects discussed here also carry over to the preparation of peptides using a Boc/Bzl strategy.

The major problems associated with management of Met reflect the susceptibility of the thioether to alkylation and oxidation (1–3). One of the merits of the Fmoc/tBu strategy, in contrast to Boc/Bzl, is that in the former strategy Met is usually introduced without recourse to a protecting group for the thioether side-chain. As documented in this chapter, a proper under-standing of acidolytic cleavage conditions and the availability of selective procedures to reverse any inadvertent oxidation are likely to lead to success in obtaining homogeneous peptides containing Met.

Management of Cys provides additional significant challenges (4–8). For some targets, Cys is required with its side-chain in the free thiol form, whereas for other targets, an even number of Cys residues pair with each other via disulphide linkage(s) to provide cystine residue(s). Disulphide bridges play an important role in the folding and structural stabilization of many natural peptides and proteins, and their artificial introduction into natural or designed peptides is a useful approach to improve biological activities/specificities and stabilities. Furthermore, use of a disulphide bridge is a preferred method to conjugate peptides to protein carriers for increasing the response in immunological studies, to link two separate chains for developing discontinuous epitopes, and to generate active site models. This chapter describes Cys protecting groups, how they are removed to provide either free thiols or disulphides directly, and various strategies and practical considerations to minimize side reactions and maximize formation of the desired products.

## 2. Methionine

The thioether side-chain of Met is subject to alkylation and oxidation side reactions, either during the synthetic process or during subsequent handling of the Met-containing peptide. Alkylation of Met by carbocations proceeds rapidly under acidic conditions to give relatively stable sulphonium salts (9); in Fmoc/*t*Bu synthesis, the risk is during the final acidolytic deprotection/ cleavage step. Suitable scavengers must then be used that compete with peptide alkylation, and hence serve to minimize by-product formation. Any of a variety of cocktails developed for complete removal of side-chain protecting groups in the Fmoc/*t*Bu strategy are also satisfactory for preventing Met alkylation; these include Reagent K (TFA–phenol–$H_2O$–thioanisole– 1,2-ethanedithiol (82.5:5:5:5:2.5), ref. 10), Reagent R (TFA–thioanisole– 1,2-ethanedithiol–anisole (90:5:3:2), ref. 11), and Reagent B (TFA–phenol– $H_2O$–triisopropylsilane (88:5:5:2), ref. 12). As has been shown for peptides synthesized by Boc/Bzl chemistry, where the risk is present at each step during the repetitive acidolytic $N^\alpha$-Boc removal, the *tert*-butylation of Met is ultimately reversible (9). Thus, crude peptides that include as components the unwanted *tert*-butyl sulphonium salts may be heated at 50–55°C for 24 h, either in the solid form or in 1 M aqueous AcOH solution, and then submitted to chromatographic purification (9).

Complete oxidation of Met to its sulphone takes place only under severe conditions, and has not been reported as a problem for synthetic peptides. However, for some sequences, partial oxidation to methionine sulphoxide [Met(O)] can occur relatively easily upon prolonged exposure to air (13). Such oxidations are possible upon handling/storage of the peptide-resin, or during work-up of the peptide once it has been cleaved from the resin; they are minimal when operations are conducted under an inert atmosphere. However, the conditions under which photolabile linkers are cleaved lead to almost quantitative Met(O) formation (14, 15). Finally, Met is susceptible to modification (along with Trp, and to a lesser extent Tyr) when peptide-resins or free peptides are treated with oxidizing reagents, for example, with the goal to prepare disulphide bridges (discussed later in Section 3.3).

While the sulphoxide can be reduced at a later stage (see following), its mere presence via partial oxidation of the thioether may complicate handling and analysis of the peptide. Sulphoxide moieties are chiral, and thus peptide samples can be resolved into at least three components, the unoxidized thioether and the two sulphoxide diastereoisomers (15).

Cases do arise when protection of Met is indicated even in concert with Fmoc chemistry, and this is achieved by intentional incorporation of a Met(O) residue (16). In fact, the corresponding course is followed widely in Boc/Bzl chemistry, where the use of Met(O) is recommended to prevent alkylation side reactions (1). The standard 'high' conditions for final cleavage of side-chain protecting groups in Boc chemistry, involving strong acids such as HF or trifluoromethylsulphonic acid (TFMSA), are such as to leave the sulphoxide unaffected, but the parent thioether is conveniently regenerated when strong acids are used in the presence of a high concentration of $(CH_3)_2S$, the first stage of recommended 'low/high' cleavage methods (17). Fmoc-Met(O)-OH has been recommended in the preparation of protected peptides in convergent strategies, not only to avoid unwanted side reactions, but also due to the beneficial effect of the extra polarity from the sulphoxide moiety for the purification steps. In addition, some synthetic strategies directed at peptides which will contain disulphide moieties and Met together may rely on oxidation methodologies to create the sulphur–sulphur bridges that are incompatible with unprotected Met; in these cases, Met(O) should be used.

When Met(O) residues are present in a peptide, either inadvertently due to partial oxidation, or because the corresponding oxidized building blocks were used during synthesis, a separate step must be carried out to reduce these to the thioethers. Several methods have been described for reduction of Met(O)-containing peptides under relatively mild conditions. When these are applied to disulphide-containing peptides, there is often a risk that already-formed bridges could be partially reduced and/or scrambled. A method using trimethylsilyl bromide (TMSBr) and 1,2-ethanedithiol (EDT) under *anhydrous* conditions is particularly suitable for the reduction of large numbers of peptides, for example as prepared by multiple synthesis and combinatorial library techniques, although it is not compatible with disulphides (18). Reduction is successful with TFA as solvent, meaning that Met(O) reduction can be conducted simultaneously with the final cleavage of peptide from support (18). Moreover, no side reactions have been observed for Trp- and Tyr-containing peptides. A different strategy defers the conversion until a later stage of the overall process, that is after work-up of the peptide is complete and it has been dissolved in an aqueous milieu. Among numerous reducing agents that were tested, N-methylmercaptoacetamide (NMA) proved to have the best kinetic parameters (19). Even so, complete reaction with NMA takes more than a day, and the reagent is apparently not compatible with the presence of disulphides (19). A further method, based on the redox system $TiCl_4/NaI$, gives rapid reduction to the thioether (20, 21). However, reaction conditions are difficult to control, and with extended time disulphide bonds are reduced and Trp residues are unstable. Finally, reduction by $NH_4I–(CH_3)_2S$ in TFA is compatible with the presence of disulphide bridges, as well as with His and Tyr (6, 16, 22–25). When the treatment was carried out on a peptide with a single free thiol, the corresponding peptide homodimer formed (25). Trp residues may dimerize through 2-indolylindoline derivatives, but this side reaction can be minimized due to the presence of $(CH_3)_2S$, and by careful control of reaction times (25).

---

**Protocol 1.** Reduction of Met(O)-containing peptides

*Equipment and reagents*

- Glass-stoppered flask
- Met(O)-containing peptide
- Anhydrous solvent: $CH_2Cl_2$, MeCN, DMF, or TFA
- Diethyl ether
- NMA
- $CCl_4$

- Screw-cap centrifuge tubes
- AcOH
- EDT
- TMSBr
- *tert*-Butyl alcohol
- $(CH_3)_2S$
- $NH_4I$

A. *TMSBr–EDT reduction (18)*

*Reduction of individual peptides*

1. Dissolve the peptide (*ca.* 20 mg ≡ 20 mM) in a suitable anhydrous solvent (10–20 ml; $CH_2Cl_2$, MeCN, DMF, or TFA can be used; in some cases, for improving solubility, 5% (v/v) of AcOH can be added) in a glass-stoppered flask.
2. Add EDT (15.7 µl/ml solvent; final concentration 0.2 M) followed by TMSBr (13 µl/ml solvent; final concentration 0.1 M) to the above stirred solution. A fluffy colourless precipitate forms after a short time.
3. After 15 min at 25°C, remove the solvent *in vacuo*.
4. Add chilled diethyl ether to the remaining residue.
5. Centrifuge the suspension of the precipitated reduced peptide, wash twice with diethyl ether.
6. Dissolve the pellet and lyophilize from *tert*-butyl alcohol–$H_2O$ (4:1).

*Simultaneous reduction of multiple peptides*

1. Place the peptides (1–2 mg ≡ 1–2 mM) in screw-cap centrifuge tubes and dissolve in TFA (1 ml/mg peptide).
2. Add EDT (15.7 µl/ml TFA) followed by TMSBr (13 µl/ml TFA).
3. Close the tubes tightly.
4. After 15 min at 25°C, add chilled diethyl ether (8 ml/tube).
5. Centrifuge and wash the precipitated peptides with diethyl ether.

*Simultaneous reductive cleavage of multiple peptides from the polymeric support*

1. Subject the peptide-resins (50 mg) to the usual protocols for the cleavage of peptides, with the exception that the cleavage cocktails (0.5 ml) must omit $H_2O$ as a scavenger.

---

2. Add EDT (8 µl) and TMSBr (6.5 µl) to the mixture 15 min before the end of the cleavage time.
3. Isolate the peptide as usual.

B. *N-Methylmercaptoacetamide reduction (19)*

1. Dissolve the peptides in 10% aqueous AcOH, to a concentration of 1–10 mg/ml ($\equiv$ 1–10 mM).
2. Add NMA (10 eq).
3. Flush the mixture with $N_2$ and leave for 12–36 h at 35–40°C.
4. Lyophilize the solution, and purify the peptide by HPLC.

C. *$NH_4I$–$(CH_3)_2S$ reduction (16, 24, 25)*

1. Dissolve the peptide in TFA to a final concentration of $\sim$ 2 mg/ml ($\equiv$ 2 mM).
2. Cool the solution to 0°C.
3. Add $(CH_3)_2S$ ($\sim$ 30 eq) and $NH_4I$ ($\sim$ 30 eq), and vigorously stir the resultant suspension for 3 h.
4. Quench the reaction by adding 4 volumes each of $H_2O$ and $CCl_4$.
5. Wash the aqueous phase with $CCl_4$ (3 $\times$).
6. Evaporate *in vacuo* volatile components from the aqueous solution, and lyophilize the solution.

# 3. Cysteine

## 3.1 Cysteine protection

The cysteine protecting groups that are most useful for an Fmoc/tBu strategy are listed in *Table 1* (see also Chapter 2, Table 4), together with indications of conditions under which they are stable, as well as preferred methods to remove them. In contrast to most other side-chain protecting groups, where cleavage provides the parent functionality, the removal of protecting groups from cysteine can yield the free thiol, thiolate metal salt (mercaptide), or disulphide, depending on the reagent(s) applied and the precise reaction conditions. The availability of several cleavage methods becomes useful in designing regioselective routes to peptides with multiple disulphide bridges (discussed later, in Section 3.3.5). The Fmoc synthesis of peptides with N-terminal Cys involves some special considerations which are covered elsewhere in this chapter (Section 3.3.4).

For many applications, the most convenient protecting groups are those that are removed concurrently with all other side-chain protecting groups and simultaneous cleavage from the support. Given that the last step in the

Fmoc/*t*Bu strategy uses nearly neat TFA, together with appropriate scavengers, suitable Cys protecting groups include *S*-9*H*-xanthen-9-yl (*S*-Xan) (26), *S*-4-methoxytriphenylmethyl (*S*-Mmt) (27), *S*-2,4,6-trimethoxybenzyl (*S*-Tmob) (28), and *S*-triphenylmethyl (*S*-Trt) (29) (ordered by decreasing relative rate of cleavage). Convenient deprotection/cleavage cocktails include Reagent K (10), Reagent R (11), and Reagent B (12). With relatively few exceptions, the deblocked Cys residues are obtained in the predominantly free thiol form as a result of these treatments. The more labile of these protecting groups have the additional advantage that their removal can be carried out selectively while the peptide is still anchored to acid-labile handle supports. Thus, *S*-Xan and *S*-Tmob can be cleaved in the presence of the same scavengers, but at substantially reduced TFA concentration, e.g. 1 and 7%, respectively, that do not cleave common anchors such as PAC and PAL (26, 28, 30). This approach provides thiol substrates for resin-bound oxidation to disulphides. Finally, the *S*-Xan, *S*-Tmob, *S*-Mmt, and *S*-Trt groups can all be oxidized directly to disulphides with appropriate reagents (Section 3.3.3).

The TFA-stable protecting groups, *S*-4-methoxybenzyl (*S*-Mob) and *S*-4-methylbenzyl (*S*-Meb), used normally in the Boc/Bzl strategy, can also be of value in Fmoc strategies (1, 5, 7, 8). The aforementioned protecting groups will survive many of the manipulations involving other Cys residues that are protected with more labile groups, and hence can provide an extra level of selectivity in regioselective schemes. Ultimately, removal of *S*-Mob or *S*-Meb with HF–anisole (9:1) or HF–*p*-cresol (9:1) at 0 °C will liberate additional thiol groups ready for chemical manipulation. Alternatively, *S*-Mob can be removed with $CH_3SiCl_3$–Ph(SO)Ph–TFA (31) to furnish a disulphide directly. In a similar approach, the properties of the *S*-*tert*-butyl (*S*-*t*Bu) group can be exploited: *S*-*t*Bu is stable to TFA, and even to short treatment with HF–*m*-cresol at 0 °C, although *S*-*t*Bu is removed with neat HF at 20 °C, and at 0 °C with HF in the presence of anisole and/or sulphur-containing scavengers such as thiophenol and thioanisole (5, 32–34). Moreover, *S*-*t*Bu is similarly converted to disulphides with $CH_3SiCl_3$–Ph(SO)Ph–TFA (31).

The *S*-acetamidomethyl (*S*-Acm) (35) group and its phenyl analogue *S*-phenylacetamidomethyl (*S*-Phacm) (36, 37) are essentially stable to acid conditions; their removal is mediated by certain metals or electrophilic reagents in properly chosen solvents. Treatment of Cys(Acm)-containing peptides with either thallium(III) trifluoroacetate [Tl(tfa)$_3$] in TFA (23) or with iodine in AcOH–$H_2O$ mixtures (38, 39) leads directly to the corresponding peptides with a disulphide bond. Alternatively, *S*-Acm removal occurs upon reaction with either mercuric acetate [Hg(OAc)$_2$] in pH 4.0 buffer (35) or with a TFA solution of silver trifluoromethanesulphonate in the presence of anisole (40, 41). For each of these latter procedures, the initial mercaptide is next treated with excess either hydrogen sulphide, β-mercaptoethanol, or dithiothreitol to remove the metal and provide the free thiol. All of the aforementioned transformations can be carried out both in solution and in the

**Table 1.** Cysteine protecting groups compatible with Fmoc/tBu solid-phase peptide synthesis [a]

| Lability | Protecting group | Structure | Stability | Removal conditions |
|---|---|---|---|---|
| **Moderate acid** | | | | |
| | Xan | | Base, Nucleophiles | Dilute TFA/scavengers, $I_2$, Tl(III) |
| | Tmob | | Base, Nucleophiles | Dilute TFA/scavengers, $I_2$, Tl(III) |
| | Mmt | | Base, Nucleophiles | Dilute TFA/scavengers, Hg(II), Ag(I), $I_2$, Tl(III), RSCl |
| | Trt | | Base, Nucleophiles | Dilute TFA/scavengers, Hg(II), Ag(I), $I_2$, Tl(III), RSCl |

83

Table 1. *Continued*

| Lability | Protecting group | Structure | Stability | Removal conditions |
|---|---|---|---|---|
| **Concentrated acid** | | | | |
| | Mob | ~S–CH2–C6H4–OCH3 | TFA, Base, RSCl, I2 | HF(0°C), Tl(III), Hg(III), Ag(I), Ph(SO)Ph–CH3SiCl3 |
| | Meb | ~S–CH2–C6H4–CH3 | TFA, Ag(I), Base, RSCl | HF(0°C), Tl(III), Ph(SO)Ph–CH3SiCl3 |
| | tBu | ~S–C(CH3)(CH3)CH3 | HF(0°C), Base, Ag(I), I2 | HF(20°C), Hg(II), NpsCl, Ph(SO)Ph–CH3SiCl3 |
| **Metal ions** | | | | |
| | Acm | ~S–CH2–NH–C(=O)CH3 | HF(0°C), TFA, Base | Hg(II), Ag(I), I2, Tl(III), RSCl, Ph(SO)Ph–CH3SiCl3 |
| | Phacm | ~S–CH2–NH–C(=O)CH2–C6H5 | HF, Base | Hg(II), Tl(III), I2, Penicillin amidohydrolase |

**Reducing agents**

| | | |
|---|---|---|
| S-*t*Bu | TFA, HF (partial), Base, RSCl | RSH, Bu$_3$P, other reducing agents |
| Pic | TFA, HBr–AcOH, Base | Electrolytic reduction, Zn–AcOH |

[structure: ~~S–S–C(CH$_3$)$_3$ with three CH$_3$ groups]

[structure: ~~S–CH$_2$–pyridine (N)]

[a] Adapted from similar table in ref. 8. The protecting group structure is drawn to include the sulphur (shown in **bold**) of the cysteine that is being protected. Unless indicated otherwise, conditions or reagents outlined under 'Removal' are intended for quantitative removal and/or oxidative cleavage; the products from these reactions may be free thiols, mercaptides, or disulphides. Acid-promoted deprotections must be carried out in the presence of appropriate scavengers that prevent reattachment of the protecting carbocations. For metal-mediated removals, the counterion and solvent are sometimes critical. Also, when metal ions are used as deprotecting agents, treatment with excess β-mercaptoethanol or hydrogen sulphide is necessary to generate the free thiol. Sometimes metal ions bind very tightly and are difficult to remove completely from the peptide. For further details on all aspects of this table, as well as literature citation, see text.

solid-phase mode (5, 7, 8). An advantage of the latter is that removal of noxious reagents and by-products can be achieved by simple filtration and washing.

In the context of developing regioselective routes for the preparation of multiple-cystine peptides, the use of *S*-Phacm is particularly promising. *S*–Phacm is removed selectively, even in the presence of Acm groups, by means of penicillin G acylase (PAH) from *Escherichia coli* (EC 3.5.1.11), an enzyme with $P_1$ specificity for the phenylacetyl residue (36). A convenient way to facilitate work-up involves the use of enzyme that is immobilized on either a dry fibre or acrylic beads. Under the enzymatic cleavage conditions (pH 7.8, 35 °C), the unstable *S*-aminomethyl intermediates decompose to the thiols, but these undergo further oxidation to form disulphides (i.e. an intra-molecular cyclic product if the product has two appropriately positioned Cys residues or a symmetrical homodimer if a single Cys) (37).

---

**Protocol 2.** Enzymatic deprotection (37)

*Reagents*
- Penicillin G acylase (PAH) immobilized on acrylic beads (Eupergit-PcA; Rohm Pharma)
- Aqueous buffer: Et$_3$N·AcOH or NH$_4$OAc
- AcOH–H$_2$O (9:1)

*Method*

1. Dissolve the peptide in either Et$_3$N·AcOH (0.02 M) or NH$_4$OAc (0.05 M) pH 7.8 aqueous buffers at 35°C, to obtain a final concentration of 0.2 mg/ml ($\equiv$ 0.2 mM).

2. Add PAH immobilized on acrylic beads (Eupergit-PcA; Rohm Pharma, Weiterstad, Germany) (1 EU/μmol of Phacm), and monitor the reaction by HPLC.

3. On completion of the reaction (usually 24 h), add ~ 3 vol. AcOH–H$_2$O (9:1).

4. Filter the mixture through a disposable pipette with a glass wool plug, and lyophilize the filtrate.

5. Purify the peptide using preparative HPLC.[a]

[a] Yields of 80–90% have been reported.

---

Mixed disulphides are usually precluded as cysteine protecting groups in the Fmoc/*t*Bu strategy, because such functions are relatively unstable on exposure to piperidine. However, the *S-tert*-butylmercapto (*S-St*Bu) group (42) is sufficiently hindered to be compatible; its removal is achieved by treatment with either β-mercaptoethanol, dithiothreitol (DTT), tri-*n*-butylphosphine (Bu$_3$P), or other reducing agents (5, 6).

The $S$-4-picolyl group has been used only rarely in solid-phase synthesis, but it might become useful for the synthesis of large peptides because its presence could confer improved solubility (43, 44). Removal is achieved either by electrolytic reduction or (preferably) by means of Zn–AcOH (44).

## 3.2 Solid-phase synthesis of cysteine-containing peptides

### 3.2.1 C-terminal cysteine peptide acids

Racemization during the esterification of protected amino acids onto hydroxymethyl functionalities of substituted benzyl anchor-resins is a well-documented problem in solid-phase peptide synthesis (1–3, 45). This side reaction is exacerbated when couplings are mediated by carbodiimides in the presence of the catalyst 4-dimethylaminopyridine (DMAP), especially in the case of cysteine (46–48). Standard methods provide as much as 10–40% of D-Cys. Safer ways to anchor C-terminal Cys involve the use of halomethyl functionalized resins, such as the 2-chlorotrityl chloride resin (48, 49), the 4-(bromomethyl)phenoxyacetamide resin (similar to Wang resin) (50), and the bromo-SASRIN resin (51, 52) (*Figure 1*). With these resins, the esterification step is a nucleophilic displacement of the halogen group by the carboxylate of the protected amino acid derivative, and racemization levels are significantly lower (0.3–2.5%, depending on the $S$-protecting group; $S$-Trt relatively more prone to racemization) (53).

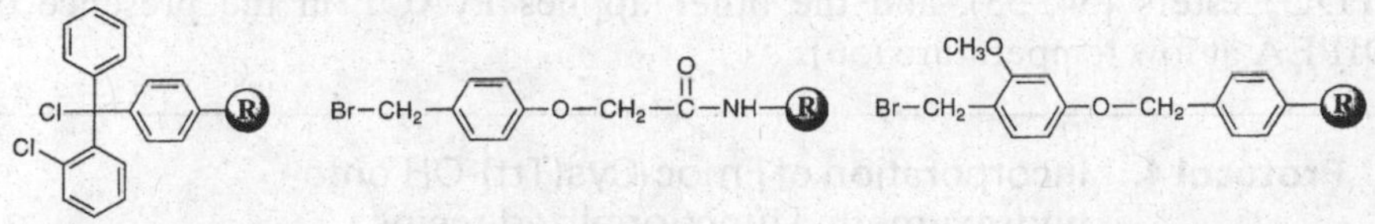

**Chlorotrityl chloride resin**    **Bromomethylphenoxy acetamide resin**    **Bromo-SASRIN resin**

**Figure 1.** Functionalized resins used for the solid-phase synthesis of peptide acids with C-terminal cysteine.

---

**Protocol 3.** Esterification of Fmoc-Cys(Trt)-OH onto 2-chlorotrityl chloride resin[a] (49, 53)

*Reagents*
- 2-Chlorotrityl chloride resin
- Fmoc-Cys(Trt)-OH
- MeOH
- Piperidine
- $CH_2Cl_2$
- DIPEA
- DMF
- Isopropanol

*Method*

1. Wash the 2-chlorotrityl chloride resin (typical loading 1.0 mmol/g) with $CH_2Cl_2$.

**Protocol 3.** *Continued*

2. Add a solution of Fmoc-Cys(Trt)-OH (3 eq with respect to Cl groups on resin) and DIPEA (3.3 eq) in a minimal volume of $CH_2Cl_2$.

3. Stir the mixture for 5 min at 25°C.

4. Add a further 6.6 eq DIPEA and stir the mixture for 2 h.

5. Terminate the reaction by adding MeOH (1 µl/1 µmol of starting chloride).

6. After 10 min stirring, filter the Fmoc-Cys(Trt)-ClTrt resin, and subject the filtered resin to the following further washings and treatments: DMF, $CH_2Cl_2$, DMF, piperidine–$CH_2Cl_2$–DMF (1:10:10, 10 min), piperidine–DMF (1:4, 15 min),[b] DMF, isopropanol, DMF, isopropanol, and MeOH.

7. Dry the resin *in vacuo*.

[a] Also refer to Chapter 9, Section 2.2.1.
[b] As a general practice, amino acyl esters to the 2-chlorotrityl chloride resin are best stored with the free amino group, after Fmoc removal.

When carrying out esterification of Fmoc-protected cysteine derivatives to hydroxymethyl functionalized resins, two methods that avoid the use of DMAP appear promising to reduce the level of racemization. One involves the use of 2,3-dihydro-2,5-diphenyl-4-hydroxy-3-oxo-thiophen-1,1-dioxide (TDO) esters (54, 55), and the other applies PyAOP in the presence of DIPEA at low temperature (56).

**Protocol 4.** Incorporation of Fmoc-Cys(Trt)-OH onto hydroxymethyl functionalized resins

*Reagents*
- Fmoc-Cys(Trt)-TDO ester
- DMF
- Fmoc-Cys(Trt)-OH
- DIPEA
- PyAOP
- $CCl_4$

A. *Use of TDO esters (54, 55)*

1. Add a solution of Fmoc-Cys(Trt)-TDO ester (3 eq with respect to hydroxyl function) and DIPEA (3 eq) in minimal amount of DMF to the resin.

2. Gently shake the suspension for 2 h at 25 °C.

3. Filter off the red solution and wash the resin thoroughly with DMF.

B. *Use of PyAOP (56)*

1. Add a solution of Fmoc-Cys(Trt)-OH (5 eq with respect to hydroxyl function) and DIPEA (10 eq) in the minimal volume of DMF to the resin at –20 °C (dry ice–$CCl_4$ bath).

2. After 5 min, add solid PyAOP (5 eq).

3. Stir the reaction mixture for 2 h at –20 °C.

4. Remove the cooling bath and stir the suspension for an additional 2 h at 25 °C.

5. Filter the suspension and wash the resin thoroughly with DMF.

Even if C-terminal Cys is safely anchored as an ester, further epimerization may occur upon chain assembly, specifically during repetitive $N^\alpha$-Fmoc removals promoted by the base piperidine (47). With p-alkoxybenzyl ester anchoring, the degree of epimerization was found to vary with the S-protecting group in the order StBu (34% after 16 h of treatment with piperidine–DMF, 1:4) > Trt (24%) >> Acm (11%), tBu (9%). However, racemization was not detected when starting with 2-chlorotrityl chloride resin, perhaps due to steric hindrance from the resin (48). Another effective strategy, confirmed experimentally to circumvent racemization in the synthesis of C-terminal Cys peptide acids, involves side-chain anchoring of Fmoc-Cys tBu ester via its free thiol to a 5-(9-hydroxyxanthen-2-oxy)valeric acid handle, which then becomes attached to amino-functionalized supports by standard coupling methods (57) (*Figure 2*). Finally, there is no racemization when the C-terminal Cys is anchored as an amide to PAL and related supports (30, 47).

### 3.2.2 Chain assembly using protected cysteine derivatives

It has recently become appreciated that cysteine derivatives can undergo substantial levels of racemization specifically at the step where they are incorporated (58–60). For example, standard protocols for coupling mediated by the recently popularized phosphonium and aminium (uronium) salts (BOP, PyAOP, HBTU, HATU) (61) typically involve 5-min pre-activation times and are conducted in the presence of suitable additives, such as HOBt or HOAt, plus a tertiary amine base, such as DIPEA or NMM. Under such conditions, the levels of racemization in a model peptide were in the range 5–33% (60). However, these levels were in general reduced by a factor of six- or sevenfold by avoiding the pre-activation step (60). Additional strategies to

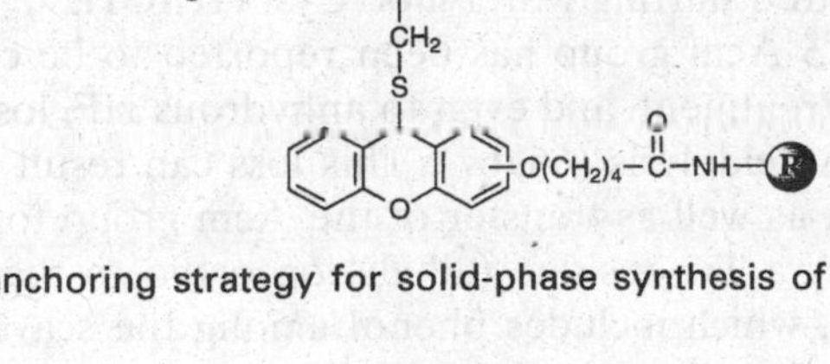

**Figure 2.** Side-chain anchoring strategy for solid-phase synthesis of peptide acids with C-terminal cysteine.

reduce racemization involved change to a hindered and/or weaker base such as 2,4,6-trimethylpyridine (TMP, collidine) or 2,6-di-*tert*-butyl-4-(dimethyl-amino)pyridine [DB(DMAP)] (62); twofold reduction in the amount of base; and change in solvent from neat DMF to the less polar $CH_2Cl_2$–DMF (1:1) (60, 63).

Coupling methods for the *safe* incorporation of cysteine with minimal racemization (< 1% per step) include aminium or phosphonium salts/HOBt or HOAt/TMP or DB(DMAP) (1:1:1) without pre-activation in $CH_2Cl_2$–DMF (1:1); DIC/HOBt (or HOAt) (1:1) with 5-min pre-activation (avoiding the pre-activation step notably increase racemization in this case); and preformed OPfp esters in $CH_2Cl_2$–DMF (1:1) (60, 63).

### 3.2.3 Side reactions

The most important family of side reactions, namely potential loss of con-figuration at various stages of solid-phase anchoring and chain assembly, has been mentioned already (Sections 3.2.1 and 3.2.2). Other problems should be considered as well. Oxidation and alkylation of the thioether of S-protected Cys is always a possibility, but such processes appear to be less severe in the case of Cys than with Met. A number of side reactions are likely to be sequence dependent; some of these involve modifications of other residues by species released during deprotection and/or oxidation.

When C-terminal cysteine peptides are synthesized, a 3-(1-piperidinyl)-alanine residue can form at the C-terminus (64). This side reaction might be explained by a base-catalysed β-elimination of the protected sulphur of the side-chain, to give a dehydroalanine intermediate which is trapped by a nucleophilic addition of the deprotection base piperidine. Occurrence of the side reaction is diagnosed by mass spectrometry, i.e. a peak of 51 Da higher in mass for peptides synthesized with Trt, and 20 Da lower for those synthesized with Acm. The amount of undesired by-product depends on the polymeric support, the side-chain protecting group, and the linker used to anchor cysteine to the resin. Although not recognized at the time, this side reaction explains a failed literature synthesis (65) of a peptide acid that started with C-terminal Fmoc-Cys(S*t*Bu)-Wang-PS resin. In a systematic study (64), a greater amount of 3-(1-piperidinyl)alanine adduct was detected with PS resins compared with PEG–PS supports, with *S*-Acm protection compared with *S*-Trt, and with C-terminal anchoring to Wang-type resins compared with PAL. In contrast, the side reaction is negligible when C-terminal peptide amides are prepared starting with Fmoc-Cys(Trt)-OH.

Although the *S*-Acm group has been reported to be completely stable to prolonged TFA treatment, and even to anhydrous HF, loss of Acm can occur during final TFA acidolysis (66, 67). This loss can result in premature disul-phide formation, as well as transfer of the Acm group to suitably positioned acceptors such as a Tyr residue (alkylation ortho to the phenolic function) (67). Reagent K, which includes phenol among the scavengers in its recipe, applied in a volume to ensure high peptide dilution once cleavage has occurred, has been reported to minimize these side reactions (67). Similarly, alkylation of the carboxamide group of Asn was reported during the treatment of the tripeptide Boc-Asn-Cys(Acm)-Pro-OBzl with HCl in diglyme (68), and the hydroxyl groups of Ser and Thr were modified on removal of the *S*-Acm group with metal ions reagents Tl(tfa)$_3$ and Hg(OAc)$_2$ (69). The latter side reaction was circumvented by adding a trivalent alcohol, i.e. glycerol, as a scavenger.

## 3.3 Formation of disulphides

### 3.3.1 General strategies and potential problems

Once the linear sequence of a cysteine-containing peptide has been assembled satisfactorily, additional steps are needed should the desired synthetic product include one or more disulphide bridge(s). Further considerations apply depending on whether the cysteine residues involved are on the same chain or on different chains, i.e. intramolecular or intermolecular bonds, respectively (the preparation of two-chain parallel symmetrical dimers bridged by two disulphides is a special case, and different from the corresponding antiparallel dimers). Intramolecular oxidations are often plagued by the unwanted formation of dimers and higher oligomers, a problem which can be mitigated but not necessarily eliminated by working under high-dilution conditions. A further option, often useful, involves disulphide formation from resin-bound precursors, taking advantage of the *pseudo-dilution* phenomenon which can favour intramolecular cyclization (1, 70, 71).

For many peptides with a single intramolecular disulphide, the most convenient way to create the needed sulphur–sulphur bridges needed is by oxidation of the corresponding linear precursors with two free sulphydryls (*Scheme 1*, Approach A). Deprotection/oxidation of the precursor with the sulphydryls protected (*Scheme 1*, Approach B) can be applied as well. Peptides with four, six, or eight cysteine residues present a statistical obstacle, since they can, in principle, give rise to 3, 15, and 105 different intramolecular isomers. Nevertheless, random co-oxidation (generalization of Approach A) under highly optimized conditions can often be successful to produce a major regioisomer with the naturally occurring pairing, since the polypeptide chains in their reduced poly(thiol) form may fold into native-like conformations that favour the proper alignments of disulphides. Alternatively, regioselective routes aimed at multiple intramolecular disulphide peptides (and intermolecular parallel symmetrical dimers) are readily contemplated; these extend the pairwise co-oxidation approaches with at least two families of selectively removable S-protecting groups. The challenge is to avoid *scrambling* side reactions, that is the breaking of already formed disulphide bridges with resultant production of non-desired isomers.

The preparation of peptides with intermolecular disulphides also requires

Fernando Albericio et al.

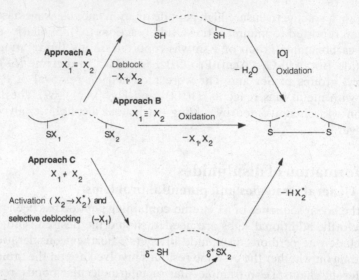

**Scheme 1.** Synthetic approaches to the formation of disulphide bridges. The residues to be linked are either on the same (intramolecular reactions; dotted line indicates intervening residues) chain or different (intermolecular reaction; dotted line indicates discontinuity) chains. $X_1$ and $X_2$ designate S-protecting groups that are stable to chain assembly conditions. $X_2$ and $X_2^*$ may or may not be the same, depending on the chemistry chosen for the directed method. Reactions can be carried out either on-resin or in solution; this facet and additional considerations are described more fully in the text.

the use of two different S-protecting groups, together with chemistry for 'directed' disulphide formation (*Scheme 1*, Approach C). While one might contemplate a simple co-oxidation of the two chains in either the free thiol or the S-protected form, such strategies are of limited reliability because of the statistical formation of both unwanted homodimers. In the directed approach, one cysteine is differentially activated (either directly due to the structure of an appropriate S-protecting group, or after selective deprotection and conversion to an activated form); a subsequent displacement reaction from a second cysteine, which has been separately deblocked, gives the disulphide bond. Conditions must be found for minimal *disproportionation*, that is conversion of the desired unsymmetrical intermolecular heterodimers to symmetrical homodimer species. Directed approaches are also sometimes beneficial for intramolecular cyclizations.

### 3.3.2 From free thiol precursors

The simplest method for the preparation of disulphide-containing peptides in concert with an Fmoc/*t*Bu synthesis strategy is represented by Approach A in *Scheme 1*. On completion of linear chain assembly using the same TFA-labile protecting group (i.e. *S*-Xan, *S*-Tmob, *S*-Mmt, or *S*-Trt) for all Cys residues, the final acidolytic cleavage/deprotection generates the precursor with free sulphydryl groups. A variety of oxidation procedures can then be investigated, and it should be stressed that the precise experimental conditions, such as pH, ionic strength, organic co-solvent, temperature, time, and concentration, are often critical (1–8, 72). High dilution is recommended in order to avoid physical aggregation and the formation of dimers, oligomers, or intractable polymers. However, the formation of such intermolecular covalently bridged by-products cannot always be avoided and 'recycling' of non-monomeric species through alternation of reduction/re-oxidation steps is also quite difficult to achieve in practice (5, 73).

For many peptide targets, it is generally helpful to pre-treat the crude material with appropriate reducing agents, such as DTT or β-mercaptoethanol. Tertiary phosphines, particularly the water-soluble tris(2-carboxyethyl)phosphine (TCEP) (74), can also be used for this purpose and they have the added advantage of being effective at somewhat acidic pH. Regardless of the reduction protocol followed, the resultant poly(thiol) linear precursor is then purified, at acidic pH, to minimize unwanted premature inter- or intramolecular disulphide formation, and to ensure that only discrete linear monomeric species, free of oligomeric or disulphide-mispaired contaminants, will be subjected to the oxidation/folding process.

---

**Protocol 5.** Pretreatment to reduce undesired disulphide bridges

*Reagents*
- Peptide
- 0.1 M NH$_4$HCO$_3$
- DTT
- 10% AcOH(aq)

- Concentrated HCl(aq)
- Buffer: 0.1 M Tris–HCl, 6 M guanidine-hydrochloride (Gdm–HCl), and 1 mM EDTA

A. *Reduction with DTT (75–77)*

1. Dissolve the crude peptide in 0.1 M NH$_4$HCO$_3$, pH 7.9–8.0, to obtain a final concentration of 1–3 mg/ml (≡ 1–3 mM).

2. Add DTT (5 eq relative to the number of desired thiol groups), and flush the mixture with N$_2$ and leave standing for 2–6 h at 25°C.[a]

3. Add 10% AcOH(aq) to adjust the reaction mixture to pH 2–3, and lyophilize the mixture.

4. Remove the excess DTT and other low molecular weight components by gel filtration.

B. *Reduction with DTT under denaturing conditions (78)*

1. Dissolve the crude peptide, to a concentration of 1–10 mg/ml (≡ 1–10 mM), in a buffer [b] containing 0.1 M Tris–HCl, 6 M guanidine hydrochloride (Gdm–HCl), and 1 mM EDTA, pH 8.5–8.7.

---

**Protocol 5.** *Continued*

2. Prepare a 0.8 M stock solution of DTT using the above buffer solution.

3. Add to the reaction mixture a portion of the DDT stock solution, to achieve a final DTT concentration of 0.03–0.1 M (20–50 eq relative to the number of thiol groups).

4. Flush the reaction mixture and leave the mixture under $N_2$ at 25°C.[a]

5. Reaction times are found to vary from 3 to 16 h.

6. On completion of reaction, add concentrated HCl(aq), to adjust the mixture to pH 2–3.

7. Carry out semi-preparative HPLC to purify the peptide.[c]

[a] Can be carried out at higher temperatures when needed.
[b] Degas and flush the buffer with argon.
[c] Sometimes, the reduced peptide is prone to rapid re-oxidation, not necessarily to the desired products. In these cases, it is recommended to carry out a rapid (< 2 min) desalting by passing the reduced peptide through a PD-10 column (Sephadex G-25M, Pharmacia), eluted with the buffer chosen for the disulphide formation/folding process.

Disulphide bridge formation from free sulphydryl precursors may also be carried out while the peptide is retained on the solid support (5, 8, 30, 65, 71, 79–81). This approach has several advantages, including elimination of problems of peptide solubility, and facile removal of oxidizing agents and solvents by filtration. In addition, the already discussed pseudo-dilution phenomenon that applies to solid-phase reactions favours intramolecular disulphide formation over intermolecular side reactions. Alternatively, chains with a single free sulphydryl can be subjected to on-resin intermolecular dimerization. In all of these cases, Cys residues which are protected by S-Xan, S-Tmob, S-Mmt, or S-StBu are first selectively deblocked by mild acid, or by thiolysis (Section 3.1). On-resin oxidation is then carried out (sometimes in the presence of a second family of paired orthogonally S-protected Cys residues) and, at an appropriate stage, the peptide is released from the support by cleavage cocktails that do not affect the formed disulphide bridge(s).

### 3.3.2.1 Air oxidation

The most convenient oxidation is carried out in the presence of atmospheric oxygen, at high dilution, and generally under slightly alkaline conditions (1–8). This widely used approach may be subject to one or more of the following limitations: (a) dimerization or worse, despite precautions to carry out reactions at low concentrations of substrate; (b) inadequate solubility for basic or hydrophobic peptides; (c) very long duration (up to 5 days) sometimes required for complete reaction; (d) difficulty in controlling oxidation, because the rate depends on trace amounts of metal ions; and (e) accumulation of side products due to oxidation of Met residues.

**Protocol 6.** Air oxidation

*Reagents*
- Peptide (free thiol form)
- Peptide-resin
- Et₃N
- Buffer[a]
- NMP

A. *Oxidation in solution (1, 5, 8, 80, 81)*

1. Dissolve the peptide (free thiol form) in an appropriate buffer[a] at a concentration of 0.01–0.10 mg/ml (≡ 0.01–0.1 mM). [b]
2. Stir the solution in an open atmosphere, or with oxygen bubbling through it.
3. Monitoring of reactions should be carried out by HPLC. Reaction times are found to vary from a few hours to a few days.

B. *On-resin oxidation (71, 79, 80, 81)*

1. Suspend the peptide-resin in a solution of Et₃N (0.02–0.175 M, 2–10 eq) in *N*-methylpyrrolidone (NMP).
2. Gently bubble air or oxygen through the suspension for 5–36 h at 25 °C.[c]

[a] Buffers used for air oxidation typically have pH values of 6.5–8.5, with the rate of oxidation higher with increasing pH (assuming that the substrate is soluble at the pH value chosen). The most widely used aqueous buffers are 0.1–0.2 M Tris–HCl or Tris–acetate, pH 7.7–8.7; 0.01 M phosphate buffers, pH 7–8; 0.2 M ammonium acetate, pH 6–7; and 0.01 M ammonium bicarbonate, pH 8. For the oxidation of certain peptides, the use of organic co-solvents (methanol, acetonitrile, dioxane, TFE) and the addition of tertiary amines (NMM, Et₃N, or DIPEA) is recommended (72, 80, 81). Gdm–HCl (2–8 M) is sometimes added to aid in solubility, and to increase the conformational flexibility of peptides—this is reported to result in overall improvements of the rates and yields of air oxidations. Addition of CuCl₂ (0.1–1 μM) has also been reported to improve certain oxidations.
[b] Concentrations as low as 1 μM and as high as 1 mM have been reported.
[c] Other polar solvents that swell the resin can also be used. Both intra- and intermolecular disulphides have been made in this way.

### 3.3.2.2 Oxidation in the presence of redox buffers

When poly(thiol) precursors are oxidized in the presence of mixtures of low molecular weight disulphides and thiols, overall rates and yields are often better than in the case of straightforward air oxidation. This is because the mechanism changes from direct oxidation (free radical intermediates) to thiol–disulphide exchange (thiolate intermediate), which facilitates the reshuffling of incorrect disulphides to the natural ones. Mixtures of oxidized and reduced glutathione, cysteine, cysteamine, or β-mercaptoethanol are commonly used. As before, high dilution of the peptide or protein substrate is necessary to maximize yields of the desired intramolecular disulphide-bridged species, and avoid formation of oligomers (5, 73).

---

**Protocol 7.** Use of redox buffers (5, 8, 73, 76, 81–84)

*Reagents*
- Poly(thiol) peptide
- Gdm–HCl or urea
- Aqueous buffers

- Redox buffer: Tris–HCl, EDTA, reduced and oxidized glutathione

*Method*

1. Prepare a buffer of 0.1–0.2 M Tris–HCl, pH 7.7–8.7, containing 1 mM EDTA, reduced (1–10 mM) and oxidized (0.1–1 mM) glutathione.[a] Typically, the optimal molar ratio of the reduced to oxidized compound is 10:1, but ratios of up to 1:1 have been reported.

2. Dissolve the previously reduced and purified poly(thiol) peptide in the above redox buffer to a concentration of 0.05–0.1 mg/ml ($\equiv$ 0.05–0.1 mM).

3. In cases where formation of physical aggregates occur, add a non-denaturing chaotropic (1–2 M Gdm–HCl or urea).

4. Allow the oxidation to proceed at 25–35°C, and monitor the progress of the oxidation by HPLC; 16 h to 2 days are common reaction times.[b]

5. Concentrate the oxidized peptide by lyophilization and purify by gel filtration on a Sephadex G-10 or G-25 column, using aqueous buffers at acidic pH.

[a] Cysteine/cystine, cysteamine/cystamine, β-mercaptoethanol/2-hydroxyethyl disulphide, reduced/oxidized dithiothreitol can be use as alternatives.
[b] If intractable precipitates form during this procedure, an alternative folding/oxidation is recommended in which the peptide is oxidized against a series of redox buffers with a slow pH gradient from 2.2 to 8 (85, 86).

---

### 3.3.2.3 Dimethylsulphoxide-mediated oxidation

Addition of dimethylsulphoxide (DMSO) is often useful for promoting oxidation of free thiols to disulphides (87, 88). The oxidant is miscible with water, so concentrations of up to 20% (v/v) can be used to promote faster reactions. Oxidation occurs over an extended pH range (pH 3–8), meaning that conditions can be found under which the substrates that undergo oxidation have improved solubility characteristics. Side reactions involving oxidation of sensitive side-chains (Met, Trp, or Tyr) have not been observed. Sometimes, complete removal of DMSO presents problems (30).

---

**Protocol 8.** DMSO mediated oxidation

*Reagents*
- Crude peptide
- $(NH_4)_2CO_3$
- TFA
- Phosphate buffer

- AcOH
- DMSO
- $CH_3CN$

A. *Oxidation under slightly acidic pH conditions (81, 88)*

1. Dissolve the crude peptide in AcOH and water (as required).
2. Dilute the solution to obtain a final peptide concentration of 0.5–1.6 mg/ml ($\equiv$ 0.5–1.6 mM), and a final concentration of 5% AcOH in water.
3. Adjust the pH to 6 with $(NH_4)_2CO_3$.
4. Add DMSO (10–20% by volume), and allow the oxidation to proceed for 5–24 h at 25°C (monitor by HPLC).
5. Dilute the final reaction mixture twofold with 0.05% aq. TFA–5% aqueous $CH_3CN$.
6. Load the solution onto a preparative reversed-phase HPLC column for DMSO removal and product purification.

B. *Oxidation under slightly basic pH conditions (81, 88)*

1. Dissolve the crude peptide in 0.01 M phosphate buffer, pH 7.5, to obtain a final concentration of $\sim$ 1 mg/ml ($\equiv$ 1 mM).
2. Add DMSO (1% by volume) and leave the reaction mixture at 25°C
3. Monitor the reaction by HPLC, and after completion ($\sim$ 3–7 h), lyophilize the mixture.

---

### 3.3.2.4 Potassium ferricyanide-mediated oxidation

Potassium ferricyanide is a relatively mild inorganic oxidizing reagent that is used widely for the conversion of bis(thiol) to disulphides (5, 6, 8). Since $K_3Fe(CN)_6$ is slightly light sensitive, reactions are best conducted in the dark. Oxidation side products are possible when Met or Trp residues are present in the substrate (89, 90).

---

**Protocol 9.** $K_3Fe(CN)_6$ mediated oxidation

*Reagents*
- Peptide (free thiol form)
- $K_3Fe(CN)_6$
- Peptide-resin
- $CH_2Cl_2$

- Buffer
- AcOH
- DMF

---

**Protocol 9.** *Continued*

A. *Oxidation in solution (91, 92)*

1. Dissolve the peptide (free thiol form) to a concentration of 0.1–1 mg/ml ($\equiv$ 0.1–1 mM), in a suitable buffer (acidic or basic, depending on the solubility characteristics of the peptide).

2. Slowly add the above peptide solution to a 0.01 M aqueous $K_3Fe(CN)_6$ solution, under nitrogen, at 25°C. [a,b]

3. On completion of reaction, adjust the pH to 5 with 50% aqueous AcOH.

4. Filter the solution, first through Celite and then (under mild suction) through a weakly basic anion exchange column (AG-3) to remove ferro- and ferricyanide ions.

5. Wash the AG-3 column with water, and apply the filtrate, along with the washes, to a weakly acidic cation exchange column or a Sephadex column equilibrated with 50% aqueous AcOH.

6. Elute to obtain the peptide. Yields of 20–50% purified product are typically obtained.

B. *On-resin oxidation (65, 81)*

1. Swell the peptide-resin in DMF for 1 h.

2. Drain off the DMF.

3. Add a homogeneous solution of 0.1–0.5 M $K_3Fe(CN)_6$ in $H_2O$–DMF (1:1 to 1:10). [c]

4. Gently agitate the suspension overnight at 25°C in the dark.

5. After 12–24 h reaction, wash the peptide-resin several times with $H_2O$, DMF, and $CH_2Cl_2$.

6. Dry the oxidized peptide-resin *in vacuo*, and cleave the peptide from the resin with an acid cocktail of choice that excludes thiol scavengers.

---

[a] The amount of oxidant used should be in 20% excess over theory, and the pH of the reaction solution should be kept constant at 6.8–7.0 by controlled addition of 10% aqueous ammonia. Addition times vary between 6 and 24 h, with purer products noted upon slower addition. For the formation of intermolecular homodimers, the peptide thiol solution must be more concentrated (> 1 mg/ml $\equiv$ 1 mM)), and addition is carried out inversely, i.e. oxidizing solution added to peptide solution.

[b] For peptides containing Trp and/or Met, which have side-chains susceptible to oxidation in the presence of excess $K_3Fe(CN)_6$, Misicka and Hruby (90) proposed a modification of the above procedure: add simultaneously the peptide solution and the oxidant solution (same concentrations as above), very slowly and at the same rate, to a reaction flask. This modification allows for peptide and ferricyanide to be at the equimolar highly-dilute concentration believed to allow for optimal intramolecular cyclization without side reactions.

[c] The overall amount of $K_3Fe(CN)_6$ used is calculated to be 10–20 eq with respect to the amount of SH groups on the peptide-resin, and the higher concentration of oxidant in aqueous solution requires correspondingly less of the DMF co-solvent.

### 3.3.2.5 Oxidation mediated by solid-phase Ellman's reagent

Ellman's reagent, 5,5'-dithiobis(2-nitrobenzoic acid) (developed originally in the context of an assay for measuring free thiol concentration under physiological conditions (93)), when bound through two sites to a suitable solid-support (PEG–PS or modified Sephadex) is an effective, mild oxidizing reagent that promotes the formation of intramolecular disulphide bridges (94) (*Figure 3*). This approach presents several advantages, including circumvention of potential problems due to solubility characteristics of substrates as well as reagent, the chance to carry out reactions under mild conditions conducive to product formation and to drive reactions to completion with excess reagent, the ready separation of reagent and concomitant isolation of product by simple filtration, and the recovery of reagent suitable for regeneration and reuse. In addition, the pseudo-dilution principle for polymer-supported reactions favours intramolecular reactions, so the level of dimer formation is reduced and often negligible. The rates and yields are dependent on the substrate's intrinsic propensity to be oxidized, as well as pH, resin loading, and the solid support. The main side reaction involves covalent adsorption of

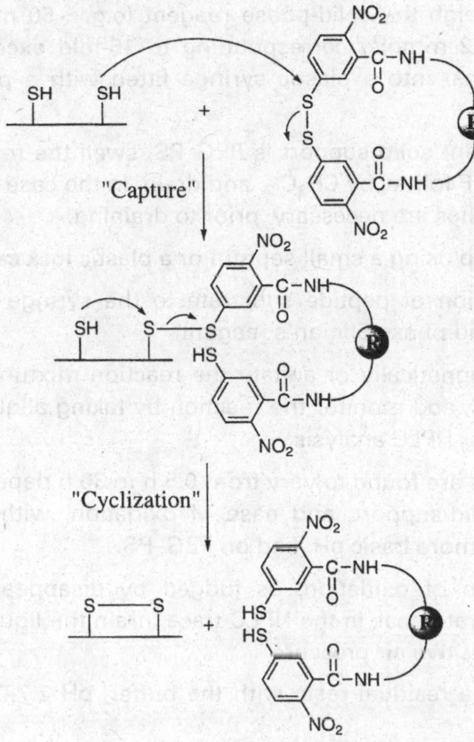

**Figure 3.** Intramoecular disulphide formation mediated by solid-phase Ellman's reagents.

the substrate to the support through intermolecular disulphide bridges; this diminishes the yield but does not affect the purity of the desired intra-molecular disulphide which is isolated. Moreover, the thiol substrate can be recovered by reduction, and recycled through the oxidation process.

---

**Protocol 10.** Oxidation mediated by solid-phase Ellman's reagent (94)

*Equipment and reagents*
- Plastic syringe fitted with a polypropylene frit
- Rotary shaker
- Peptide (free thiol form)
- Buffer–CH$_3$CN–CH$_3$OH

- CH$_2$Cl$_2$
- DMF
- Solid-phase Ellman's reagent: 5,5′-dithiobis (2-nitrobenzoic acid) (DTNB)-derivatised solid support

*Method*

1. Dissolve the peptide (free thiol form), to a concentration of 0.1–1 mg/ml ($\equiv$ 0.1–1 mM), in a suitable aqueous buffer pH 2.7–7.0 – CH$_3$CN–CH$_3$OH (2:1:1).

2. Separately, weigh the solid-phase reagent (e.g. ~50 mg/ml peptide solution, at 0.2 mmol/g corresponding to 15-fold excess based on DTNB functions) into a plastic syringe fitted with a polypropylene frit.

3. When the parent solid support is PEG–PS, swell the resin in CH$_2$Cl$_2$, wash with DMF follow by CH$_2$Cl$_2$, and drain. In the case of Sephadex, only DMF washes are necessary, prior to draining.

4. Plug syringe tip using a small septum or a plastic lock cap.

5. Add the solution of peptide substrate to the syringe already con-taining the solid-phase Ellman's reagent.

6. Gently stir magnetically or agitate the reaction mixture on a rotary shaker at 25°C, and monitor the reaction by taking aliquots from the liquid phase for HPLC analysis.

7. Reaction times are found to vary from 0.5 h to 30 h depending on the buffer pH, solid-support, and ease of oxidation, with significantly faster rates in more basic pH, and on PEG–PS.

8. On completion of oxidation, as judged by disappearance of the reduced substrate peak in the HPLC trace, drain the liquid phase into a vial using positive air pressure.

9. Wash (3 ×) the residual resin with the buffer, pH 2.7–7.0 – CH$_3$CN–CH$_3$OH (2:1:1).

10. Combine the washes with the drained liquid phase.

---

100

> **11.** Lyophilize to give the oxidized peptide. If the peptide in the free thiol form was pure, no additional purification of the oxidized peptide is required.

### 3.3.3 From protected thiol precursors

Disulphide bridges can also be formed directly from certain S-protected cysteine residues, without the intermediacy of free bis(thiol) intermediates that might undergo unwanted premature oxidation (*Scheme 1*, Approach B). Both of the cysteine residues to be paired are protected with the same kind of protecting group, and the approach can be generalized in experiments aimed at the regioselective construction of multiple disulphide bridges.

Suitable oxidative reagents include iodine (38), Tl(tfa)$_3$ (23), and alkyltrichlorosilanes–sulphoxide (31). The deprotection/oxidation step can be carried out either in solution, once the free peptide has been cleaved from the resin, or while the protected peptide is still on the resin. In a variation on the theme, silver-mediated deprotection can be followed immediately by a mild oxidizing environment, e.g. acidic DMSO (40, 41, 95). As specified in the subsections that follow, caution is required when peptide substrates contain sensitive residues, especially Met and Trp (38, 96).

### 3.3.3.1 Iodine oxidation

Iodine-mediated oxidation has been applied in conjunction with S-Xan, S-Tmob, S-Mmt, S-Trt, and S-Acm protection (5, 8, 26–28, 38). Optimal solvents include AcOH–H$_2$O (4:1), MeOH–H$_2$O (4:1), and DMF–H$_2$O (4:1) (39). The main caution with iodine is to avoid over-oxidation of the thiol functionality to the corresponding sulphonic acid, as well as to prevent or minimize modification of other sensitive amino acid side-chains (Tyr, Met, Trp) (5, 8, 38). It is advisable to monitor reactions, e.g. by HPLC; once complete oxidation is verified, quenching should be carried out so as to lessen the extent of any possible side reactions. Other oxidizing agents, such as cyanogen iodide (97) or N-iodosuccinimide (98), can be used.

---

**Protocol 11.** Iodine oxidation

*Reagents*
- Peptide
- AcOH
- Iodine
- CCl$_4$

- Peptide-resin (protected with S-Acm, S-Xan, S-Tmob, or S-Trt)
- DMF

A. *Oxidation in solution (38, 39, 81, 99)*

**1.** Dissolve the peptide in AcOH–H$_2$O (4:1), to a final concentration of about 2 mg/ml ($\equiv$ 2 mM).

---

**Protocol 11.** *Continued*

2. Add iodine (10 eq) to the above solution in one portion.

3. Stir the reaction mixture for 10 min to 1 h at 25°C.

4. On complete oxidation, revealed by HPLC monitoring, quench the reaction by diluting to twice the volume with $H_2O$, and extract the iodine with $CCl_4$ (5–6 times, equal volume each time).

5. Lyophilize the aqueous phase, and purify the product by preparative HPLC.

B. *On-resin oxidation (26, 81)*

1. Swell the peptide-resin (protected with *S*-Acm, *S*-Xan, *S*-Tmob, or *S*-Trt) in DMF.

2. Add iodine (10–20 eq) to the suspension.

3. Gently stir the mixture at 25°C for 1–4 h.

4. Wash, drain, and dry *in vacuo* the oxidized peptide-resin. Oxidation efficiency by this procedure is in the range 60–80%.

### 3.3.3.2 *Thallium(III) trifluoroacetate oxidation*

Thallium(III) trifluoroacetate [$Tl(tfa)_3$] is a mild oxidant that sometimes gives better yields and purities of desired disulphide products, with respect to methods using $I_2$ (23). TFA is the solvent of choice for reasons both of solubilization and chemistry, but DMF is an acceptable solvent for on-resin oxidations in conjunction with TFA-labile anchoring linkages (71). His and Tyr survive exposure to $Tl(tfa)_3$, but Met and Trp are compatible only if protected (23, 96). Thus, the only way that the use of Tl for disulphide formation in the presence of Trp can be reconciled with the Fmoc/*t*Bu strategy is when Trp is protected with $N^{in}$-Boc (100) and solid-phase oxidation is applied (the Boc group is removed during TFA cleavage, so oxidation cannot be carried out in solution; also the $N^{in}$-For (1, 5–7, 101) group favoured in conjunction with Boc/Bzl chemistry is lost under the conditions of $N^{\alpha}$-Fmoc removal). When starting with bis(*S*-Acm) sequences, only a slight excess of $Tl(tfa)_3$ should be used. Bis(*S*-Trt) sequences cannot be oxidized at all with this reagent, while bis(Tmob) sequences tolerate a range of reagent excess (28). The major limitation of this reagent is its high toxicity. As with other metals, thallium can be difficult to remove entirely from sulphur-containing peptides.

**Protocol 12.** $Tl(tfa)_3$ oxidation

*Reagents*

- Peptide (*S*-Acm or *S*-Tmob)
- $Tl(tfa)_3$
- Bis(*S*-Acm) peptide-resin
- DMF

- TFA
- Anisole
- Diethyl ether
- $CH_2Cl_2$

**A.** *Oxidation in solution (30)*

1. Dissolve the peptide in TFA–anisole (19:1), to a final concentration of about 1.1 mg/ml ($\equiv$ 1.1 mM).
2. Chill the solution to 0 °C.
3. Add Tl(tfa)$_3$ (1.2 eq) and stir the resultant mixture for 5–18 h at 4 °C.
4. Concentrate the reaction mixture, and add diethyl ether ($\sim$3.5 ml for each $\mu$mol peptide) to precipitate the crude product.
5. Triturate the product with diethyl ether for 2 min, followed by centrifugation and removal of the ether by decantation.
6. Repeat the trituration/centrifugation cycle three times, to ensure the complete removal of thallium salts. Yields of 35–50% of crude cyclic peptide have been reported.

**B.** *On-resin oxidation (30, 71)*

1. Swell the bis(S-Acm) peptide-resin in DMF.
2. Drain, and add Tl(tfa)$_3$ (1.5–2 eq) in DMF–anisole (19:1, $\sim$0.35 ml/25 mg resin).
3. After 2–18 h at 0 °C, wash the resin with DMF and CH$_2$Cl$_2$ to remove the excess Tl reagent. Yields of up to 80–95% have been reported.

### 3.3.3.3 Alkyltrichlorosilane–sulphoxide oxidation

Mixtures of alkyltrichlorosilanes and sulphoxides can cleave several S-protecting groups (e.g. *S*-Acm, *S*-*t*Bu, *S*-Mob, *S*-Meb) resulting in the direct formation of disulphides (31, 102–106). The reactions are fast, and are reported to be compatible with pre-existing disulphides formed in an orthogonal scheme. Depending on the protecting groups present in the substrate, certain chlorosilane–sulphoxide combinations are more effective than others; the most generally effective milieus apply diphenylsulphoxide [Ph(SO)Ph] with either CH$_3$SiCl$_3$ or SiCl$_4$ (102). The method is compatible with most amino acid side-chains, although not with Trp, because the side-chain indole can become fully chlorinated under the oxidizing conditions (77).

**Protocol 13.** Alkyltrichlorosilane–sulphoxide oxidation (33, 77, 102, 104–106)

*Reagents*
- Peptide-resin (*S*-Acm, *S*-*t*Bu, *S*-Mob, or *S*-Meb protected)
- TFA
- CH$_3$SiCl$_3$
- Ph(SO)Ph
- Anisole
- NH$_4$F
- Diethyl ether
- AcOH

**Protocol 13.** *Continued*

*Method*

1. Dissolve the peptide (*S*-Acm, *S*-*t*Bu, *S*-Mob, or *S*-Meb protected) in TFA, to a concentration of 1–10 μg/ml ($\equiv$ 1–10 μM).

2. Add sequentially Ph(SO)Ph (10 eq), $CH_3SiCl_3$ (100–250 eq), and anisole (100 eq).

3. Allow the reaction to proceed for 10–30 min at 25°C.

4. Add solid $NH_4F$ (300 eq) in order to quench the reaction.

5. Precipitate the crude product by addition of a large excess of dry diethyl ether.

6. Collect the precipitate by centrifugation.

7. Purify the product on a Sephadex G-15 column, eluting with 4 N aqueous AcOH.

8. Lyophilize the collected fractions, and the product can be further purified by preparative HPLC. Yields of 20–40% have been reported.

### 3.3.3.4 *Silver trifluoromethanesulphonate–DMSO–aqueous HCl deprotection/oxidation*

The use of metal reagents such as silver trifluoromethanesulphonate to remove *S*-Acm and *S*-Mob groups in acid media has been alluded to earlier (Section 3.1). Fujii and co-workers have developed a two-step procedure to prepare disulphides based on the idea that the mercaptide intermediate can then be oxidized directly, for example by treatment with DMSO (40, 41, 95). The oxidation is carried out in the presence of aqueous HCl, leading to concomitant precipitation of AgCl. These conditions are reported to avoid oxidative decomposition of Trp residues, and to not affect pre-existing disulphides formed in an orthogonal scheme (41).

**Protocol 14.** AgOTf and subsequent DMSO–aqueous HCl deprotection/oxidation (40, 41)

*Reagents*

- Peptide (*S*-Acm, *S*-Mob protected)
- Anisole
- Diethyl ether
- 1 M HCl(aq)
- TFA
- AgOTf
- DMSO

*Method*

1. Dissolve the peptide (*S*-Acm, *S*-Mob protected) in TFA–anisole (99:1), to a concentration of 1 mg/ml ($\equiv$ 1 mM).

2. Add AgOTf (100 eq) and leave for 2 h at 4°C.

3. Add ice-cold diethyl ether (15 volumes) and collect the resulting precipitate.

4. Wash the precipitated product with ice-cold diethyl ether (3 times).

5. Dissolve the peptide product in DMSO–1 M HCl(aq) (1:1), to achieve a final peptide concentration of 0.1 mM.

6. Allow the oxidation reaction to proceed for 7 h at 25 °C.

7. Remove the AgCl precipitate by filtration.

8. Dilute the filtrate with $H_2O$ and apply to preparative reversed-phase HPLC to purify the product. Yields by this procedure are in the range 50–60%.

### 3.3.4 Directed disulphide formation

Directed methods of disulphide bridge formation (*Scheme 1*, Approach C) offer control whenever two different peptide chains must be linked (1, 4, 5, 8, 107). (The applications of directed methods for intramolecular disulphide formation can be gleaned from the original publications (32, 108–110) and are not covered here.) One chain will have a Cys residue with a free thiol, prepared by methods already described (Section 3.1), whereas in the other chain the Cys residue needs to be blocked by a suitably activated form, such as *S*-3-nitro-2-pyridinesulphenyl (*S*-Npys) (111–113), *S*-2-pyridinesulphenyl (*S*-Pyr) (33, 34, 114), *S*-methoxycarbonylsulphenyl (*S*-Scm) (115, 116), or *S*-(*N*-methyl-*N*-phenylcarbamoyl)-sulphenyl (*S*-Snm) (110, 116). This latter need immediately raises a problem in the context of the Fmoc/*t*Bu strategy for chain assembly, because the mentioned dual-purpose S-protecting/activating groups are labile to bases such as piperidine used for repetitive $N^\alpha$-Fmoc removal. Thus, if the plan is to incorporate a protected Cys residue without further transformations, it must be the N-terminal residue (fortunately, a convenient position when the goal is conjugation to a protein carrier). Boc-Cys(Npys)-OH (111, 112) and Boc-Cys(Snm)-OH (110, 116) can be used in this way.

Generation of internal activated Cys residues, as needed for directed methods of disulphide formation, involves multi-stage procedures whereby a different S-protecting group is used to incorporate the Cys into the linear sequence. Accordingly, *S*-Trt, *S*-Acm, *S*-Tmob, or *S*-Xan protected peptide-resins, as well as peptides in solution containing those same S-protecting group or a free thiol, can be treated with appropriate sulphenyl chlorides as follows: (1) methoxycarbonylsulphenyl chloride in chloroform and/or methanol will give *S*-Scm (107, 115, 116); (2) consecutive reactions in $CH_2Cl_2$ with chloro-carbonylsulphenyl chloride, followed by *N*-methylaniline, will give *S*-Snm (110, 116); (3) Pyr-Cl and Npys-Cl in acetic acid will give respectively *S*-Pyr (117) and *S*-Npys (111, 112, 118) (*Figure 4*). Sulphenyl chlorides are highly reactive electrophilic reagents which are sometimes difficult to manipulate;

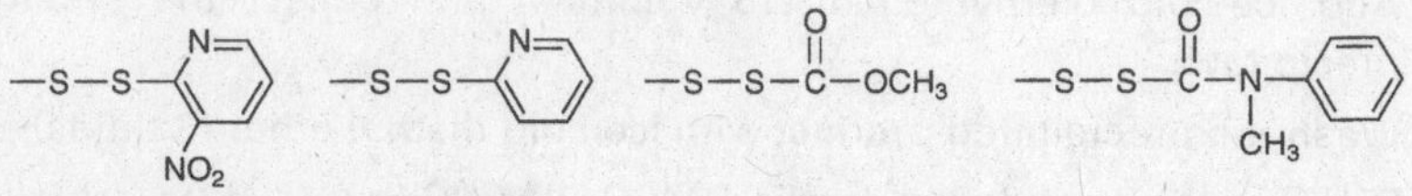

**Figure 4.** Dual S-protecting/activating groups for cysteine residues.

they will cause irreversible modification of Trp, and it is not possible to achieve selective activation with multiple Cys. An attractive and mild alternative is to prepare the peptide where the Cys is obtained in its free thiol form, and then carry out reaction with a reagent such as 2,2′-dithiobispyridine (34, 114, 119, 120) or 2,2′-dithiobis(5-nitropyridine) (DTNP) (121) to give respectively *S*-Pyr and *S*-Npys. Reaction with DTNP is particularly fast, requiring less than a minute between pH 3.5 and 6.5, and proceeding at an acceptable rate in aqueous AcOH (121). The aforementioned transformations are usually achieved with a purified peptide, but in some cases, the aromatic disulphides have been used to quench the thiols directly as they are liberated in an appropriate S-deprotective step (34, 114). Another reagent which can be used to activate a Cys-containing peptide preparatory to directed disulphide formation is bis(*tert*-butyl)azodicarboxylate (Boc–N=N–Boc), which is used to generate an isolable sulphenyl hydrazide intermediate (122, 123) (*Figure 5*).

Directed reactions to form disulphide bridges, applying the various activated forms already described, are often rapid, facile, and readily driven to completion due to the electronic characteristics of the leaving group. As described in the leading references, reactions can be conducted in solution or on-resin, in aqueous or organic milieus, and under acidic or mildly basic conditions.

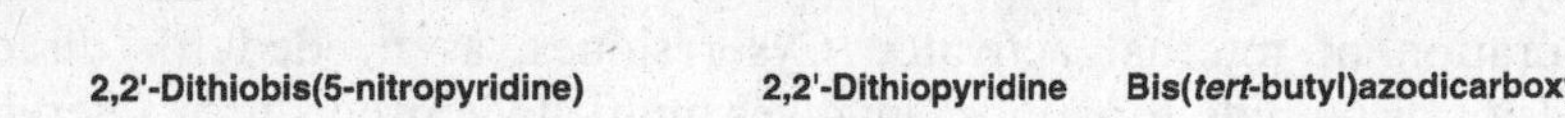

**2,2′-Dithiobis(5-nitropyridine)**    **2,2′-Dithiopyridine**    **Bis(*tert*-butyl)azodicarboxylate**

**Figure 5.** Reagents used for the activation of cysteine residues.

---

**Protocol 15.    Directed disulphide formation**

*Reagents*

- Peptide (free thiol form)
- AcOH
- Diethyl ether
- DTT
- Acetate-buffered saline

- 2,2′-Dithiobis(5-nitropyridine) (DTNP)
- TFA
- CHCl₃
- Keyhole limpet haemocyanin (KLH)
- NaOH

---

- *n*-Propanol–0.1 M Tris
- 2,2'-Dithiopyridine
- 10% aq. sodium carbonate solution
- Ar-saturated DMF
- $NH_4CO_3$

- Tri-*n*-butylphosphine
- MeCN–0.1 M Tris
- Boc–N=N–Boc
- Diethyl ether

### A. Disulphide formation through activation with DTNP (121) and conjugation to a carrier protein (113)

1. Dissolve DTNP (3–5 eq) in a minimal amount of $AcOH–H_2O$ (3:1).

2. Add the peptide with the free thiol function in one portion with vigorous stirring.

3. After 4–6 h reaction, add AcOH (3 volumes) and lyophilize.

4. Extract the resultant solid with 0.1% aqueous TFA, sonicate, centrifuge, and finally lyophilize the supernatant to obtain a powder.

5. Wash with diethyl ether–$CHCl_3$ (7:3) to remove residual organic reagents and by-products.

6. Separately, partially reduce the carrier protein (keyhole limpet haemocyanin, KLH) by adding DTT (1.16 mg) to a solution of KLH (7.5 mg) in acetate-buffered saline (ABS; 0.1 M sodium acetate, 0.1 M sodium chloride, pH 4.5; 0.5 ml).

7. Allow the mixture to react for 1 h.

8. Purify on a Sephadex G-25 column (1 × 19 cm) by eluting with ABS, and UV monitoring at 220 nm.

9. Pool the protein fractions (3–4 ml).

10. Add to the protein solution, the Cys(Npys)-peptide (5–10 mg, lyophilized powder) from step 5.

11. Adjust the pH to 5.0 with 1 M NaOH.

12. After stirring overnight, adjust the solution to pH 7.0 by addition of 1 M aqueous NaOH, and stir for another 3 h.

13. Dialyse the solution against 10 mM $NH_4CO_3$ (3 × 200 ml). The peptide:protein ratios are determined from acid hydrolysates of the conjugates.

### B. Disulphide formation through activation with 2,2'-dithiopyridine (114)

1. Dissolve the peptide with the free thiol function in *n*-propanol–0.1 M Tris pH 8 (1:1).

2. Add tri-*n*-butylphosphine (1.1 eq) and allow reaction to proceed for 10 h under Ar atmosphere.

3. Add a solution of 2,2'-dithiopyridine (40 eq) in *n*-propanol–0.1 M Tris pH 8 (1:1).

4. Stir the resultant mixture for 10 min.

**Protocol 15.** *Continued*

5. Purify by chromatography on Sephadex G-10, eluting with 0.1 M AcOH to give the activated peptide.

6. Dissolve the second peptide, containing a free thiol, in AcOH–$H_2O$ (1:9).

7. Add to a solution of the activated peptide (from step 5) in MeCN–0.1 M Tris pH 8 (1:1).

8. Adjust the pH to 6 with 10% aqueous sodium carbonate solution.

9. After 30 min, concentrate the reaction mixture and purify the product disulphide by preparative HPLC.

C. *Disulphide formation through activation with Boc–N=N–Boc (122)*

1. Add dropwise a solution of a peptide (with a free thiol function) in Ar-saturated DMF to a solution of Boc–N=N–Boc (1 eq) in Ar-saturated DMF, under vigorous stirring.

2. Allow the reaction to proceed for 12 h at 25°C,

3. Remove the bulk of the solvent by evaporation *in vacuo*.

4. Triturate the residue with diethyl ether.

5. Collect the precipitate, wash with diethyl ether, and dry *in vacuo*.

6. Dissolve the above peptide product (containing an activated thiol; 1 eq) in Ar-saturated DMF.

7. Add dropwise the activated peptide solution to a solution of peptide (with a free thiol function; 1.4 eq) in Ar-saturated DMF, with stirring at 25°C.

8. After 24 h, repeat steps 3–5 to obtain a precipitate and purify further by Sephadex chromatography.

### 3.3.5 Regioselective formation of disulphides

Regioselective formation of disulphides can be tackled by applying some of the already described methods (1–8). The general approach, when the disulphides are intramolecular, involves graduated deprotection (often ortho-gonal) and/or co-oxidations of pairwise cysteine residues, as specified by the original protection scheme. If an intermolecular disulphide is part of the target, a directed method is used to link two chains. Although these principles are reasonably self-evident, their implementation in practice retains a substantial measure of art and empiricism. Selectivities in the Cys deprotection chemistries, compatibilities with sensitive residues, stabilities of already-formed disulphides to subsequent transformations, and solubility considerations are among the factors that influence success. For more complicated targets, the yields (if reported at all) tend to be quite low. The following paragraphs provide a brief overview, but readers interested in more information should consult the aforementioned general reviews and original literature contributions cited therein.

For the preparation by the Fmoc/*t*Bu strategy of peptides containing two disulphides, the best combination appears to be an acid-labile protecting group (*S*-Trt, *S*-Tmob, or *S*-Xan) for two of the Cys, and *S*-Acm for the remaining two Cys (5, 30, 81). The formation of the first disulphide bridge occurs either in solution or on-resin after selective acidolysis has given a bis(thiol), bis(Acm) intermediate. In a subsequent step, deprotection/ oxidation of Acm is carried out to form the second bridge. An alternative two-dimensional orthogonal strategy uses acid-stable, thiolysable *S*-*St*Bu to protect the two Cys residues that are joined first, and again *S*-Acm for the remaining two Cys (66, 122). In addition, the *S*-Phacm/*S*-Acm combination has been demonstrated for a simple parallel cystine dimer (37).

The extra selectivity to prepare peptides containing three disulphides can be provided by incorporation of more acid-stable groups such as *S*-Mob or *S*-*t*Bu. Deprotection/oxidation upon treatment with alkyltrichlorosilanes– sulphoxides, or related procedures, may give the last disulphide; while strong acid, for example HF at 0°C or TFMSA at 25°C in the presence of carefully chosen scavengers, provides a free thiol which may be manipulated further in favourable cases (34, 80, 124).

It may be adequate to adopt an intermediate partially regioselective strategy with the goal to obtain peptides with a multiple disulphide array. The idea is to use one class of protecting groups to control the pairing of one particular set of Cys residues, and the second class to block all of the remaining Cys. Thus, one might hope that establishment of the initial disulphide bridge might bias the oxidation process that creates the subsequent bridges. Literature precedents are available when this was done in concert with chain assembly by either Boc/Bzl (125, 126) or Fmoc/*t*Bu (127) chemistry.

# References

1. Barany, G. and Merrifield, R. B. (1979). In *The peptides—analysis, synthesis, biology* (ed. E. Gross and J. Meienhofer), Vol. 2, pp. 1–284. Academic Press, New York.
2. Fields, G. B., Tian, Z., and Barany, G. (1992). In *Synthetic peptides: a user's guide* (ed. Grant, G. A.), pp. 259–345. W. H. Freeman, New York.
3. Lloyd-Williams, P., Albericio, F., and Giralt, E. (1997). *Chemical approaches to the synthesis of peptides and proteins.* CRC, Boca Raton, Florida.
4. Büllesbach, E. E. (1992). *Kontakte* (Darmstadt), **1**, 21.
5. Andreu, D., Albericio, F., Solé, N. A., Munson, M. C., Ferrer, M., and Barany, G. (1994). In *Methods in molecular biology, Vol. 35: Peptide synthesis protocols* (ed. M. W. Pennington and B. M. Dunn), pp. 91–169. Humana Press, Totowa, NJ.
6. Kiso, Y. and Yajima, H. (1995). In *Peptides—synthesis, structures, and applications* (ed. B. Gutte), pp. 39–91. Academic Press, San Diego.

7. Moroder, L., Besse, D., Musiol, H. J., Rudolph-Böhner, S., and Siedler, F. (1996). *Biopolymers*, **40**, 207.
8. Annis, I., Hargittai, B., and Barany, G. (1997). *Methods in Enzymology*, **289**, 198.
9. Noble, R. L., Yamashiro, D., and Li, C. H. (1976). *J. Am. Chem. Soc.*, **98**, 2324, and references cited therein.
10. King, D. S., Fields, C. G., and Fields, G. B. (1990). *Int. J. Pept. Prot. Res.*, **36**, 255.
11. Albericio, F., Kneib-Cordonier, N., Biancalana, S., Gera, L., Masada, R. I., Hudson, D., and Barany, G. (1990). *J. Org. Chem.*, **55**, 3730.
12. Solé, N. A. and Barany, G. (1992). *J. Org. Chem.*, **57**, 5399.
13. Hofmann, K., Haas, W., Smithers, M. J., Wells, R. D., Wolman, Y., Yanaihara, N., and Zanetti, G. (1965). *J. Am. Chem. Soc.*, **87**, 620.
14. Hammer, R. P., Albericio, F., Gera, L., and Barany, G. (1990). *Int. J. Pept. Prot. Res.*, **36**, 31.
15. Lloyd-Williams, P., Albericio, F., and Giralt, E. (1991). *Int. J. Pept. Prot. Res.*, **37**, 58.
16. Ferrer, T., Nicolás, E., and Giralt, E. (1999). *Lett. Pept. Sci.*, **6**, 165.
17. Tam, J. P. and Merrifield, R. B. (1987). In *The Peptides* (ed. S. Udenfriend and J. Meienhofer), Vol. 9, pp. 185–248, Academic Press, New York.
18. Beck, W. and Jung, G. (1994). *Lett. Pept. Sci.*, **1**, 31.
19. Houghten, R. A. and Li, C. H. (1979). *Anal. Biochem.*, **98**, 3646.
20. Mapelli, C. and Leftheris, K. (1992). In *Peptides 1992. Proceedings of the Twenty-Second European Peptide Symposium* (ed. C. H. Schneider and A. N. Eberle), pp. 207–208. Escom, Leiden, The Netherlands.
21. Pennington, M. W. and Byrnes, M. (1995). *Peptide Research*, **8**, 39.
22. Izeboud, E., and Beyerman, H. C. (1978). *Recueil des Travaux Chimiques des Pays-Bas*, **97**, 1.
23. Fujii, N., Otaka, A., Funakoshi, S., Bessho, K., Watanabe, T., Akaji, K., and Yajima, H. (1987). *Chem. Pharm. Bull.*, **35**, 2339, and references cited therein.
24. Nicolás, E., Vilaseca, M., and Giralt, E. (1995). *Tetrahedron*, **51**, 5701.
25. Vilaseca, M., Nicolás, E., Capdevila, F., and Giralt. E. (1998). *Tetrahedron*, **54**, 15273.
26. Han, Y. and Barany, G. (1997). *J. Org. Chem.*, **62**, 3841.
27. Barlos, K., Gatos, D., Chatzi, O., Koch, N., and Koutsogianni, S. (1996). *Int. J. Pept. Prot. Res.*, **47**, 148.
28. Munson, M. C., García-Echeverría, C., Albericio, F., and Barany, G. (1992). *J. Org. Chem.*, **57**, 3013.
29. Zervas, L. and Theodoporopoulos, D. M. (1956). *J. Am. Chem. Soc.*, **78**, 1359.
30. Munson, M. C. and Barany, G. (1993). *J. Am. Chem. Soc.*, **115**, 10203.
31. Akaji, K., Tatsumi, T., Yoshida, M., Kimura, T., Fujiwara, Y., and Kiso, Y. (1991). *J. Chem. Soc., Chem. Commun.*, 167, and references cited therein.
32. Ploux, O., Chassaing, G., and Marquet, A. (1987). *Int. J. Pept. Prot. Res.*, **29**, 162.
33. Akaji, K., Fujino, K., Tatsumi, T., and Kiso, Y. (1992). *Tetrahedron Lett.*, **33**, 1073, and references cited therein.
34. Maruyama, K., Nagata, K., Tanaka, M., Nagasawa, H., Isogai, A., Ishizaki, H., and Suzuki, A. (1992). *J. Prot. Chem.*, **11**, 1.
35. Veber, D. F., Milkowski, J. D., Varga, S. L., Denkewalter, R. G., and Hirschmann, R. (1972). *J. Am. Chem. Soc.*, **94**, 5456.

36. Hermann, P. and Schillings, T. (1993). In *Peptides 1992: Proceedings of the Twenty-Second European Peptide Symposium* (ed. C. H. Schneider and A. N. Eberle), pp. 411–412. Escom, Leiden, The Netherlands, and references cited therein.
37. Royo, M., Alsina, J., Giralt, E., Slomczynska, U., and Albericio, F. (1995). *J. Chem. Soc., Perkin Trans. 1*, 1095.
38. Kamber, B., Hartmann, A., Eisler, K., Riniker, B., Rink, H., Sieber, P., and Rittel, W. (1980). *Helv. Chim. Acta*, **63**, 899.
39. Ruiz-Gayo, M., Albericio, F., Royo, M., García Echeverría, C., Pedroso, E., Pons, M., and Giralt, E. (1989). *Anales de Química*, **85C**, 116.
40. Tamamura, H., Otaka, A., Nakamura, J., Okubo, K., Koide, T., Ikeda, K., Ibuka, T., and Fujii, N. (1995). *Int. J. Pept. Prot. Res.*, **45**, 312.
41. Tamamura, H., Matsumoto, F., Sakano, K., Ibuka, T., and Fujii, N. (1998). *Chem. Commun.*, 151.
42. Weber, U. and Hartter, P. (1970). *Hoppe-Seyler's Zeitschrift für Physiologische Chemie*, **351**, 1384.
43. Gosden, A., MacRae, R., and Young, G. T. (1977). *J. Chem. Res. (Synopsis)*, 22.
44. Ramage, R. (1999). Personal communication.
45. Grandas, A., Jorba, X., Giralt, E., and Pedroso, E. (1989). *Int. J. Pept. Prot. Res.*, **33**, 386.
46. Sieber, P. (1987). *Tetrahedron Lett.*, **28**, 6147.
47. Atherton, E., Hardy, P. M., Harris, D..E., and Mattews, B. H. (1991). In *Peptides 1990: Proceedings of the Twenty-First European Peptide Symposium* (ed. E. Giralt and D. Andreu), pp. 243–244. Escom, Leiden, The Netherlands.
48. Fujiwara, Y., Akajii, K., and Kiso, Y. (1994). *Chem. Pharm. Bull.*, **42**, 724.
49. Barlos, K., Chatzi, O., Gatos, D., and Stavropoulos, G.. (1991). *Int. J. Pept. Prot. Res.*, **37**, 513.
50. Bernatowicz, M. S., Kearney, T., Neves, R. S., and Köster, H. (1989). *Tetrahedron Lett.*, **30**, 4341.
51. Mergler, M., Nyfeler, R., Tanner, R., Gosteli, J., and Grogg, P. (1988). *Tetrahedron Lett.*, **29**, 4005.
52. Mergler, M., Gosteli, J., Nyfeler, R., Tanner, R., and Grogg, P. (1989). In *Peptides 1988: Proceedings of the 20th European Peptide Symposium* (ed. G. Jung and E. Bayer), pp. 133–135. Walter de Gruyter, New York.
53. Chiva, C., Vilaseca, M., Giralt, E., and Albericio, F. (1999). *J. Pept. Sci.*, **5**, 131.
54. Kirstgen, R., Sheppard, R. C., and Steglich, W. (1987). *J. Chem. Soc., Chem. Commun.*, 1870.
55. Kirstgen, R., Olbrich, A., Rehwinkel, H., and Steglich, W. (1988). *Justus Liebigs Annalen der Chemie*, 437.
56. Kates, S. A., Diekmann, E., El-Faham, A., Ionescu, D., McGuinness, B. F., Triolo, S. A., Albericio, F., and Carpino, L.A. (1996). In *Techniques in protein chemistry VII* (ed. D. R. Marshak), pp. 515–523. Academic Press, New York.
57. Han, Y., Vagner, J., and Barany, G. (1996). In *Innovation and perspectives in solid phase synthesis and combinatorial libraries: peptides, proteins and nucleic acids, small molecule organic chemical diversity* (ed. R. Epton) , pp. 385–388. Mayflower Worldwide Ltd., Birmingham, UK.
58. Musiol, H. J., Siedler, F., Quarzago, D., and Moroder, L. (1994). *Biopolymer*, **34**, 1553.

59. Kaiser, T., Nicholson, G., Kohlbau, H. J., and Voelter, W. (1996). *Tetrahedron Lett.*, **37**, 1187.
60. Han, Y., Albericio, F., and Barany, G. (1997). *J. Org. Chem.*, **62**, 4307.
61. Albericio, F. and Carpino, L. A. (1997). *Methods in Enzymology*, **289**, 104.
62. Carpino, L. A., Ionescu, D., and El-Faham, A. (1996). *J. Org. Chem.*, **61**, 2460.
63. Angell, Y. M., Han, Y., Albericio, F., and Barany, G. (1998). In *Peptides— chemistry and biology. Proceedings of the Fifteen American Peptide Symposium* (ed. J. P. Tam and P. T. P. Kaumaya), pp. 339–340. Kluwer–ESCOM, Dordrecht, The Netherlands.
64. Lukszo, J., Patterson, D., Albericio, F., and Kates, S. A. (1996). *Lett. Pept. Sci.*, **3**, 157.
65. Eritja, R., Ziehler-Martin, J. P., Walker, P. A., Lee, T. D., Legesse, K., Albericio, F., and Kaplan, B. E. (1987). *Tetrahedron*, **43**, 2675.
66. Atherton, E., Sheppard, R. C., and Ward, P. (1985). *J. Chem. Soc., Perkin Trans. 1*, 2065.
67. Engebretsen, M., Agner, E., Sandosham, J., and Fischer, P. M. (1997). *Int. J. Pept. Prot. Res.*, **49**, 341.
68. Mendelson, W. L., Tickner, A. M., Holmes, M. M., and Lantos, I. (1990). *Int. J. Pept. Prot. Res.*, **35**, 249.
69. Lamthanh, H., Roumestand, C., Deprun, C., and Ménez, A. (1993). *Int. J. Pept. Prot. Res.*, **41**, 85.
70. Mazur, S. and Jayalekshmy, P. (1978). *J. Am. Chem. Soc.*, **101**, 677.
71. Albericio, F., Hammer, R. P., García-Echevarría, C., Molins, M. A., Chang, J. L., Munson, M. C., Pons, M., Giralt, E., and Barany, G. (1991). *Int. J. Pept. Prot. Res.*, **37**, 402.
72. Royo, M., Contreras, M. A., Giralt, E., Albericio, F., and Pons, M. (1998). *J. Am. Chem. Soc.*, **120**, 6639.
73. Jaenicke, R. and Rudolph, R. (1989). In *Protein structure: a practical approach* (ed. T. E. Creighton), pp. 191–223. IRL Press, Oxford, and references cited therein.
74. Burns, J. A., Butler, J. C., Moran, J., and Whitesides, G. M. (1991). *J. Org. Chem.*, **56**, 2648.
75. Cleland, W. W. (1964). *Biochemistry*, **3**, 480.
76. Hantgan, R. R., Hammes, G. G., and Scheraga, H. A. (1974). *Biochemistry*, **13**, 3421.
77. Akaji, K., Tatsumi, T., Yoshida, M., Kimura, T., Fujiwara, Y., and Kiso, Y. (1992). *J. Am. Chem. Soc.*, **114**, 4137.
78. Ferrer, M., Woodward, C., and Barany, G. (1992). *Int. J. Pept. Prot. Res.*, **40**, 194.
79. Munson, M. C., Lebl, M., Slaninová, J., and Barany, G. (1993). *Peptide Research*, **6**, 155.
80. Kellenberger, C., Hietter, H., and Luu, B. (1995). *Peptide Research*, **8**, 321.
81. Barany, G., Angell, Y. M., Annis, I., Chen, L., Gross, C. M., and Hargittai B. (1998). In *Proceedings of the Fifth International Symposium on Solid-Phase Synthesis and Combinatorial Chemical Libraries* (ed. R. Epton), pp. 85–88. Mayflower Scientific Ltd., UK.
82. Karim Ahmed, A., Schaffer, S. W., and Wetlaufer D. B. (1975). *J. Biol. Chem.*, **250**, 8477.
83. Rothwarf, D. M. and Scheraga, H. A. (1993). *Biochemistry*, **32**, 2680.
84. Ruoppolo, M. and Freedman, R. B. (1995). *Biochemistry*, **34**, 9380.

85. Sabatier, J.-M., Darbon, H., Fourquet, P., Rochat, H., and Van Rietschoten, J. (1987). *Int. J. Pept. Prot. Res.*, **30**, 125.
86. Kuhelj, R., Dolinar, M., Pungercar, J., and Turk, V. (1995). *Eur. J. Biochem.*, **229**, 533.
87. Wallace, T. J. (1964). *J. Am. Chem. Soc.*, **86**, 2018.
88. Tam, J. P., Wu, C.-R., Liu, W., and Zhang, J.-W. (1991). *J. Am. Chem. Soc.*, **113**, 6657.
89. Sieber, P., Eisler, K., Kamber, B., Riniker, B., Rittel, W., Märki, F., and De Gasparo, M. (1978). *Hoppe-Seyler's Zeitschrift für Physiologische Chemie*, **359**, 113.
90. Misicka, A. and Hruby, V. J. (1994). *Polish J. Chem.*, **68**, 893.
91. Rivier, J., Kaiser, R., and Galyean, R. (1978). *Biopolymers*, **17**, 1927, and references cited therein.
92. Gray, W. R., Luque, F. A., Galyean, R., Atherton, E., Sheppard, R. C., Stone, B. L., Reyes, A., Alford, J., McIntosh, M., Olivera, B. M., Cruz, L. J., and Rivier, J. (1984). *Biochemistry*, **23**, 2796.
93. Ellman, G. L. (1959). *Archives of Biochemistry and Biophysics*, **82**, 70.
94. Annis, I., Chen, L., and Barany, G. (1998). *J. Am. Chem. Soc.*, **120**, 7226.
95. Tamamura, H., Otaka, A., Nakamura, J., Okubo, K., Koide, T., Ikeda, K., and Fujii, N. (1993). *J. Chem. Soc., Chem. Commun.*, 283.
96. Edwards, W. B., Fields, C. G., Anderson, C. J., Pajeau, T. S., Welch, M. J., and Fields, G. B. (1994). *J. Med. Chem.*, **37**, 3749.
97. Bishop, P. and Chmielewski, J. (1992). *Tetrahedron Lett.*, **33**, 6263.
98. Shih, H. (1993). *J. Org. Chem.*, **58**, 3003.
99. Chen, L., Bauerová, H., Slaninová, J., and Barany, G. (1996). *Peptide Research*, **9**, 114.
100. White, P. (1992). In *Peptides: chemistry, and biology. Proceedings of the 12th American Peptide Symposium* (ed. J. A. Smith and J. E. Rivier), pp. 537–538, Escom, Leiden, The Netherlands.
101. Previero, A., Antonia, M., Previero, C., and Cavadore, J. C. (1976). *Biochimica Biophysica Acta*, **147**, 453.
102. Koide, T., Otaka, A., Suzuki, H., and Fujii, N. (1991). *Synthetic Letters*, 345.
103. Akaji, K., Nishiuchi, H., and Kiso, Y. (1995). *Tetrahedron Lett.*, **36**, 1875.
104. Otaka, A., Koide, T., Shide, A., and Fujii, N. (1991). *Tetrahedron Lett.*, **32**, 1223.
105. Koide, T., Otaka, A., and Fujii, N. (1993). *Chem. Pharm. Bull.*, **41**, 1030.
106. Adeva, A., Camarero, J. A., Giralt, E., and Andreu, D. (1995). *Tetrahedron Lett.*, **36**, 3885.
107. Brois, S. J., Pilot, J. F., and Barnum, H. W. (1970). *J. Am. Chem. Soc.*, **92**, 7629.
108. Mott, A. W., Somczynska, U., and Barany, G (1986). In *Forum peptides le Cap d'Agde 1984* (ed. B. Castro and J. Martinez), pp. 321. Les Impressions Dohr, Nancy, France.
109. Ridge, R. J., Matsueda, G. R., Haber, E., and Matsueda, R. (1982). *Int. J. Pept. Prot. Res.*, **19**, 490.
110. Chen, L., Zoulíková, I., Slaninová, J., and Barany, G. (1997). *J. Med. Chem.*, **40**, 864.
111. Matsueda, R., Kimura, T., Kaiser, E.T., and Matsueda, G. R. (1981). *Chemistry Letters*, **6**, 737.

112. Bernatowicz, M. S., Matsueda, R., and Matsueda, G. R. (1986). *Int. J. Pept. Prot. Res.*, **28**, 107.
113. Albericio, F., Andreu, D., Giralt, E., Navalpotro, C., Pedroso, E., Ponsati, B., and Ruiz-Gayo, M. (1989). *Int. J. Pept. Prot. Res.*, **34**, 124.
114. Ruiz-Gayo, M., Royo, M., Fernández, I., Albericio, F., Giralt, E., and Pons, M. (1993). *J. Org. Chem.*, **58**. 6319.
115. Kamber, B. (1973). *Helv. Chim. Acta*, **56**, 1370.
116. Schroll A. L., and Barany, G. (1989). *J. Org. Chem.*, **54**, 244, and references cited therein.
117. Castell, J. V. and Tun-Kyi, A. (1979). *Helv. Chim. Acta*, **62**, 2507.
118. Simmonds, R. G., Tupper, D. E., and Harris, J. R. (1994). *Int. J. Pept. Prot. Res.*, **43**, 363.
119. Carlsson, J., Drevin, H., and Axen, R. (1978). *Biochemistry J.*, **173**, 723.
120. Mant, C. T., Kondejewski, L. H., Cachia, P. J., Monera, O. D., and Hodges, R. S. (1996). *Methods in Enzymology*, **289**, 426.
121. Rabanal, F., DeGrado, W. F., and Dutton, P. L, (1996). *Tetrahedron Lett.*, **37**, 1347.
122. Wünsch, E., Moroder, L., Göhring-Romani, S., Musiol, H.-J., Göhring, W., and Bovermann, G. (1988). *Int. J. Pept. Prot. Res.*, **32**, 368.
123. Wünsch, E. and Romani, S. (1982). *Hoppe-Seyler's Zeitschrift für Physiologische Chemie*, **363**, 449.
124. Büllesbach, E. E. and Schwabe, C. (1991). *J. Biol. Chem.*, **266**, 754.
125. Tam, J. P. and Lu, Y.-A. (1997). *Tetrahedron Lett.*, **38**, 5599.
126. Shimonishi, Y., Hidaka, Y., Koizumi, M., Hane, M., Aimoto, S., Takeda, T., Miwatani, T., and Takeda, Y. (1987). *FEBS Lett.*, **215**, 165.
127. Spetzler, J. C., Rao, C., and Tam, J. P. (1994). *Int. J. Pept. Prot. Res.*, **43**, 351.

# 5

# Difficult peptides

MARTIN QUIBELL and TONY JOHNSON

## 1. Introduction

Solid phase peptide synthesis (SPPS), first proposed by R. B. Merrifield in 1962 (1), has evolved over three decades into a tremendously powerful method for the preparation of peptides and small proteins (2, 3). An absolute prerequisite for successful syntheses in all solid phase schemes is that reactions which accumulate solid supported products, and by the very nature of the technique contaminating by-products, must proceed cleanly and efficiently. During the earlier years of SPPS this optimal situation was not always achieved, primarily due to contaminated reagents and ill-defined polymers in combination with poorly flexible protection strategies (4, 5). As the methods of SPPS gained popularity and more widespread application, reagents and protection strategies were improved and refined (6, 7). However, reports of notable successful syntheses were accompanied by then unexplained failures, which have since been collectively termed 'difficult peptides' (8, 9). This chapter describes how an intrinsic understanding behind the occurrence of 'difficult peptides' has accumulated, leading to a general synthetic solution—the utilization of a backbone amide protection strategy.

## 2. Difficult peptides—an overview

### 2.1 Background

Within a few years of the introduction of SPPS, it was recognized that the assembly of some peptide sequences posed a special synthetic problem. The main feature evident during these syntheses was a sudden decrease in reaction kinetics, leading to incomplete amino-acylation by activated amino acid residues. The unreacted sites were readily detected by the Kaiser test for free amine (10); however, couplings showed no significant improvement even upon repeated or prolonged reaction. Efficient reactions are known to occur within a fully solvated peptide–polymer matrix, where reagent penetration is rapid and unhindered (11). This optimal situation no longer exists during the assembly of a difficult peptide, where the normally accessible solid phase reaction matrix becomes partially inaccessible during assembly. This situation arises suddenly, typically 6–12 residues into the synthesis, and may then persist for a number of cycles before easing, or in extreme cases remain throughout the completion of the assembly. The crude products are particularly poor if slower coupling β-branched residues (isoleucine, valine, threonine) are introduced after the onset of synthetic difficulties. The principles underlying the occurrence of difficult peptide sequences have for many years been the focus of intense debate and research (8, 9, 12). An intrinsic feature of the numerous ideas proposed is that *aggregation* occurs resulting in poor solvation within the peptide–polymer matrix.

In a fully solvated system, the reactive peptide amino termini are well separated and readily accessible, along with polymer chains and cross-links that are extended and mobile. This is eloquently represented in *Figure 1a*, along with scenarios in which partial collapse or aggregation could occur leading to hindrance of the amino termini. In scenario *b*, the once solvated peptide chains have self-associated, possibly through intramolecular hydrogen bonding, minimizing interaction between the reaction medium and the amino termini and limiting the accessibility of reagents. Scenario *c* depicts poor solvation of the polymer backbone, again hindering reagent access to the peptide chains. Scenario *d* depicts peptide chain association, but now via *interchain* as opposed to *intrachain* bonding as in *b*. This effectively produces a more highly cross-linked matrix that is less mobile or accessible than shown in the optimal situation *a*.

During the early 1970s it was suggested that the occurrence of difficult peptides possibly arose due to an incompatibility between the apolar polystyrene resins, polar peptide chains, and apolar reaction media, such as dichloromethane, used for SPPS as depicted in *Figure 1b* or *1c* (13). This prompted the investigation of alternative polymer supports and reaction media to the

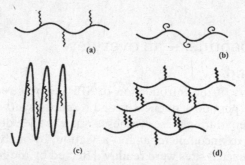

**Figure 1.** Scenarios depicting the possible solvation or aggregation states of a polymer-bound peptide sequence. (a) Peptide and polymer chains (bold) are fully solvated. (b) Peptide chains are intramolecularly aggregated whilst polymer is solvated. (c) Peptide chains are well solvated whilst the polymer backbone is poorly solvated. (d) Peptide chains are intermolecularly aggregated whilst polymer is solvated, effectively increasing the overall level of cross-linking in the matrix. (Reproduced from ref. 7 with permission.)

original Merrifield system. Since peptides are polyamides, a major advance in compatibility was the introduction by Sheppard and co-workers (14) of a polyamide-based polymer support along with the use of dipolar aprotic amide solvents. These new protocols enabled the successful preparation of the deca-peptide sequence from the acyl carrier protein (ACP)-(65–74), which was not prepared readily by standard solid phase techniques (15, 16). However, this breakthrough in the synthesis of difficult peptides did not provide the complete solution since some sequences still remained prone to aggregation (17). It is now accepted that a major contributing factor to the phenomenon of difficult peptides originates in the intrinsic properties of the peptide sequence itself. A wealth of physical evidence has been accumulated (see later) to show that aggregation primarily occurs through the self-association of polymer-bound peptide sequences by *intermolecular* hydrogen bonding as depicted in *Figure 1d*. Thus, a general solution to difficult peptide syntheses requires inhibition of the root cause of the problem, intermolecular backbone hydrogen bonding.

## 2.2 Identifying the effects of aggregation

Following the introduction of polyamide supports, Sheppard and co-workers advocated a milder orthogonal Fmoc/*t*-butyl chemistry for SPPS in place of the somewhat harsh Boc/benzyl protocols of the original Merrifield scheme (18, 19). This new chemistry, along with the advent of a flow stable physically supported polyamide gel, led to the widely adopted continuous flow Fmoc-polyamide technique for SPPS (20–22). This proved to be a pivotal contribution in the understanding of the difficult peptide phenomenon through the real-time spectrophotometric monitoring of a fluorene derived chromophore, released at each $N^\alpha$-Fmoc deprotection cycle, into the solvent stream.

The unhindered release of fluorene derivatives from a fully solvated peptide–polymer matrix results in a characteristic 'bell-shaped' elution profile (20) (*Figure 2a*). Syntheses that exhibit such profiles throughout their assembly are generally of an excellent crude quality. In the case of difficult peptide sequences, the slowing of reaction kinetics at the onset of aggregation is readily observed as a broadening of the release of fluorene derivatives from the reaction matrix (*Figure 2b*). In severe instances, the monitored release of

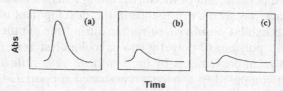

**Figure 2.** Typical Fmoc deprotection profiles (measured at 275 nm) observed during continuous flow SPPS. (a) Unhindered release of fluorene chromophore from a fully solvated peptide–polymer matrix. (b) Release from a partially aggregated peptide–polymer. (c) Release from a fully aggregated peptide–polymer matrix.

fluorene shows a dramatically flattened profile, indicating slow and possibly incomplete removal of the $N^{\alpha}$-Fmoc group (*Figure 2c*), occasionally seen as Fmoc-terminated products in the crude peptide (23).

Numerous non-invasive methods for the identification of difficult peptide sequences have been described, each adding to an overall understanding of the phenomenon. A 'swellographic' method, which is based on the actual physical bed volume of a swollen gel peptide–resin matrix (24–27) provides compelling visual evidence for aggregation. All the popular polymeric gel supports used for peptide synthesis absorb organic solvents and swell to many times their dry bed volume. As a resin-bound peptide sequence is extended, the swollen bed volume of the matrix increases (12). However, in the synthesis of a difficult peptide sequence, the swollen bed volume suddenly decreases by up to 50%, reflecting the increased cross-linking introduced through intermolecular hydrogen bonding between the peptide chains.

The fundamental physiochemical properties of difficult peptide sequences have been investigated using infrared (28–31) and NMR analysis of polymer-bound peptide sequences. Infrared measurement of peptide–polymer gel matrices clearly show the formation of β-secondary structure on transition of a fully solvated to an aggregated sequence. On aggregation, clear bands at 1220, 1240, and 1300 cm$^{-1}$ in the amide III region and 1665 cm$^{-1}$ in the amide I region, characteristic of a β-sheet secondary structure, are observed. A similar effect has been described by Narita *et al.* (32) correlating protected peptide solubility in solution with structure. With Fmoc/polyamide SPPS techniques, the ortho-substituted phenyl ring of residual Fmoc may be observed at 1025 cm$^{-1}$ on deprotection of a difficult peptide sequence (29). Mapelli and Swerdloff (33) have described the $^{13}$C NMR of a gel polymer-bound ACP-(65–74), showing a broadening of the NMR line widths upon aggregation.

## 2.3 Effect of resins, solvents, and additives on aggregation

Numerous modified SPPS protocols have been suggested to improve the quality of difficult peptide syntheses. A common theme among the majority of these protocols is an attempt to limit the extent of β-sheet secondary structure formation by altering the solvation properties of the polymer, the polymer-bound protected peptide, and/or the surrounding reaction media.

The last decade has seen a resurgence in the development of new polymer supports, which exhibit *improved solvation* during the synthesis of difficult peptides. A graft polymer of polyethylene glycol–polystyrene (PEG–PS) has shown impressive improvements in the synthesis of a difficult polyalanine sequence when compared to a polystyrene-based support (34). Meldal (35) has described the use of a polyethylene glycol–polydimethylacrylamide (PEGA) based polymer in an improved preparation of ACP-(65–74). An important factor to consider in the performance of any polymer support is the overall functional loading and the *distribution* of that loading, highlighted in an investigation by Tam and Lu (27). The use of low-substitution resins (typically 0.1–0.4 mmol/g) may be advantageous since a high density of peptide chains may mimic an increased degree of cross-linking and poor solvation. However, Tam and Lu conclude that many such commercial resins may not be evenly functionalized and can contain areas of highly clustered functionality resulting in very poor solvation.

Alternatively, attempts have been made to improve the solvation properties of the reaction media, rather than those of the polymer support alone. Virtually all SPPS, regardless of the type of polymer support, is now performed in dipolar aprotic amide solvents originally suggested by Sheppard and co-workers (14). Additionally, numerous additives that may facilitate the disruption of secondary structure formation through the potential elimination of peptide-chain intermolecular hydrogen bonds have been investigated. The addition of chaotropic salts such as potassium thiocyanate (36) or lithium chloride (37), typically at a concentration of 0.4 M, or detergent (38) at 1% v/v has been suggested. Improved solvation has been described upon the addition of trifluoroethanol (39), hexafluoroisopropanol (40), N-methyl-pyrrolidinone (41), and dimethylsulphoxide (42) to reaction media.

Each of the aforementioned modifications has undoubtedly provided improved crude products in a number of such syntheses. However, improved solvation of the polymer–peptide/reaction media, although extremely desirable, only aims to limit the effects and extent of β-sheet secondary structure formation. The potential for intermolecular backbone hydrogen bonding, the fundamental cause of the phenomenon, still exists.

## 3. Predicting difficult peptide sequences

In order to effectively assembly difficult peptides, more complex protocols aimed at the disruption of secondary structure formation need to be introduced *prior* to the onset of synthetic difficulties. Ideally, this could be achieved if a predictive method describing where and when β-structure formation will occur, based simply on examination of the primary sequence, was available. Unfortunately, there is no complete solution available to date, although a number of groups have developed reasonably successful guidelines. Current predictive methods are derived from four main investigations:

1. Statistical analysis of data gathered during the SPPS of hundreds of peptides (43–46);
2. Ab initio calculations based upon Chou and Fasman parameters (47) from known protein structures (28);
3. Correlation of protected peptide solubility in solution with structure (32, 48, 49);
4. The effects of host–guest substitutions within a given difficult peptide framework (50, 51).

A statistical analysis method has been used to ascribe an *aggregation parameter* for each amino acid (44, 45) based on the change in swollen volume of a peptidyl-resin on couplings in 87 unrelated sequences. Amino acids such as proline, arginine, and histidine were rarely present towards the carboxy terminus of sequences that showed subsequent aggregation and were accordingly given low indices. Conversely, amino acids such as alanine, isoleucine, methionine, and lysine were often present and went on to give difficult peptide problems. The method predicts difficult peptide syntheses with reasonable success, through a run of high aggregation parameter residues in the 6–9 carboxy-terminal residue region of an assembly. The predictive methods of Van Woerkom and Van Nispen (43) are based on the analysis of 696 couplings from 88 sequences. Residues that are difficult to couple with, and the effects of chain length on these couplings, are described.

Narita *et al.* (48) developed a predictive method which can potentially identify soluble protected peptide intermediates for solution-based fragment condensations. A fully solvated peptide–polymer is considered to contain a 'soluble' peptide chain, whilst an aggregated matrix contains a partially soluble or 'insoluble' chain (12). Thus, the findings of Narita *et al.* might be used to predict solid phase difficult peptide syntheses if one considers the addition of each amino acid during assembly as the formation of a new protected segment and access that segment accordingly. The single most striking conclusion from the work of Narita *et al.* was the onset of an unordered structure in solution of a segment containing an appropriately positioned tertiary amide bond. Such segments showed substantially improved solubility (32, 49).

The host–guest substitution investigations of Sheppard and co-workers were based on the introduction of guest residues into a host sequence (**I**) with a strong tendency towards aggregation, followed by continuous flow SPPS monitoring of the $N^\alpha$-Fmoc fluorene-derived chromophore released at each cycle (51). After introduction of the two guest residues, the syntheses were continued by addition of alanine residues until aggregation was observed through broad Fmoc deprotection profiles.

$$\text{H-(Ala)}_n\text{-X-X-(Ala)}_3\text{-Val-polydimethylacrylamide} \qquad \textbf{(I)}$$

*Figures 3a–d* show the deprotection profile peak heights as a function of chain length and guest residue and a number of conclusions were clearly apparent. The hydrophobic guest residues alanine, valine, isoleucine, leucine, and phenylalanine showed a strong tendency to aggregate on the addition of only one or two more alanine residues (*Figure 3a*). A similar result based on aggregation parameters was obtained by Krchnak *et al.* (44, 45). The side-chain protecting group of trifunctional residues, e.g. ranging from the bulky hydrophobic trityl to the relatively polar acetamidomethyl for a series of cysteine derivatives, showed little effect on the propensity for aggregation

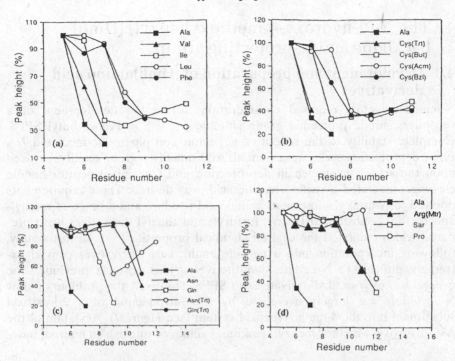

**Figure 3.** Deprotection peak heights as a function of host residue (-X-X-) and chain length in the SPPS of the known difficult peptide test sequence H-(Ala)$_n$-X-X-(Ala)$_3$-Val-poly-dimethylacrylamide (I). (a) General hydrophobic residues. (b) Effect of side-chain protection of cysteine. (c) Effect of trityl protection of side-chain carboxamide of glutamine and asparagine. (d) Effect of arginine and tertiary amino acids proline and sarcosine. (Reproduced from ref. 51 with permission from Munksgaard International Publishers Ltd., Copenhagen.)

(*Figure 3b*). An exception was the side-chain carboxamide protection of glutamine and possibly asparagine by trityl, which showed a beneficial effect in delaying aggregation (*Figure 3c*). The introduction of arginine delayed the onset of aggregation significantly, as did the introduction of histidine (data not shown) (*Figure 3d*). The most striking result was the introduction of the tertiary amide residue such as proline or sarcosine which abolished the propensity for aggregation until a further six alanine residues were added (*Figure 3d*), this being the minimum length at which polyalanine itself aggregated (*Figure 3a*).

The host–guest results, along with the segment solubility results of Narita *et al.* (48, 49) and Blackmeer *et al.* (52), suggested that the difficult peptide problem might be eliminated by the introduction of a reversibly *N*-alkylated amino acid residue *prior* to aggregation.

# 4. The *N*-(2-hydroxy-4-methoxybenzyl) (Hmb) backbone amide-protecting group

## 4.1 Development and preparation of (Hmb)amino acid derivatives

Numerous factors required consideration during the development of a secondary amide protecting group suitable for use in Fmoc/*t*-butyl SPPS. Complete stability to the repetitive acylation and piperidine-mediated $N^{\alpha}$-Fmoc deprotection cycles was essential. Additionally, a group readily removed upon the acidolytic side-chain deprotection and C-terminal peptide-handle cleavage, mediated by trifluoroacetic acid, was desired. These requirements pointed to *N*-aralkyl-substituted amino acid residues and a range of methyl- and methoxy-substituted benzyl, furfuryl, and thienyl derivatives were prepared which exhibited the desired chemical properties (53). Unfortunately, following incorporation into a peptide chain, such derivatives proved extremely difficult to N-acylate quantitatively. This obstacle precludes the *general* use of *N*-aralkyl derivatives in SPPS. However, the problem of slow N-acylation was largely overcome by the introduction of a 2-hydroxyl substituent into the 4-methoxybenzyl system (see *Figure 4*). Acylation of the *N*-(2-hydroxy-4-methoxybenzyl) residue is achieved through an internal base-

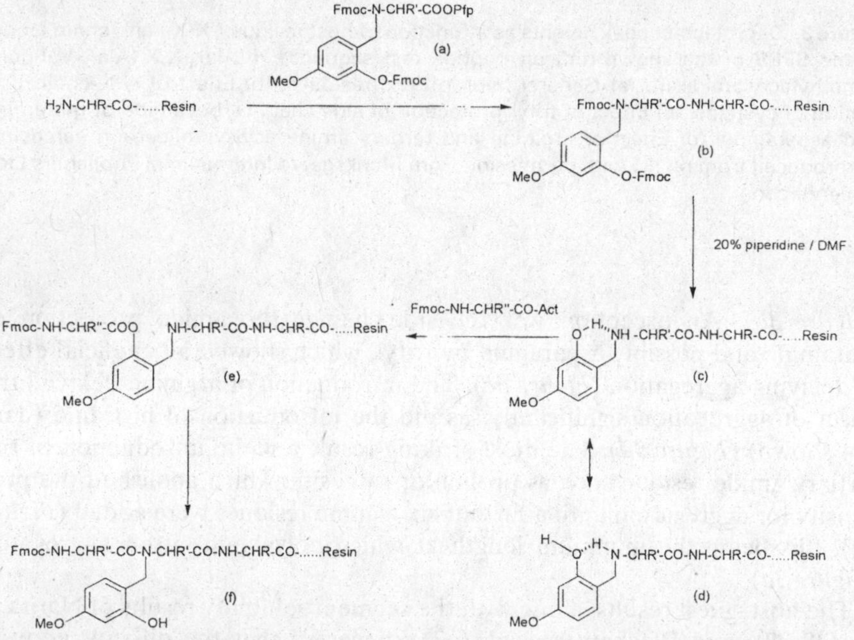

**Figure 4.** Scheme for the incorporation of *N,O*-bis(Fmoc)-*N*-(Hmb)-protected pentafluorophenyl esters (a) into a peptide chain. Key hydrogen-bonded tautomer (c) catalyses esterification to (e) which undergoes O →N acyl transfer.

catalysed esterification providing *4e*, favoured via tautomer *4c*, followed by an intramolecular O →N acyl transfer giving *4f*. The acyl transfer mechanism overcomes the severe steric hindrance encountered during the direct acylation of the *N*-aralkyl secondary amino group and although coupling of the in-coming activated residue is relatively slow compared to standard reactions, full incorporation may be achieved in the vast majority of cases.

The Hmb-substituted amino acids of non-functional and side-chain *t*-butyl/ Boc-protected derivatives are readily prepared from commercially available 2-hydroxy-4-methoxybenzaldehyde and amino acid residues by reductive amination (*Protocol 1*). To date the following Hmb residues have been prepared: Ala, Arg(Pmc), Asn(Trt), Asp(O*t*Bu), Cys(Acm), Glu(O*t*Bu), Gln(Trt), Gly, Ile, Leu, Lys(Boc), Phe, Ser(*t*Bu), Thr(*t*Bu), Tyr(*t*Bu), and Val.

---

**Protocol 1.** Preparation of *N*-(2-hydroxy-4-methoxybenzyl)amino acids

*Reagents*

- Amino acid
- 2-Hydroxy-4-methoxybenzaldehyde
- Sodium borohydride
- Concentrated HCl
- Phosphorous pentoxide
- KOH
- Ethanol
- 1 M NaOH
- MeOH

*Method*

1. Dissolve the amino acid (10 mmol) with stirring in water (10 ml) containing potassium hydroxide (10 mmol). Note that some of the more hydrophobic side-chain protected derivatives, e.g. *O-t*-butyl tyrosine, require the addition of ethanol to achieve full dissolution.

2. Add a solution of 2-hydroxy-4-methoxybenzaldehyde (10 mmol) in ethanol (10 ml) in a single portion and stir the resultant bright yellow solution at room temperature for 30 min.

3. Add a solution of sodium borohydride (10.1 mmol) in water (5 ml containing a few drops of 1 M NaOH) dropwise over 15 min, resulting in effervescence and some colour loss.

4. Heat the mixture briefly on a water bath at 60°C until the yellow solution becomes essentially colourless (requires no longer than 10 min), and then cool slowly to room temperature.

5. Acidify the cooled solution to pH 4 with concentrated hydrochloric acid. If product precipitation is not observed, concentrate the acidified mixture *in vacuo* to approximately half-volume.

6. Leave standing at 4°C overnight, collect the precipitate and wash carefully with cold 20% aq. MeOH (3 × 50 ml), water (2 × 50 ml), and finally vacuum dry over phosphorous pentoxide.

---

---

**Protocol 2.** Preparation of *N,O*-bis(Fmoc)-*N*-(2-hydroxy-4-methoxybenzyl)amino acids

*Reagents*

- *N*-(2-Hydroxy-4-methoxybenzyl)amino acid
- 1,4-Dioxane
- Diethyl ether
- $Na_2SO_4$
- $CHCl_3$

- Sodium carbonate decahydrate
- Fmoc-Cl
- 0.5 M HCl
- MeOH

*Method*

1. Add the *N*-(2-hydroxy-4-methoxybenzyl)amino acid (5 mmol) to an ice-cold stirred solution of sodium carbonate decahydrate (16 mmol, 3.2 excess) in water (50 ml). Add 1,4-dioxane (30–50 ml) to ensure dissolution.

2. Add dropwise a solution of Fmoc-Cl (10.5 mmol, 2.1 excess) in dioxane (10 ml) over 1 h.

3. After a further 1–2 h reaction, monitored by reverse-phase HPLC, the original clear solution should thicken with the precipitated product.

4. Quench the reaction mixture with diethyl ether (100 ml), and extract the aqueous layer with further ether (2 × 100 ml).[a]

5. Wash the combined ethereal extracts with 0.5 M HCl (2 × 100 ml), dry over $Na_2SO_4$, and evaporate *in vacuo* to an oily solid.

6. Dry the solid *in vacuo* overnight.

7. Purify the crude derivative using Kieselgel 60 (Merck) (20 g silica/g crude product), initially packed and sample loaded in $CHCl_3$. Elute the column with $CHCl_3$ until the UV positive material no longer elutes.

8. Gradually change the elution solvent up to 5% MeOH in $CHCl_3$.[b]

9. Pool the desired fractions and evaporate *in vacuo* to a crystalline solid. Isolated yields are generally around 20% based on the starting (Hmb)amino acid.

[a] These ethereal layers contain the sodium salt of the desired bis(Fmoc) product.
[b] In general the bis(Fmoc) derivatives elute at $R_f = 0.3$ (5% MeOH in $CHCl_3$).

---

A key point to note in the synthesis of the *N,O*-bis(Fmoc)-*N*-(2-hydroxy-4-methoxy benzyl)amino acid derivatives is the balance of water to dioxane in the reaction mixture. Improved yields are obtained if the sodium salt of the desired product precipitates in the reaction mixture on its formation, since the Fmoc group present on the 2-hydroxyl as a carbonate is removed slowly by aq. $Na_2CO_3$. Thus, if the proportion of dioxane present in the reaction mixture is sufficient to keep the desired product is solution, then predominately the monoFmoc species are isolated. The bisFmoc acid derivatives of (Hmb)Ala, (Hmb)Asp(O$^t$Bu), (Hmb)Gly, (Hmb)Leu, (Hmb)Lys(Boc), (Hmb)Phe, and (Hmb)Val along with a selection of the pentafluorophenyl esters are available from Novabiochem.

---

**Protocol 3.** Preparation of *N,O*-bis(Fmoc)-*N*-(2-hydroxy-4-methoxybenzyl)amino acid pentafluorophenyl esters

*Reagents*
- *N,O*-bis(Fmoc)-*N*-(2-hydroxy-4-methoxy benzyl)amino acid
- DCC
- THF
- Pentafluorophenol
- CHCl$_3$

*Method*

1. Dissolve *N,O*-bis(Fmoc)-*N*-(2-hydroxy-4-methoxybenzyl)amino acid (10 mmol) and pentafluorophenol (11 mmol) in dry THF (40 ml) with stirring and ice-cooling.
2. Add dropwise a solution of DCC (10.5 mmol) in dry THF (10 ml) over 5 min, leave at 0°C for 30 min.
3. Stir overnight at room temperature.
4. Filter off the precipitated dicyclohexylurea, wash with THF (2 × 10 ml), and discard.
5. Evaporate *in vacuo* the combined THF solution to give an oily solid.
6. Purify the crude ester using Kieselgel 60 (Merck) (10 g silica/g crude product) packed with CHCl$_3$ and load the sample in CHCl$_3$. Elute the column with CHCl$_3$ until UV positive material no longer elutes.
7. Combine the appropriate fractions and evaporated to dryness *in vacuo*. Triturate the residual material with hexane.
8. Collect the purified ester by filtration and dry *in vacuo*. Isolated yields are typically > 90% based on starting acid.

---

## 4.2 Incorporation of *N*-(2-hydroxy-4-methoxybenzyl)amino acid residues

The incorporation of (Hmb)amino acid residues is conveniently achieved through 1-hydroxybenzotriazole activation of the *N,O*-bis(Fmoc)-*N*-(Hmb)-protected pentafluorophenyl esters (*Figure 4a*). A number of groups have recently described the successful use of the *N*$^\alpha$-Fmoc-*N*-(Hmb)-amino acid derivatives (54, 55) in the synthesis of difficult peptides. However, we have previously found that the monoFmoc derivatives of residues other than glycine or alanine couple slowly to resin-bound peptide chains (56). The slow coupling of these species is attributed to the formation of a relatively low

**Table 1.** (Fmoc-AA-)$_2$O/DCM coupling times (in hours) to terminal (Hmb)AA···resin [a,b]

| Incoming residue | N-terminal (Hmb)amino acid residue | | | | | | |
|---|---|---|---|---|---|---|---|
| | Gly | Ala | Leu | Asp(O$^t$Bu) | Phe | Gln(Trt) | Val |
| Gly | 0.25 | 0.25 | 0.5 | 1 | 1 | 3 | 3 |
| Ala | 0.25 | 0.25 | 1 | 2 | 2 | 4 | 5 |
| Leu | 0.5 | 1 | 3 | 4 | 5 | 24 | 24 |
| Asp(O$^t$Bu) | 1 | 2 | 4 | 8 | 24 | 24 | 24 |
| Phe | 1 | 4 | 6 | 24 | 24 | 24 | 24 |
| Val | 1 | –[c] | –[c] | –[c] | –[c] | –[c] | –[c] |

[a] Conditions: 10 equiv. of anhydride in a minimum of dichloromethane. Note that some anhydrides are not very soluble, e.g. Phe, Cys(Acm), to which a maximum of 10% v/v DMF may be added to aid solubility. Acylations still proceed adequately if the anhydride remains gelatinous.
[b] Readily monitored by the Kaiser amine test (10).
[c] Slow reactions; not reaching completion even upon prolonged coupling.

reactivity 4,5-dihydro-8-methoxy-1,4-benzoxazepin-2(3H)-one, a seven-membered lactone, upon activation of the monoFmoc derivative (54, 57). In general, we have found that the fully protected, pentafluorophenyl ester activated derivatives give superior results. As standard, the N,O-bis(Fmoc)-N-(Hmb)-protected pentafluorophenyl esters are coupled at a threefold excess. Even the most hindered additions, e.g. Fmoc-N-(FmocHmb)Ile-OPfp/HOBt onto H-Ile—PepsynKA®, proceed quantitatively with an overnight reaction.

Quantitative N-acylation of the terminal Hmb residue, even with the aid of the acyl transfer mechanism, requires special attention. As illustrated in *Figure 4*, an initial base-catalysed esterification occurs which is favoured through tautomer *4c*. Optimal acylation occurs via symmetric anhydride (7) (*Protocol 4*), urethane-N-carboxy anhydride (58), or pre-formed acid fluoride (59) in the apolar aprotic solvent dichloromethane, conditions which preserve the crucial hydrogen-bonded structure *4c*. Typical acylation times are detailed in *Table 1*.

---

**Protocol 4.** General preparation of symmetric anhydrides for Hmb residue acylation

*Equipment and reagents*
- 50 ml Falcon tube
- Fmoc-amino acid derivative
- DIC
- Resin-bound N-terminal (Hmb)amino acid
- DCM
- DMF
- MeOH
- Ether

*Method*

1. Dissolve the Fmoc-amino acid derivative (20 eq excess to resin loading) in DCM (5 ml/mmol amino acid) with stirring and ice-cooling

---

in a 50 ml Falcon tube. If the amino acid appears particularly insoluble, then add DMF (500 μl) to aid dissolution. Note that some residues will still appear only partially soluble, e.g. Phe, Ala, Cys(Acm).

2. Add DIC (10 eq) in DCM (1 ml) over a few minutes—initially the mixture generally becomes much more soluble.

3. Stir the mixture at 0 °C for 30 min.

4. Add the resin-bound *N*-terminal (Hmb)amino acid to the anhydride solution.[a]

5. Either dilute the reaction volume with additional DCM or evaporate briefly *in vacuo* until the resin mixture is just covered. Seal the Falcon tube to minimize solvent evaporation and leave the reaction mixture for an appropriate length of time.[b]

6. Collect the resin by filtration and wash the resin with DMF (5 vol.) followed by methanol (2 vol.) and ether (2 vol.).

[a] A number of anhydrides appear fairly insoluble, but experience has shown that acylation does occur.
[b] Refer to *Table 1*.

## 4.3 Site selection for Hmb backbone protection

The choice of site for Hmb protection is based on the application of a set of standard considerations of the difficult peptide being synthesized:

1. Protection must be implemented *prior* to aggregation to be effective.

2. Backbone protection need only be introduced a maximum of each sixth residue (51, 60, 61). As the synthesis extends this becomes more flexible. Thus for a 20 residue sequence containing no proline residues, a maximum of 2 or 3 Hmb residues are required located approximately at residues 5–8 and 12–15.

3. The amide bond to be protected needs to be compatible with Hmb chemistry. The main restriction here is that incoming β-branched residues (Val, Ile, Thr) do not couple quantitatively to Hmb residues other than glycine. All other combinations may be protected.

4. Aspartyl bonds prone to base-catalysed transformations, e.g. Asp(O$^t$Bu)-Asn or Asp(O$^t$Bu)-Gly, require protection (62).

5. Given a choice of position for Hmb protection, dependent on the desired sequence, the chemically easiest bond to synthesize is protected (see following example).

The details of synthetic improvements upon the introduction of Hmb protection into a wide range of difficult peptides have been published, illustrating the decisions (55, 60, 61, 63–66).

## 4.4 Improved difficult peptide syntheses through Hmb backbone protection

The following examples illustrate the dramatic improvement obtained in difficult peptide syntheses on the introduction of only one or two judiciously placed Hmb residues and exemplifies the decision processes undertaken when choosing the optimal backbone amide bond for protection.

### 4.4.1 Example 1. An HIV-related octadecapeptide sequence

The syntheses were performed on 0.1 mmol/g Pepsyn KA® resins (Novabiochem) under standard conditions (7) using a Crystal® automated synthesizer (Novabiochem). The 18-mer target (**II**) which constituted the C-terminal section of a larger 44-residue target was originally prepared without side-chain protection for $Asn^{3,5}$ or $Gln^{11}$ and differential protection of the cysteine residues as $Cys^4(Trt)$ and $Cys^{13}(Acm)$.

$$H\text{-}Leu^1\text{-}Ile\text{-}Asn\text{-}Cys(Trt)\text{-}Asn^5\text{-}Thr(^tBu)\text{-}Ser(^tBu)\text{-}Val^8\text{-}Ile\text{-}Thr(^tBu)\text{-}Gln^{11}$$
$$Ala\text{-}Cys(Acm)\text{-}Pro\text{-}Lys(Boc)\text{-}Val\text{-}Ser(^tBu)\text{-}Phe^{18}\text{-}Pepsyn\ KA^{®} \qquad \textbf{(II)}$$

$N^\alpha$-Fmoc deprotection monitoring showed a dramatic broadening at $Val^8$, which is situated six residues from the β-structure inhibiting tertiary residue $Pro^{14}$. Subsequent additions of the remaining seven residues were extremely slow, especially $Thr(^tBu)^6$ and $Cys(Trt)^4$, and could not be driven to completion even on repeated couplings (as judged by the Kaiser test; ref. 10). The resultant crude product was of a poor quality containing only 33% (by HPLC integration) of the desired material along with numerous impurities (*Figure 5a*).

The synthesis was repeated using trityl side-chain protection of the $Gln^{11}$ carboxamide (*cf.* results illustrated in *Figure 3c*). The broadening of deprotection profiles was inhibited by three additional residues from $Val^8$ to $Asn^5$. However, synthetic difficulties persisted throughout the remainder of the assembly, with only a modest improvement in the final crude product.

With the above observed delay of aggregation onset by the introduction of $Gln(Trt)^{11}$, the addition of a single appropriately placed Hmb residue was predicted to inhibit aggregation throughout the remainder of the assembly. Since the point of Hmb introduction needs to be prior to aggregation difficulties (4.3 rule 1), this defined residues $Asn^5$—$Thr(^tBu)^{10}$ as possibilities, taking into account the presence of $Pro^{14}$. Since the use of an incoming β-branched residue onto an Hmb residue (other than Gly) is not quantitative (4.3 rule 3), the possibilities are limited to $Cys(Trt)^4$ onto $(Hmb)Asn^5$ or $Ser(^tBu)^7$ onto $(Hmb)Val^8$. The optimal point of introduction for backbone protection was chosen as $Val^8$ (4.3 rule 5) which was predicted to delay aggregation for at least a further six residues which would complete the majority of the assembly. $(Hmb)Val^8$ was coupled via 3 eq Fmoc-$N$-(FmocHmb)Val-OPfp/HOBt onto H-$Ile^9$—Pepsyn KA® in 6 h. This was

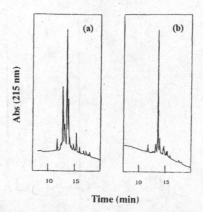

Time (min)

**Figure 5.** HPLC profiles of total crude products from the assembly of octadecapeptide **II**. (a) Standard assembly without side-chain protection of $Gln^{11}$, aggregation onset at $Val^8$. (b) Introduction of $Gln(Trt)^{11}$ and $(Hmb)Val^8$, no aggregation during assembly. Conditions: Aquapore RP-300 $C_8$ (250 × 4.6 mm); 0–20% B in A linear gradient over 25 min, 1.5 ml/min, 215 nm UV detection, where solvent A = 0.1% aq. TFA and solvent B = 90% acetonitrile/10% solvent A. (Reproduced from ref. 61 with permission from Munksgaard International Publishers Ltd., Copenhagen.)

followed by 10 eq $(FmocSer(^tBu)^7)_2O$ (prepared via *Protocol 4* in DCM) for 24 h and a repeated coupling of the anhydride. No synthetic difficulties were observed during the remainder of the assembly and a dramatically improved quality of product was obtained (*Figure 5b*).

### 4.4.2 Example 2. A myelin sheath-related dodecapeptide sequence

The syntheses were performed on 0.2 mmol/g PEG–PS resin (Novabiochem) under standard conditions (7) using a Pioneer® automated synthesizer (Perceptive Biosystems). The 12-mer target (**III**) was an in-house request for a rheumatoid arthritis related project.

$$H-Leu^1-Val-Val-Asn(Trt)-Lys(Boc)-Ile-Arg(Pmc)^7-Ala^8-Thr(^tBu)-$$
$$Phe-Lys(Boc)-Ser(^tBu)-PEG-PS \qquad (III)$$

The standard synthesis produced a crude product that contained a main correct component along with four significant impurities (*Figure 6a*). The five components were identified by electrospray mass spectrometry as (referring to *Figure 6a*): $R_t$ = 6.2 min, *des*$Leu^1$; 6.8 min, *des*$Val^2$ or *des*$Val^3$; 7.8 min, correct material; 13.8 min, Fmoc-$Leu^1$-·····-$Ser^{12}$-$NH_2$ + *des*$Val^2$ or *des*$Val^3$; 14.7 min, Fmoc-protected full-length material. Since the Pioneer® synthesizer does not detail the point of aggregation through extended deprotection profiles, we needed to predict where the synthetic difficulties began and implement appropriate Hmb protection. Since one of the main impurities in the crude product is *des*$Val^2$ or des$Val^3$, one can conclude that aggregation probably started on deprotection of $Val^3$ or $Asn(Trt)^4$. If aggregation began at

129

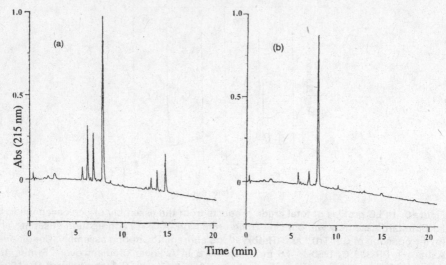

**Figure 6.** HPLC profiles of total crude products from the assembly of decadecapeptide **III**. (a) Standard assembly without backbone protection. (b) Introduction of (Hmb)Ala[8], no aggregation during assembly. Conditions: Phenomenex $C_4$ (250 × 4.6 mm); 0–90% B in A linear gradient over 25 min, 1.5 ml/min, 215 nm UV detection, where solvent A = 0.1% aq. TFA and solvent B = 90% acetonitrile/10% solvent A.

Lys(Boc)[5], then one would expect to see a significant *des*Asn[4] product. This later position for aggregation onset, at residue 3 or 4, is consistent with the inhibiting effect exhibited by Arg(Pmc)[7] (*cf. Figure 4d*, Arg(Pmc) behaves in a similar manner to Arg(Mtr)). Thus Hmb protection needs to be implemented between Lys(Boc)[5] and Ala[8] in order to be effective for the remainder of the synthesis. The three available options are between Asn(Trt)[4] and Lys(Boc)[5], Lys(Boc)[5] and Ile[6], or Arg(Pmc)[7] and Ala[8]. Each of these options is reasonable, thus the decision point is simply based on the bond with the least steric constraints and the easiest (and therefore least precious) Hmb residue to prepare.

The assembly was repeated using (Hmb)Ala[8], coupled via 3 eq Fmoc-*N*-(FmocHmb)Ala-OPfp/HOBt onto H-Thr(^tBu)[9]—PEG–PS in 4 h. This was followed by 10 eq FmocArg(Pmc)[7])$_2$O (prepared via *Protocol 4* in DCM) for 12 h. No synthetic difficulties were observed during the remainder of the assembly and a substantially improved quality of product was obtained (*Figure 6b*).

### 4.4.3 Example 3. Acyl carrier protein, ACP-(65–74)

This decapeptide sequence (**IV**) is a classic difficult peptide synthesis that has been used to assess the performance of new coupling procedures (8), additives (67), resin supports (35), backbone protection (60), etc.

$$\text{H-Val}^{65}\text{-Gln-Ala-Ala}^{68}\text{-Ile-Asp-Tyr-Ile-Asn-Gly}^{74}\text{-OH} \qquad (\textbf{IV})$$

130

A standard Fmoc continuous flow synthesis of ACP-(65–74), on Pepsyn KA®, exhibits aggregation upon the deprotection of the penultimate Gln[66] residue (51, 60). Addition of the final residue Val[65] is now very hindered and occurs to around 85% under conditions that would normally give quantitative incorporation in a non-aggregating synthesis. The crude product contains an earlier eluting *des*Val[65] impurity along with the desired full-length material (*Figure 7a*). The choices for positioning of an Hmb residue range between Ala[67] and Tyr[71]. The Asp[70]–Tyr[71] bond is not particularly prone to aspartimide formation (4.3 rule 4), thus the protection of this bond is not obligatory. The option of Ile[69] onto (Hmb)Asp[70] is not viable (4.3 rule 3). Chemically, the easiest bond to form (4.3 rule 5) is between the residues Ala[67] and Ala[68]; this occurs before the point of aggregation and was chosen as the optimal bond for protection.

The assembly was repeated using (Hmb)Ala[68], coupled via 3 eq Fmoc-*N*-(FmocHmb)Ala-OPfp/ HOBt onto H-Ile[69]—Pepsyn KA® in 2 h. This was followed by 10 eq (FmocAla[67])$_2$O (prepared via *Protocol 4* in DCM) for 1 h. No synthetic difficulties were observed during the remainder of the assembly and a substantially improved quality of product was obtained (*Figure 7b*).

## 4.5 Use of Hmb protection to increase solution solubility

Peptide insolubility occurs when the combined inter- and intrapeptide ionic and hydrophobic interactions outweigh favourable peptide–solvent interactions.

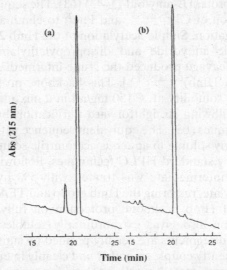

**Figure 7.** HPLC profiles of total crude products from the assembly of decapeptide **IV**. (a) Standard assembly without backbone amide protection. (b) Introduction of (Hmb)Ala[68], no aggregation during remaining assembly. Conditions: Aquapore RP-300 C$_8$ (250 × 4.6 mm); 10–50% B in A linear gradient over 25 min, 1.5 ml/min, 215 nm UV detection, where solvent A = 0.1% aq. TFA and solvent B = 90% acetonitrile/10% solvent A. (Reproduced from ref. 60 with permission from The Royal Society of Chemistry.)

As described in the preceding discussions, when this situation arises within a resin-bound peptide chain, aggregation occurs. When this occurs in solution, individual peptide chains associate leading to precipitation, the major interaction being interchain hydrogen bonding, mediated primarily via the secondary amide bonds along the peptide backbone. As detailed earlier, Narita and co-workers (32, 48, 49, 68) have shown that these precipitates are an ordered β-structure by FT-IR. We have shown through light microscopy and X-ray diffraction that the hydrophobic β-amyloid-(34–42) C-terminal forms ordered precipitates in solution (69).

The key to solubilizing such sequences is the elimination of ordered β-structures. Narita *et al.* (49) and Blaakmeer *et al.* (52) have described the chain dissociating and powerful solubilizing effects gained through protection of an amide bond in solution with the 2,4-dimethoxybenzyl (Dmob)-protecting group. Retention of the structurally similar Hmb amide-protecting group shows the same solubilizing effect on an otherwise free peptide in solution. However, the Hmb group was initially designed to be labile to trifluoroacetic acid. Thus, an Fmoc/*t*-butyl protection-based solid phase synthesis cannot be directly cleaved to the free peptide whilst still retaining Hmb protection. This may, however, be achieved by attenuation of the acid lability of Hmb, in a fully reversible manner, by simple acetylation of the 2-hydroxyl functionality (70).

This protocol was successfully applied in the preparation and purification of the poorly soluble protein β-amyloid-(1–43) (63). The sequence was assembled with Hmb protection of $Gly^{25,29,33,37}$ and $Phe^{20}$ to eliminate the potential for resin-bound aggregation. Simple acetylation of the Hmb 2-hydroxyl moieties, mediated by acetic anhydride and diisopropylethylamine in DMF, and subsequent TFA cleavage produced the crude intermediate β-amyloid-[1–43, $(AcHmb)Phe^{20}$, $(AcHmb)^{Gly25,29,33,37}$]. This backbone-protected intermediate exhibited excellent solubility at > 150 mg/ml in a mixture of 0.1% aq. TFA/acetonitrile (1:1) allowing straightforward purification by standard reverse-phase HPLC techniques (63). The equivalent sequence without backbone protection is very poorly soluble in aqueous acetonitrile solutions and extremely difficult to purify by standard HPLC techniques. Following purification, the AcHmb-protected intermediate was treated with 5% hydrazine-hydrate in DMF to de-O-acetylate, restoring the Hmb group and TFA lability.

The retention of Hmb backbone protection on fully protected peptide segments in solution also gives extraordinarily soluble materials (71, 72). Backbone-protected segments are readily purified by standard reverse-phase HPLC and subsequently couple smoothly and cleanly in either DMF or DCM due to the highly concentrated coupling solutions prepared.

# References

1. Merrifield, R. B. (1962). *Fed. Proc. Fed. Am. Soc. Exp. Biol.*, **21**, 412.
2. Sheppard, R. C. (1988). *Chem. Br.*, 557, and references cited therein.

3. Fields, G. B. and Noble, R. L. (1990). *Int. J. Peptide Protein Res.*, **35**, 161.
4. Kent, S. B. H. (1984). In *Peptides: structure and function* (ed. V. J. Hruby and D. H. Rich), p. 99. Pierce Chemical Co., Rockford, IL.
5. Kent, S. B. H. and Merrifield, R. B. (1983). *Int. J. Pept. Prot. Res.*, **22**, 57.
6. Gross, E. and Meienhofer, J. (ed.) (1981). *The peptides, analysis, synthesis, biology*, Vol. 3. Academic Press, New York.
7. Atherton, E. and Sheppard, R. C. (1989). In *Solid phase peptide synthesis: a practical approach*, p. 1. Oxford University Press, UK.
8. Kent, S. B. H. (1985). In *Peptides, structure and function; Proceedings of the 9th American Peptide Symposium* (ed. C. M. Deber., V. J. Hruby, and K. D. Kopple), p. 407. Pierce Chemical Co., Rockford, IL.
9. Atherton, E, and Sheppard, R. C. (1985). In *Peptides, structure and function; Proceedings of the 9th American Peptide Symposium* (ed. C. M. Deber., V. J. Hruby, and K. D. Kopple), p. 415. Pierce Chemical Co., Rockford, IL.
10. Kaiser, E., Colescott, R. L., Bossinger, C. D., and Cook, P. I. (1970). *Analyt. Biochem.*, **34**, 595.
11. Merrifield, R. B. and Littau, V. (1969). In *Peptides 1968*, p. 179. North-Holland Publishing Co., Amsterdam.
12. Kent, S. B. H., Alewood, D., Alewood, P., Baca, M., Jones, A., and Schnollzer, M. (1992). In *Innovations and perspectives in solid phase synthesis; 2nd International Symposium* (ed. R. Epton), p. 1. Intercept Ltd., Andover, UK.
13. Sheppard, R. C. (1973). In *Peptides 1971; Proceedings of the 7th European Peptide Symposium* (ed. H. Nesvadba), p. 111, North-Holland Publishing Co., Amsterdam.
14. Atherton, E., Clive, D. L. J., and Sheppard, R. C. (1975). *J. Am. Chem. Soc.*, **97**, 6584.
15. Hancock, W. S., Prescott, D. J., Vagelos, P. R., and Marshall, G. R. (1973). *J. Org. Chem.*, **38**, 774.
16. Atherton, E. and Sheppard, R. C. (1974). In *Peptides 1974* (ed. W. Wolmer), p. 123. Wiley, New York.
17. Nguyen, O. and Sheppard, R. C. (1989). In *Proceedings of the 20th European Peptide Symposium* (ed. G. Jung), p. 151. Walter de Gruyter, Berlin, Germany.
18. Atherton, E., Fox, H., Harkiss, D., Logan, C. J., Sheppard, R. C., and Williams, B. J. (1978). *J. Chem. Soc., Chem. Commun.*, 537.
19. Atherton, E., Logan, C. J., and Sheppard, R. C. (1981). *J. Chem. Soc., Perkin. Trans. 1*, 538.
20. Dryland, A. and Sheppard, R. C. (1986). *J. Chem. Soc.. Perkin Trans 1*, 125.
21. Atherton, E. and Sheppard, R. C. (1986). *J. Chem. Soc.. Chem. Commun.*, 165.
22. Dryland, A. and Sheppard, R. C. (1988). *Tetrahedron*, **44**, 859.
23. Due Larsen, B. and Holm, A. (1994). *Int. J. Pept. Prot. Res.*, **43**, 1.
24. Rodionov, I. L., Baru, M. B., and Ivanov, V. T. (1992). In *Innovations and perspectives in solid phase synthesis; 2nd International Symposium* (ed. R. Epton), p. 449. Intercept Ltd., Andover, UK.
25. Rodionov, I. L., Baru, M. B., and Ivanov, V. T. (1992). In *Innovations and perspectives in solid phase synthesis, 2nd International Symposium* (ed. R. Epton), p. 455. Intercept Ltd., Andover, UK.
26. Beyermann, M., Klose, A., Wenschuh, H., and Bienert, M. (1993). In *Peptides 1992; Proceedings of the 22nd European Peptide Symposium* (ed. C. H. Schneider and A. N. Eberle), p. 263. Escom, Leiden, Netherlands.

27. Tam, J. P. and Lu, Y.-A. (1995). *J. Am. Chem. Soc.*, **117**, 12058.
28. De L. Milton, R. C., Milton, S. C. F., and Adams, P. A. (1990). *J. Am. Chem. Soc.*, **112**, 6039.
29. Due Larsen, B., Holm, A., Christensen, D. H., Werner, F., and Faurskov, N. (1992). In *Innovations and perspectives in solid phase synthesis; 2nd International Symposium* (ed. R. Epton), p. 363. Intercept Ltd., Andover, UK.
30. Toniolo, C., Bonara, G. M., Mutter, M., and Maser, F. (1983). *J. Chem. Soc., Chem. Commun.*, 1298, and references cited therein.
31. Narita, M., Honda, S., Umeyama, H., and Ogura, T. (1988). *Bull. Chem. Soc. Jpn.*, **61**, 1201.
32. Narità, M., Ishikawa, K., Nakano, H., and Isokawa, S. (1984). *Int. J. Pept. Prot. Res.*, **24**, 14.
33. Mapelli, C. and Swerdloff, M. D. (1991). In *Peptides 1990; Proceedings of the 21st European Peptide Symposium* (ed. E. Giralt and D. Andreu), p. 316. Escom, Leiden, Netherlands.
34. Barany, G., Sole, N. A., Van Abel, R. J., Albericio, F., and Selsted, M. E. (1992). In *Innovations and perspectives in solid phase synthesis; 2nd International Symposium* (ed. R. Epton), p. 29. Intercept Ltd., Andover, UK.
35. Meldal, M. (1992). *Tetrahedron Lett.* **33**, 3077.
36. Klis, W. A. and Stewart, J. M. (1990). In *Peptides: chemistry, structure and biology; Proceedings of the 11th American Peptide Symposium* (ed. J. E. Rivier and G. R. Marshall), p. 904. Escom, Leiden, Netherlands.
37. Thaler, A., Seebach, D., and Cardinaux, F. (1991). *Helv. Chim. Acta.*, **74**, 628.
38. Zhang, L., Goldammer, C., Henkel, B., Zuhl, F., Jung, G., and Bayer, E. (1994). In *Innovations and perspectives in solid phase synthesis; 3rd International Symposium* (ed. R. Epton), p. 711. Mayflower Worldwide Ltd., Birmingham, UK.
39. Yamashiro, D., Blake, J., and Li, C. H. (1976). *Tetrahedron Lett.*, **18**, 1469.
40. Milton, S. C. F. and De L. Milton, R. C. (1990). *Int. J. Pept. Prot. Res.*, **36**, 193.
41. Fields, G. B. and Fields, C. G. (1991). *J. Am. Chem. Soc.*, **113**, 4202.
42. Hyde, C. B., Johnson, T., and Sheppard, R. C. (1992). *J. Chem. Soc.. Chem. Commun.*, 1573.
43. Van Woerkom, W. J. and Van Nispen, J. W. (1991). *Int. J. Pept. Prot. Res.*, **38**, 103.
44. Krchnak, V., Flegelova, Z., and Vagner, J. (1993). *Int. J. Pept. Prot. Res.*, **42**, 450.
45. Krchnak, V. and Vagner, J. (1992). In *Innovations and perspectives in solid phase synthesis; 2nd International Symposium* (ed. R. Epton), p. 419. Intercept Ltd., Andover, UK.
46. Steinschneider, A. (1994). *Int. J. Pept. Prot. Res.*, **44**, 49.
47. Chou, P. Y. and Fasman, G. D. (1978). In *Advances in enzymology* (ed. A. Meister), Vol. 47, p. 45. Wiley, New York.
48. Narita, M., Ishikawa, K., Chen, J.-Y., and Kim, Y. (1984). *Int. J. Pept. Prot. Res.*, **24**, 580.
49. Narita, M., Fukunaga, T., Wakabayashi, A., Ishikawa, K. and Nakano, H. (1984). *Int. J. Pept. Prot. Res.*, **24**, 306.
50. Merrifield, R. B. (1988). Paper presented at the EURCHEM Conference on *New trends in peptide synthesis*, Port Camargue, France.
51. Bedford, J., Hyde, C., Johnson, T., Jun, W., Owen, D., Quibell, M., and Sheppard, R. C. (1992). *Int. J. Pept. Prot. Res.*, **40**, 300.

52. Blaakmeer, J., Tijsse-Klasen, T., and Tesser, G. I. (1991). *Int. J. Pept. Prot. Res.*, **37**, 556.
53. Johnson, T., Quibell, M., and Sheppard, R. C. (1995). *J. Pept. Sci.*, **1**, 11.
54. Nicolas, E., Pujades, M., Bacardit, J., Giralt, E., and Albericio, F. (1997). *Tetrahedron Lett.*, **38**, 2317.
55. Zeng, W., Regamey, P.-O., Rose, K., Wang, Y., and Bayer, E. (1997). *Int. J. Pept. Prot. Res.*, **49**, 273.
56. Quibell, M. and Johnson, T. (1993). Unpublished observations.
57. Offer, J., Johnson, T., and Quibell, M. (1997). *Tetrahedron Lett.*, **38**, 9047.
58. Fuller, W. D., Cohen, M. P., Shabankarech, M., Blair, R. K., Goodman, M., and Naider, F. R. (1990). *J. Am. Chem. Soc.*, **112**, 7414.
59. Kaduk, C., Wenschuh, H., Beyermann, M., Forner, K., Carpino, L. A., and Bienert, M. (1995). *Lett. Pept. Sci.*, **2**, 285.
60. Johnson, T. J., Quibell, M., Owen, D., and Sheppard, R. C. (1993). *J. Chem. Soc., Chem. Commun.*, **4**, 369.
61. Hyde, C., Johnson, T. J., Owen, D., Quibell, M., and Sheppard, R. C. (1994). *Int. J. Pept. Prot. Res.*, **43**, 431.
62. Quibell, M., Owen, D., Packman, L. C., and Johnson, T. J. (1994). *J. Chem. Soc., Chem. Commun.*, 2343.
63. Quibell, M., Turnell, W. G., and Johnson, T. (1995). *J. Chem. Soc., Perkin Trans. 1*, 2019.
64. Johnson, T., Packman, L. C., Hyde, C. B., Owen, D., and Quibell, M. (1995). *J. Chem. Soc,. Perkin Trans. 1*, 719.
65. Simmonds, R. G. (1996). *Int. J. Pept. Prot. Res.*, **47**, 36.
66. Packman, L. C., Quibell, M., and Johnson, T. (1994). *Pept. Res.*, **7(3)**, 125.
67. Hyde, C., Johnson, T., and Sheppard, R. C. (1992). *J. Chem. Soc., Chem. Commun.*, **21**, 1573.
68. Narita, M., Ogura, T., Sato, K., and Honda, S. (1986). *Bull. Chem. Soc. Jpn.*, **59**, 2433.
69. Quibell, M., Johnson, T., and Turnell, W. G. (1994). *Biomed. Pept. Proteins Nucleic Acids,* **1**, 3.
70. Quibell, M., Turnell, W. G., and Johnson, T. (1994). *Tetrahedron Lett.*, **35**, 2237.
71. Quibell, M., Packman, L. C., and Johnson, T. (1995). *J. Am. Chem. Soc.*, **117**, 11656.
72. Quibell, M., Packman, L. C., and Johnson, T. (1996). *J. Chem. Soc., Perkin Trans. 1*, 1227.

# 6

# Synthesis of modified peptides

SARAH L. MELLOR, DONALD A. WELLINGS,
JEAN-ALAIN FEHRENTZ, MARIELLE PARIS, JEAN MARTINEZ,
NICHOLAS J. EDE, ANDREW M. BRAY, DAVID J. EVANS
and G. B. BLOOMBERG

## 1. C-terminal modifications

### 1.1 Peptidyl *N*-alkyl amides

Several important hormones such as oxytocin, secretin, and LHRH are known
to be peptidyl amides (1–3). In addition to these, other peptidyl amides such
as indolicidin and the protegrins have been shown to exhibit potent anti-
microbial activity (4, 5). The *in vivo* production of such compounds is via
endogenous enzymatic cleavage of propeptides (6), making their synthesis by
genetic engineering notoriously difficult. Furthermore, to facilitate the survival
of synthetic peptidyl amides *in vivo*, an obvious defence against the action of
carboxypeptidases is the N-alkylation of the carboxylic amide terminus. Such
secondary amides would be expected to exhibit vastly different solubility and
transport properties to primary amides, thus their chemical synthesis is of
immense importance.

#### 1.1.1 Synthesis of peptidyl *N*-alkyl amides

The solid phase synthesis of peptidyl amides and peptidyl *N*-alkyl amides is
centred around two main strategies:

1. Ammonolysis/aminolysis of resin-bound esters (*Figure 1a*; see Section 1.2).
2. Use of resin-bound primary amines, which may in turn be chemically
   modified to generate novel secondary amine functionalized linkers for the
   synthesis of peptidyl *N*-alkyl amides (*Figure 1b*).

#### 1.1.2 Use of 9-(fluorenylmethoxycarbonylamino)xanthen-3-yloxymethyl polystyrene and its chemical modification to resin-bound secondary amines

Early examples of the use of resin-bound amines for the solid synthesis of
peptidyl amides involve the use of linkers such as benzhydrylamine or benzyl-
amine. Following peptide assembly, these linkers require highly acidic (e.g.

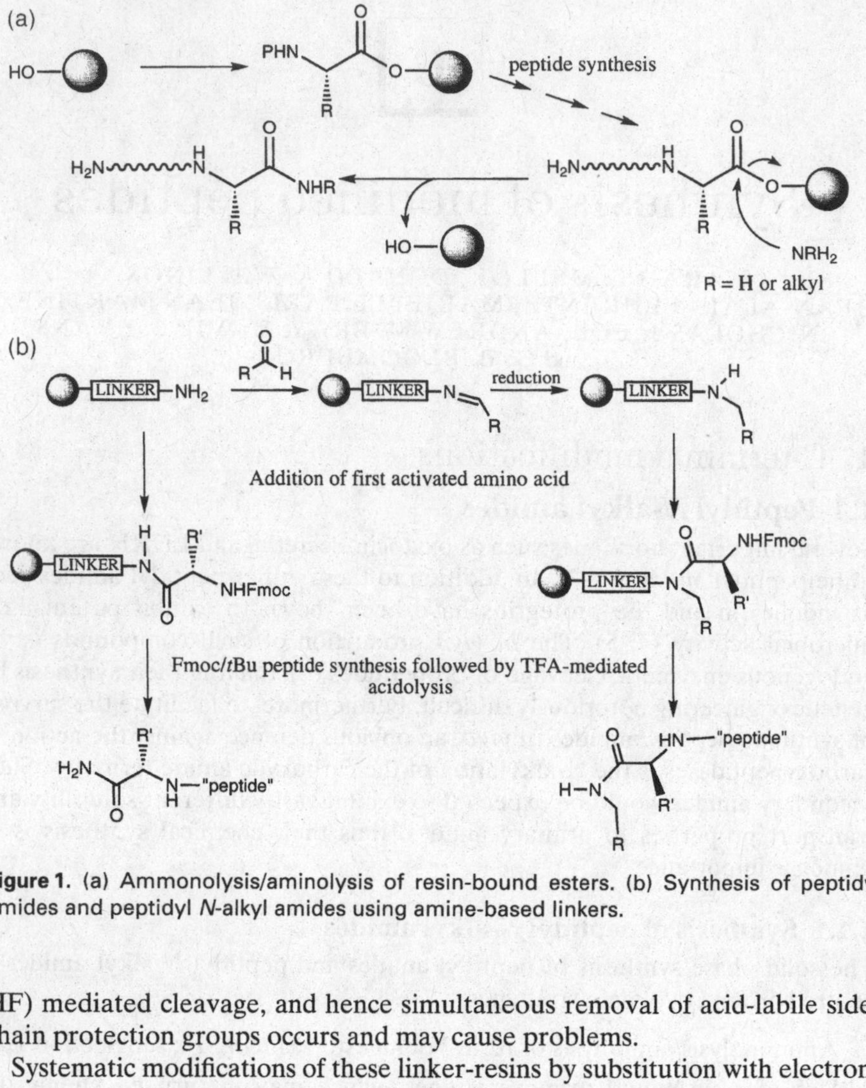

(a)

(b)

**Figure 1.** (a) Ammonolysis/aminolysis of resin-bound esters. (b) Synthesis of peptidyl amides and peptidyl *N*-alkyl amides using amine-based linkers.

HF) mediated cleavage, and hence simultaneous removal of acid-labile side-chain protection groups occurs and may cause problems.

Systematic modifications of these linker-resins by substitution with electron-donating substituents have resulted in the generation of numerous linker-resins with increased acid sensitivity. Notably, the 4-(2′,4′-dimethoxyphenylamino-methyl)phenoxy derivatized (Rink) resin **1** (7), which is cleavable by 95% v/v TFA, and the 5-(2-fluorenylmethoxycarbonylaminomethyl-3,5-dimethoxy)-phenoxyvaleric acid (PAL) linker **2** (8) acidolysed by 75% v/v TFA.

Greater acid lability has been achieved using the xanthenyl derivatized resin, 9-(fluorenylmethoxycarbonylamino)xanthen-3-yloxymethyl polystyrene (Sieber amide) resin **3** (9, 10), which besides being cleavable by 1% v/v TFA, holds the added advantage of readily undergoing reductive *N*-alkylation to afford resin-bound secondary amines for the synthesis of peptidyl *N*-alkyl

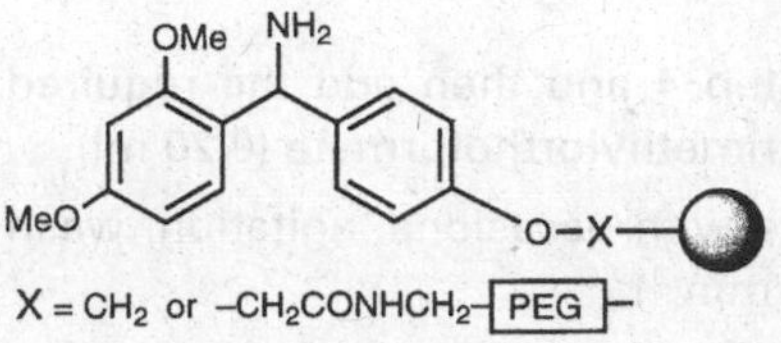

X = CH₂ or —CH₂CONHCH₂—[PEG]—

**1**

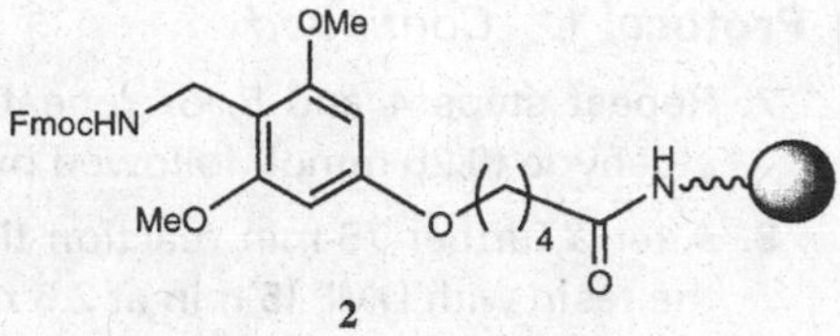

**2**

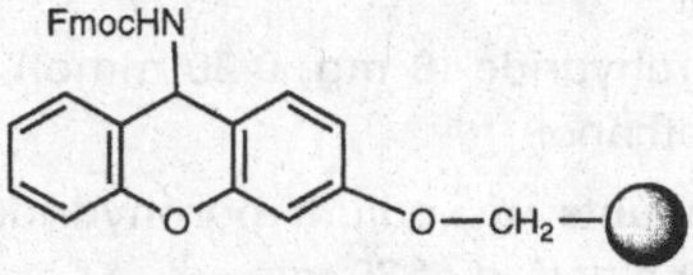

**3**

amides (11). In addition to the method detailed below (*Protocol 1*), a number of alternative approaches have recently been reported (12–14).

---

**Protocol 1.** Synthesis and use of *N*-alkylated 9-aminoxanthenyl (Sieber) resin

### Reagents

- 9-(Fmoc-amino)xanthen-3-yloxymethyl polystyrene (Sieber amide) resin
- DCM
- 20% v/v Piperidine in DMF
- DMF
- Glacial acetic acid
- Trimethylorthoformate

- Aldehyde
- Sodium borohydride
- Methanol
- *N*-Fmoc-amino acid
- HOAt
- HATU
- DIPEA

### A. Synthesis of N-alkylated 9-aminoxanthenyl (Sieber) resin

**Prior to undertaking the reaction, ensure that all glassware is oven-dried.**

1. Pre-swell the 9-(Fmoc-amino)xanthen-3-yloxymethyl polystyrene (Sieber amide) resin (0.05 mmol, *ca.* 100 mg) in DCM for 1 h in a peptide synthesizer column.

2. Treat the resin with 20% v/v piperidine in DMF (10 min at 2.5 ml min⁻¹ or 4 × 5 min batchwise).

3. Wash the resin with DMF (15 min at 2.5 ml min⁻¹).

4. Remove any excess DMF from the resin, and ensure that the resin is suspended in 0.75 ml of DMF.

5. Add the required aldehyde (0.25 mmol) followed by glacial acetic acid (15 µl, giving a 2% v/v solution in DMF) and allow the reaction to proceed for 45 min with occasional agitation.

6. Wash the resin briefly with DMF (2 min at 2.5 ml min⁻¹).

---

**Protocol 1.** *Continued*

7. Repeat steps 4 and 5, or repeat step 4 and then add the required aldehyde (0.25 mmol) followed by trimethylorthoformate (0.20 ml).

8. After a further 75-min reaction time with occasional agitation, wash the resin with DMF (5 min at 2.5 ml min$^{-1}$).

9. Remove any excess DMF, and ensure that the resin is suspended in 1.50 ml of DMF.

10. Add sodium borohydride (8 mg, 0.20 mmol), followed immediately by 0.20 ml of methanol.

11. Add further amounts of sodium borohydride (12 mg, 0.30 mmol) portionwise over a period of 30 min.

12. Gently agitate (by stirring) the reaction mixture for a further 90 min.

13. Thoroughly wash the resin with DMF (20 min at 2.5 ml min$^{-1}$).

B. *Acylation of N-alkylated 9-aminoxanthenyl (Sieber) resin*

The synthesized 9-(*N*-alkyl)aminoxanthenyl resin is then acylated with an *N*-Fmoc-protected amino acid.

1. Remove DMF by applying a positive pressure to the resin-bed contained in the column.

2. In a **dry** 5-ml round-bottomed flask or an appropriate sample-bottle, dissolve the *N*-Fmoc-amino acid (0.25 mmol) in DMF (0.8 ml), and add to the solution HOAt (0.1 mmol), HATU (0.25 mmol), and DIPEA (0.5 mmol).

3. Leave the resultant solution for 2 min, and then add to the resin.

4. Gently agitate the suspension for 18 h.

5. Wash the resin with DMF (15 min at 2.5 ml min$^{-1}$).

6. The derivatized resin may be used immediately for peptide synthesis using standard Fmoc/*t*Bu procedures[a] or collected using a Buchner funnel, and washed with dichloromethane (20 ml) and hexane (10 ml) before being dried *in vacuo* for storage.

[a] Following peptide assembly, the protected peptidyl *N*-alkyl amide may be cleaved from the linker-resin by treatment with 1–2% TFA/0.1% triethylsilane in DCM (5 × 2 min, 2–3 ml for each treatment). If a fully deprotected peptidyl *N*-alkyl amide is required, it is recommended that a two-step procedure is applied: (a) cleave the peptide from the solid support by exposure to 5% TFA/0.1% triethylsilane in DCM (10 ml) for 10 min, filter under gravity to remove resin, wash with 5% TFA in DCM (20 ml), and evaporate *in vacuo* the filtrate to dryness; (b) treat the residue material with a standard 90% TFA cocktail for complete acidolysis of side-chain protecting groups.

Numerous resin-bound alkyl amines have been generated using the reductive alkylation technique. Their subsequent acylation with Fmoc-phenylalanine has been used to assess the effectiveness of the reaction. Using

**Table 1.** N-acylation efficiencies and estimated product purities for *N*'-Fmoc-phenylalanyl-*N*-alkyl aminoxanthen-3-yloxymethyl polystyrene resin

| Alkyl group | % Efficiency | % Purity |
|---|---|---|
| 3-Phenylpropyl | 94 | 92 |
| 3,3-Dimethylbutyl | 99 | 98 |
| Propyl | 99 | 98 |
| 3-Cyclohexylpropyl | 96 | 94 |
| isoButyl | 99 | 98 |
| Benzyl | 55 | 98 |

**Table 2.** Reaction efficiencies and estimated product purities for acylation of *N*-isobutylaminoxanthen-3-yloxymethyl polystyrene resin

| Amino acid | % Efficiency | % Purity |
|---|---|---|
| Fmoc-Asp(O$^t$Bu)-OH | 100 | 92 |
| Fmoc-Ser(O$^t$Bu)-OH | 88 | 97 |
| Fmoc-Pro-OH | 67 | 94 |
| Fmoc-Val-OH | 50 | 94 |

'Fmoc-substitution' as a measure of reaction efficiency and RP–HPLC as an estimate of purity, it was established that the reaction generally generates products of high purity. However, the acylation efficiency, although generally good, is decreased markedly if the resin-bound alkyl amine is particularly sterically hindered (*Table 1*). Likewise, comparing the acylation of resin-bound isobutylamine using a variety of *N*-Fmoc-amino acids, it was also observed that acylation efficiency decreases as the acylating species becomes more bulky (*Table 2*). These considerations must therefore be part of any synthetic strategies for the solid phase construction of peptidyl *N*-alkyl amides.

# 1.2 Use of the 4-hydroxymethylbenzoic acid linkage agent for the synthesis of C-terminal modified peptides

## 1.2.1 Introduction

Since the original concept for the use of linkage agents coupled separately to polyacrylamide resin was proposed by Sheppard and co-workers (15–17) it has become commonplace to use custom-designed moieties to allow C-terminal modifications during cleavage of assembled peptides from the solid support. 4-Hydroxymethylbenzoic acid **4** is one of the most versatile of the linkage agents described by Atherton *et al.* (18) due mainly to its susceptibility to nucleophilic attack combined with an inertness to strong acids. The acylated linkage agent is cleaved readily with ammonia (19) and was therefore

$$HOCH_2 - \text{\unicode{x2394}} - COOH$$

**4**

originally used for the preparation of peptide amides. The flexibility is enhanced by the ability to remove side-chain protection before or after cleavage of the peptide from the resin, hence side-chain protected peptide amides can be prepared for subsequent fragment assembly. In a complimentary fashion, the linkage can be cleaved with hydrazine to produce hydrazides for subsequent activation as azides, for coupling in fragment assembly. The benzyl ester functionality is stable to piperidine used for removal of the Fmoc group, making the linkage agent equally applicable to Boc or Fmoc chemistry.

The use of 4-hydroxymethylbenzoic acid for routine preparation of peptide amides has been somewhat superseded by acid-labile linkage agents that liberate the amide with simultaneous cleavage of the peptide from the resin (7, 20). For preparation of peptide amides on a large scale, however, 4-hydroxymethylbenzoic acid is still favoured economically.

In more recent years, the need to create molecular diversity for peptide libraries has resulted in somewhat of a renaissance for the 4-hydroxymethylbenzoic acid linkage agent. The ability to cleave the peptide-to-resin ester with many nucleophiles including hydroxide, alkoxides, and primary amines has added to the library workers arsenal.

However, 4-hydroxymethylbenzoic acid is not without its drawbacks as a linkage agent. Peptides containing C-terminal proline are particularly prone to diketopiperazine formation (21) due to cleavage of the dipeptide from the resin by its own internal nucleophile. This can often be overcome by insertion of a 'Boc' amino acid as the second amino acid in the chain. Acidolytic removal of the Boc group affords fugitive protection of the amine as a salt, which can be neutralized *in situ* during the next amino acid coupling. The reactivity of other sites in the peptide towards nucleophiles should not be overlooked when assessing the applicability of the linker to assembly of a specific peptide (e.g. rearrangement of aspartic acid residues to aspartimide is not uncommon).

### 1.2.2 Coupling 4-hydroxymethylbenzoic acid to the solid support

The linkage agent can be coupled to amino-functionalized resin as the preformed trichlorophenyl ester using the protocols described by Atherton and Sheppard (22) or activated *in situ* using diisopropylcarbodiimide and 1-hydroxybenzotriazole as described in *Protocol 2*. Although 4-hydroxymethylbenzoic acid can be used with many amino-functionalized supports, it is best used in conjunction with polyacrylamide where the solvation properties of the resin allow more facile recovery of the peptide from the pores of the support following cleavage. As with most linkage agents, it is recommended that an internal reference amino acid such as norleucine is coupled to the resin prior to the linker (23). This allows the use amino acid analysis to assess resin loading and ultimately the extent of cleavage.

---

**Protocol 2.** Attachment of 4-hydroxymethylbenzoic acid to the support

*Equipment and reagents*

- Solid phase peptide synthesis reactor[a]
- 4-Hydroxymethylbenzoic acid (HMBA)
- Aqueous sodium hydroxide solution (1 M):DMF (1:1 v/v)
- Reagents for Kaiser test (see Chapter 3)

- Amino functionalized resin
- DIC
- HOBt
- DMF

*Method*

1. Dissolve HMBA (3 eq) and HOBt (6 eq) in the minimum amount of DMF required to swell the amino-functionalized resin.
2. Transfer the solution to the amino-functionalized resin (1 eq) contained in the solid phase reactor.[a]
3. Stir the mixture (by nitrogen bubbling) and add DIC (4 eq).
4. Allow the reaction to proceed for 2 h before removing a small sample (~2–5 mg) for analysis. Wash the sample with DMF (3 × 1 ml) by filtration or decantation. Submit the sample to the Kaiser test. A positive result (blue coloration of the solution and/or resin beads) indicates that the reaction is not complete.
5. If the reaction is not complete allow the reaction to proceed overnight.
6. If the reaction is not complete after overnight reaction, draw off the reagent mixture and replenish the reaction as described in steps 1–3 above.
7. When the reaction is complete, subject the resin to 10 × 2 min washes with the resin bed volume[b] of DMF.
8. Stir the resin with a solution of aqueous sodium hydroxide solution (1.0 M):DMF (1:1 v/v) for 15 min.[c]
9. Subject the resin to 5 × 2 min washes with the resin bed volume of DMF/H$_2$O (1:1 v/v) followed by 10 × 2 min washes with the resin bed volume of DMF.
10. The resin is now ready for coupling of the C-terminal amino acid in the peptide sequence.

[a] A manually operated solid phase peptide synthesis reactor of the type described by Atherton and Sheppard (24) is suitable for most simple operations.
[b] Various resin types have different swelling characteristics. The bed volume of the wetted resin is a good estimate for the volumes of the wash solvents required.
[c] The wash with aqueous sodium hydroxide removes any unwanted polymerization of the 4-hydroxymethylbenzoic acid on the resin following the coupling.

---

### 1.2.3 Coupling the C-terminal amino acid to 4-hydroxymethylbenzoyl resin

Precautions should be taken to avoid racemization of the C-terminal amino acid during the acylation reaction. Coupling with the carboxylate salt of the first amino acid to the halogenated linker is now popularly employed and many other methods have also been described to reduce enantiomer formation.

The procedure described in these protocols, using DIC for activation and a catalytic amount of the hypernucleophile 4-dimethylaminopyridine (DMAP) for ester bond formation, reduces racemization down to amounts that are not detectable if the conditions described in *Protocol 3* are strictly adhered to.

---

**Protocol 3.** Coupling the C-terminal amino acid to 4-hydroxymethylbenzoyl resin

*Equipment and reagents*
- Solid phase peptide synthesis reactor[a]
- 4-hydroxymethylbenzoyl resin
- Fmoc-amino acid (Fmoc-AA-OH)
- Acetic anhydride (Ac₂O)
- DIC
- DMAP
- DMF

*Method*

1. Place the 4-hydroxymethylbenzoyl resin (1 eq) in the solid phase reactor, followed by the Fmoc-AA-OH to be coupled (3 eq).

2. Add sufficient DMF to make the resin just mobile to nitrogen agitation.

3. Add DIC (4 eq), following this with the dropwise addition of a solution of DMAP dissolved in DMF (0.1 eq, *ca.* 50 mmol/l).

4. Allow the reaction to proceed for 60 min, draw off the reaction solution and subject the resin to 2 × 2 min washes with the resin bed volume of DMF.

5. Recouple the amino acid as in steps 1–3.

6. Subject the resin to 5 × 2 min washes with the resin bed volume[b] of DMF.

7. To acetylate any remaining hydroxyl functions, add sufficient DMF to make the swollen resin just mobile to nitrogen agitation.

8. Add Ac₂O (6 eq) and DMAP (0.1 eq).

9. After 1 h draw off the reaction solution and wash the resin with 10 resin bed volumes of DMF.

---

144

**10.** Use the Fmoc-amino acid derivatized resin for peptide assembly, or wash with DCM (20 ml) and methanol (10 ml), and dry *in vacuo* for 2 h prior to storage.

[a] A manually operated solid phase peptide synthesis reactor of the type described by Atherton and Sheppard (24) is suitable for most simple operations.
[b] Various resin types have different swelling characteristics. The bed volume of the wetted resin is a good estimate for the volumes of the wash solvents required.

## 1.2.4 Side-chain deprotection following peptide–resin assembly

Standard procedures for Fmoc-based peptide chain elongation are described elsewhere in this text (Chapter 3). To produce the required peptide as the unprotected molecule, side-chain deprotection is best carried out prior to cleavage of the peptide from the resin. Selection of the appropriate mixture of trifluoroacetic acid and scavengers for side-chain deprotection is described in Chapter 3, Section 10.

---

**Protocol 4.** Side-chain deprotection of the peptide–resin

*Equipment and reagents*
- Solid phase peptide synthesis reactor[a]
- Trifluoroacetic acid/scavenger cocktail (Deprotection mixture)
- Diethyl ether
- TFA
- DCM
- DMF

*Method*

1. Place the peptidyl resin in the solid phase reactor and add approximately twice the swollen resin bed volume of deprotection mixture.
2. Agitate the mixture and allow the reaction to proceed for the required amount of time.[b]
3. Draw off the deprotection mixture and subject the resin to 3 × 2 min washes with the resin bed volume of TFA.
4. Subject the resin to 2 min washes with the resin bed volume of DCM (3 times) and Et$_2$O (3 times).
5. Air dry the side-chain deprotected peptidyl resin overnight prior to cleavage.

[a] A manually operated solid phase peptide synthesis reactor of the type described by Atherton and Sheppard (24) is suitable for most simple operations.
[b] The reaction time and deprotection mixture is described in Chapter 3, Section 10.

---

## 1.2.5 Cleavage of peptide from the resin using gaseous ammonia

The cleavage procedure for removal of the peptide from the resin as the amide will differ slightly depending on the nature of the solid support.

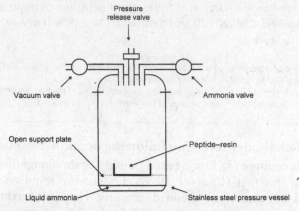

**Figure 2.** Pressure vessel layout.

Polystyrene-based matrices require wetting with an organic solvent to allow complete cleavage and often require washing with organic solvents such as DMF to completely extract the peptide from the resin. Peptides can be cleaved from polyacrylamide based supports in the gas phase due to the ease of penetration of the pores of the peptide–resin matrix with ammonia. The advantage of this approach is that the peptide can often be washed off the resin using water or aqueous solvents. The procedure described in *Protocol 5* is equally applicable to other gaseous primary amines such as methylamine and ethylamine.

The reaction should be carried out in a suitable pressure vessel with a similar layout to that described in *Figure 2*. A key feature of this equipment is the pressure release valve to prevent over-pressurizing the vessel.

---

**Protocol 5.** Cleavage of peptide from the resin with gaseous ammonia

*Equipment and reagents*
- Stainless steel pressure vessel[a]
- Peptide–resin for cleavage
- Ammonia cylinder
- Dry-ice
- Water
- DMF

*Method*
1. Place the peptide–resin in the pressure vessel within a suitable container (e.g. crystallizing dish).
2. After ensuring that the ammonia valve is closed, open the vacuum valve and evacuate the vessel.[b]
3. Close off the vacuum valve, open the ammonia valve and open the ammonia supply.

---

4. Cool the base of the vessel in dry-ice for 20–30 min to allow ammonia to condense in the pressure vessel.

5. Close off the ammonia valve and ammonia supply.

6. Remove the vessel from the dry-ice and allow the reaction to proceed overnight at room temperature.

7. Cool the base of the vessel in dry-ice for 20–30 min to allow ammonia to condense in the base of the pressure vessel.

8. Purge the ammonia from the system by passing nitrogen into the vessel through one valve and scrubbing the exhaust via the other valve.

9. Remove the mixture of peptide and resin from the pressure vessel and transfer to a glass sinter.

10. Subject the resin to 3 × 5 min washes of an appropriate solvent[c] for the peptide.

11. Recover the peptide from the solvent by standard procedures.

[a] A simple pressure vessel with a 100 psi pressure release valve is suitable. These can be purchased from Parr Ltd.
[b] A high vacuum is not required.
[c] For protected peptides DMF is often suitable. Aqueous MeCN is suitable for most deprotected peptides.

## 1.2.6 Cleavage of peptide from the resin using hydrazine

Peptide hydrazides are readily converted to the corresponding acylazides by the action of alkyl nitrites and under appropriate conditions these activated species can be among the least susceptible to racemization. Cleavage of the peptide from the resin as the hydrazide is most useful with protected molecules, which can subsequently be used in fragment condensation to assemble longer peptides.

**Protocol 6.** Cleavage of peptide from the resin with hydrazine hydrate

*Equipment and reagents*
- Solid phase peptide synthesis reactor[a]
- Rotary evaporator
- DMF
- Diethyl ether
- Peptide–resin for cleavage
- Hydrazine hydrate
- Water

*Method*
1. Place the peptide–resin in a solid phase peptide synthesis reactor.[a]
2. Add the minimum amount of DMF required to make the resin mobile to $N_2$ agitation.

147

**Protocol 6.** *Continued*

3. Add hydrazine hydrate (equivalent to 5% v/v of the DMF added above) and allow the reaction to proceed for 15 min.

4. Filter off the hydrazine hydrate solution which now contains the peptidyl hydrazide and subject the resin to 3 × 2 min washes with the resin bed volume[b] of DMF.

5. Reduce the combined hydrazine hydrate solution and DMF washes to an oil by rotary evaporation.

6. Precipitate protected peptidyl hydrazide by addition of water, or in the case of side-chain deprotected molecules precipitate by addition of Et$_2$O.

7. Recover the peptide by filtration and dry under vacuum.

---

[a] A manually operated solid phase peptide synthesis reactor of the type described by Atherton and Sheppard (24) is suitable for most simple operations.
[b] Various resin types have different swelling characteristics. The bed volume of the wetted resin is a good estimate for the volumes of the wash solvents required.

### 1.2.7 Cleavage of peptide from the resin by trans-esterification with alcohols

Cleavage of the peptide from the resin by trans-esterification can be achieved readily with most alcohols. The resultant C-terminal-protected molecules can be used in fragment condensation to assemble longer peptides. The protocol described here involves cleavage with methanol in the presence of a tertiary base. The methyl ester produced can subsequently be removed by saponification.

**Protocol 7.** Cleavage of peptide from the resin by trans-esterification with methanol

*Equipment and reagents*

- Solid phase peptide synthesis reactor[a]
- Rotary evaporator
- DMF
- Diethyl ether
- Peptide–resin for cleavage
- Methanol
- Water
- DIPEA

*Method*

1. Place the peptide–resin in a solid phase peptide synthesis reactor.[a]

2. Add the minimum amount of DMF/MeOH (1:1 v/v) required to make the resin mobile to N$_2$ agitation.

3. Add DIPEA (equivalent to 10% v/v of the DMF/MeOH added above) and allow the reaction to proceed for 1 h.

4. Filter off the DMF/MeOH solution that now contains the peptide methyl ester and subject the resin to 3 × 2 min washes with the resin bed volume[b] of DMF.

5. Reduce the combined DMF/MeOH solution and DMF washes to an oil by rotary evaporation.

6. Precipitate protected peptides by addition of water, or in the case of side-chain deprotected molecules precipitate by addition of Et$_2$O.

7. Recover the peptide by filtration and dry under vacuum.

[a] A manually operated solid phase peptide synthesis reactor of the type described by Atherton and Sheppard (24) is suitable for most simple operations.
[b] Various resin types have different swelling characteristics. The bed volume of the wetted resin is a good estimate for the volumes of the wash solvents required.

## 1.3 Peptide hydroxamic acids

### 1.3.1 Introduction

Peptide hydroxamates or hydroxamic acids have been shown to be potent enzyme inhibitors. Their target enzymes, the matrix metalloproteinases (MMPs), are a family of endopeptidases with a catalytic domain that houses a zinc atom as a coenzyme. The ability of the hydroxamate functionality to chelate such metal ions (25) and thus disrupt coenzyme activity is believed to be the reason for inhibitory activity.

A wide variety of matrix metalloproteinases have been shown to be susceptible to inhibition by a number of specific peptide hydroxamate based inhibitors, for example foroxymithine (*Figure 3a*) has been shown to inhibit angiotensin-converting enzyme (26), and propioxatin A (*Figure 3b*) acts against enkephalinase B (27), an enzyme which hydrolyses the body's naturally occurring analgesic agents. Furthermore, matlystatin B (*Figure 4*) inhibits

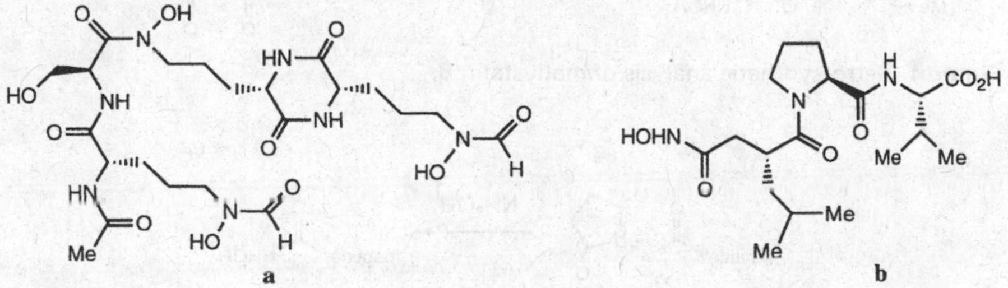

**Figure 3.** (a) Foroxymithine and (b) propioxatin A.

type IV collagenases (28), recently found to be secreted in large amounts by metastatic tumour cells. Thus, inhibition of such enzymes has important consequences for the relief of pain and the control of a variety of life-threatening diseases, ensuring that peptide hydroxamates play a pivotal role in therapeutic strategies.

## 1.3.2 Synthesis of peptide hydroxamates

Much of the research carried out on peptide hydroxamates has involved their isolation from micro-organisms or biological tissues, but to isolate sufficient quantities of pure material for biological testing by this method can involve manipulating large volumes of solvent and can be difficult and time-consuming.

Specific examples of the total synthesis of a number of peptide hydroxamates in solution have been reported, for example matlystatin B (28) (*Figure 4*), but a more generic method involves the reaction of peptidyl succinimide ester with hydroxylamine hydrochloride (*Figure 5*).

*Solid phase strategy*

The solid phase synthesis of peptide hydroxamates encompasses two main strategies. The first, the transamination of a resin-bound peptidyl ester using hydroxylamine hydrochloride, is restricted to a limited number of examples, e.g. tepoxalin (29) (*Figure 6*). The second strategy involves attaching an N-protected hydroxylamine to the resin via the hydroxyl functionality, deprotecting and using the new hydroxylamine linker as a handle for peptide synthesis (*Figure 7*).

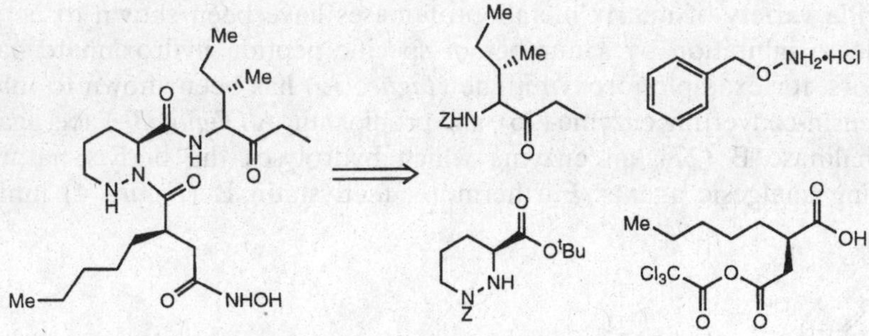

**Figure 4.** Retro-synthetic analysis of matlystatin B.

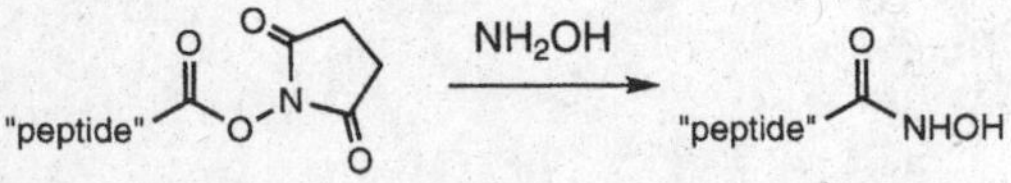

**Figure 5.** Generation of a peptidyl hydroxamic acid from a succinimide ester.

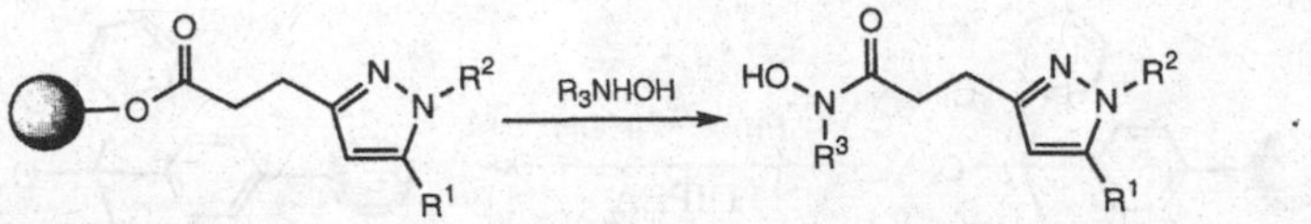

**Figure 6.** Solid phase synthesis of tepoxalin.

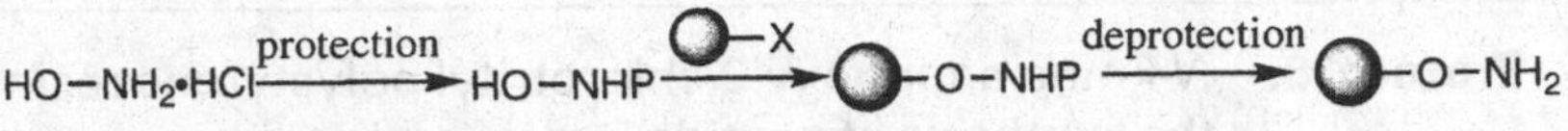

**Figure 7.** Strategy for the generation of resin-bound hydroxylamine.

Two different N-protected hydroxylamines have been used to accomplish the synthesis of these novel hydroxylamine-functionalized resins. *N*-Hydroxyphthalimide has been employed in two ways: first, as a nucleophile to displace either a resin-bound chloride ion (30) or a resin-bound mesylate ion (31) (*Figure 8a*); and second, to generate the active phosphonium species required to perform a Mitsunobu reaction on a hydroxyl-functionalized resin (32) (*Figure 8b*). In both cases, the resin-bound *N*-hydroxyphthalimide ester that is generated is subsequently treated with hydrazine to afford resin-bound hydroxylamine.

To negate the need to use the harsh and potentially hazardous deprotection reagent hydrazine, a superior alternative, which is particularly compatible for use in Fmoc/tBu solid phase synthesis, is *N*-Fmoc-hydroxylamine, which can be used to effect a nucleophilic displacement reaction of 2-chlorotrityl chloride resin (33) (*Figure* 9). The resultant *N*-Fmoc-aminooxy-2-chlorotrityl polystyrene resin can be used for the synthesis of peptide hydroxamates as described in Protocol 8.

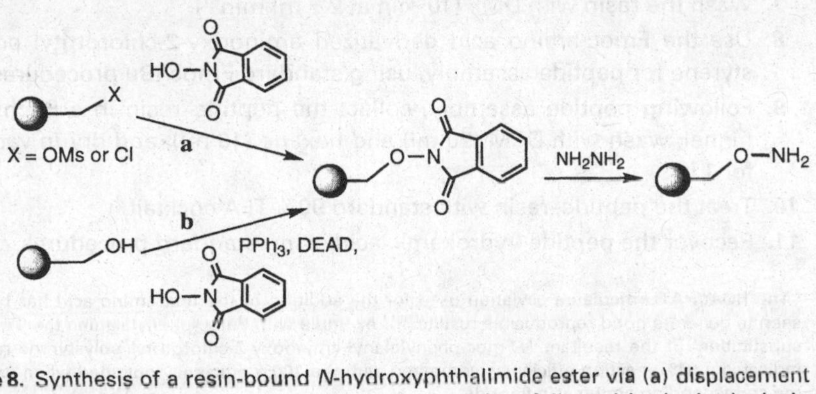

**Figure 8.** Synthesis of a resin-bound *N*-hydroxyphthalimide ester via (a) displacement of a mesylate or chloride ion or (b) a Mitsunobu reaction, followed by hydrazinolysis to generate a resin-bound hydroxylamine.

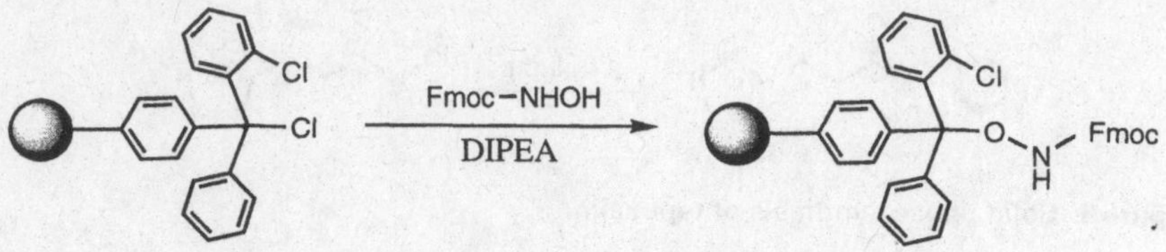

**Figure 9.** Synthesis of *N*-Fmoc-aminooxy-2-chlorotrityl polystyrene resin.

---

**Protocol 8.** *N*-Fmoc-aminooxy-2-chlorotrityl polystyrene resin for the synthesis of peptide hydroxamic acids

*Reagents*

- *N*-Fmoc-aminooxy-2-chlorotrityl polystyrene resin
- Piperidine in DMF (20% v/v)
- Fmoc-amino acid
- HATU
- DCM
- DMF
- HOAt
- DIPEA

*Method*

1. Place *N*-Fmoc-aminooxy-2-chlorotrityl polystyrene resin (0.05 mmol) in a peptide synthesizer column, and pre-swell in DCM for 24 h.

2. Treat the resin with 20% v/v piperidine in DMF (10 min at 2.5 ml min$^{-1}$, or using batchwise treatment), followed by DMF (15 min at 2.5 ml min$^{-1}$), after which excess DMF is expelled from the column.

3. In a dry 5-ml round-bottomed flask or an appropriate glass sample bottle, dissolve the Fmoc-amino acid (0.4 mmol) in 1.0 ml DMF, and add HOAt (0.4 mmol), HATU (0.4 mmol), and DIPEA (0.8 mmol).[a]

4. Allow the activation to proceed for 1 min.

5. Add the reaction mixture to the resin.

6. Gently agitate the reaction suspension for 24 h at room temperature.

7. Wash the resin with DMF (10 min at 2.5 ml min$^{-1}$).

8. Use the Fmoc-amino acid derivatized aminooxy-2-chlorotrityl polystyrene for peptide assembly, using standard Fmoc/*t*Bu procedures.

9. Following peptide assembly, collect the peptide–resin in a Buchner funnel, wash with DCM (20 ml) and hexane (10 ml), and dry *in vacuo* for 1 h.

10. Treat the peptide–resin with standard 90% TFA cocktail.

11. Recover the peptide hydroxamic acid using standard procedures.

---

[a] The HOAt/HATU-mediated acylation used for the addition of the first amino acid has been seen to generate good reproducible results, for example with Fmoc-phenylalanine the 'Fmoc-substitution' of the resultant *N*-Fmoc-phenylalanyl-aminooxy-2-chlorotrityl polystyrene resin indicated a 94% reaction efficiency, compared with the 100% efficiency obtained when using the corresponding amino acid fluoride.

## 1.4 Peptide aldehydes by solid phase synthesis

### 1.4.1 Introduction

Preparation of peptide aldehydes has been reported in numerous papers because these compounds present two main interests. First, they can be used for further chemistry (formation of reduced bond, Wittig reactions, ligation, etc.), and second, they have been found to be potential inhibitors of several classes of enzymes such as serine proteases (34, 35), prohormone convertases (36), cysteinyl proteases (34, 37, 38), and aspartyl proteases (39, 40). These inhibitory properties result from the tetrahedral hydrated C-terminus aldehydic function, which mimics the transition state of the substrate hydrolysis.

Various methods for the solution synthesis of peptide aldehydes have been described: (i) the corresponding peptide alcohol can be oxidized to the aldehyde (41, 42)—if the desired sequence does not contain other alcohol functions; (ii) reduction of the corresponding methyl esters using diiso-butylaluminium hydride (43); (iii) arginal analogue syntheses, important in studies of their anticoagulant, antithrombotic properties, have been reported, including the use of δ-lactam (35, 44, 45) and semicarbazone derivatives (34). A particular synthesis of an inhibitor of interleukin-1β converting enzyme was described by Chapman (37), in which the C-terminus aldehydic residue of this compound is an aspartyl residue: the aspartyl aldehyde moiety was protected as the corresponding *O*-benzylacylal which could be coupled and then hydrogenolysed to afford the desired compound. Moreover, a recent report mentioned the use of thiazolidines as aldehyde precursors (46). Until recently, only one publication (47) concerned solid phase synthesis, this method being derived from a solution preparation of aldehydic peptides (48).

We have focused our attention on the solid phase synthesis of such compounds and described our results here. Alternative routes for the preparation of peptide aldehydes and side-chain protected peptide aldehydes in solid phase synthesis are described. Three new linkers that are stable under classical Fmoc or Boc strategies have been developed to obtain the peptide aldehyde from the solid support. One of these linkers was conceptualized on the basis of the Weinreb amide (49) and the other on the basis of phenolic esters (50). Both strategies required the reduction with hydrides of the peptide-linker-resin to release the peptidic aldehydic function. The use of these two different approaches was demonstrated by the synthesis of N-protected α-amino-aldehydes and peptide aldehydes. The third approach used the ozonolysis reaction for the generation of the desired aldehyde. This concept requires a linker incorporating a double bond in the α-position of the asymmetric carbon of the C-terminal residue that will be cleaved by ozone to produce the carbonyl function.

### 1.4.2 Weinreb amide-based linker

Among all the described preparations of N-protected amino aldehydes and peptide aldehydes, the reduction of Weinreb amides is used most widely. This method has been applied successfully to the synthesis of N-protected α-amino-aldehydes and peptide aldehydes using Z, Boc and Fmoc chemistry. On the basis of this stable intermediate, it was reasonable to apply this strategy in the solid phase synthesis of peptide aldehydes. This was done by the synthesis of a linker (51) that incorporates the methoxyamino functionality. This linker, *N*-(*tert*-butoxycarbonyl)-*N*-methoxy-3-aminopropionic acid, was coupled to a solid support (i.e. MBHA resin), allowing after Boc-deprotection peptide elongation by either classical Boc or Fmoc strategy on the solid support.

This methodology was validated by the synthesis of Boc-Ala-H and Boc-Phe-H; these compounds were purified on a silica gel column according to the procedure described by Ho and Ngu (52) to prevent racemization during purification. Their optical purities were in accordance with those reported in the literature (53).

The peptide aldehydes were then synthesized on the solid support as outlined in *Figure 10*. The aldehydic tripeptide Boc-Phe-Val-Ala-H was synthesized as a control peptide for comparison of the methods and for the epimerization study. Dipeptides Z-Val-Phe-H and Z-Val-Ala-H were described as potential inhibitors of aspartic proteases specifically as HIV protease inhibitors (39), Ac-Leu-Leu-Lys-H or Ac-Leu-Val-Lys-H were reported as leupeptin analogues (34). The peptide aldehydes were released from the resin by treatment with AlLiH$_4$ followed by hydrolysis. The tripeptide aldehydes Boc-Leu-Leu-Lys(2Cl-Z)-H and Ac-Leu-Val-Lys(2Cl-Z)-H were obtained according to the same procedure. As described previously (54), owing to the presence of several amide functions, the equivalent amount of LiAlH$_4$ has to be increased with the length of the peptide.

In all reactions, no over-reduction was observed, confirming the formation of the stable metal–chelate intermediate described by Nahm and Weinreb (49). Furthermore, during the reduction, no aldehyde derivative could be

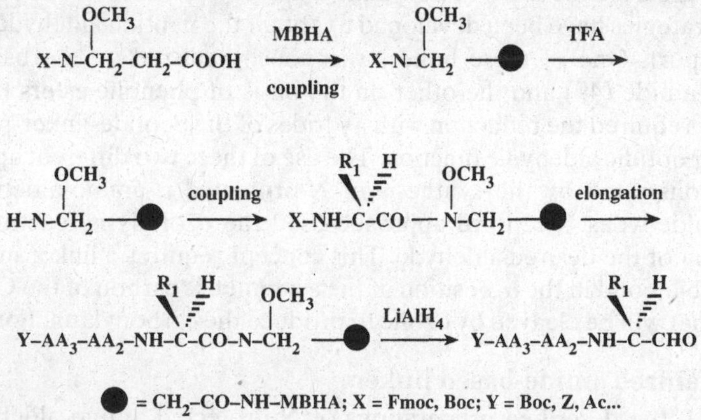

**Figure 10.** Synthesis of peptide aldehydes via the Weinreb amide linker.

detected in the supernatant of the reaction mixture before quenching. After hydrolysis, peptide aldehydes were detected by TLC. The crude products were studied by reverse-phase HPLC and by $^1$H NMR. Examination of the $^1$H NMR spectra revealed the presence of a single aldehydic proton signal, indicating the absence of epimerization and the presence of some trace impurities (55).

---

**Protocol 9.** Preparation of peptide aldehydes by reduction of a polymer-supported Weinreb amide

*Reagents*
- Boc-peptide-Weinreb amide resin,[a] dried thoroughly
- Argon
- LiAlH$_4$
- Saturated aqueous NaHCO$_3$
- DCM
- Anhydrous DMF
- 1 M aqueous KHSO$_4$
- Saturated aqueous NaCl

*Method*

1. Place peptidyl resin[a] (0.6 mmol) in a round-bottomed flask equipped with a magnetic stirrer, and add sufficient THF to cover. Flush flask with argon, seal, and place in an ice-bath.

2. Weigh LiAlH$_4$ (0.114 g, 3 mmol) into a dry sample tube. Quickly open flask and wash the LiAlH$_4$ into the reaction mixture with dry THF. Flush flask with argon, seal, and stir for 40 min.

3. Add 1 M aq. KHSO$_4$ dropwise to reaction mixture until no more gas is evolved.

4. Filter the mixture through a sintered glass funnel, collecting the filtrates.

5. Wash the resin with DCM (2 times).

6. Dilute the filtrate with DCM and wash, using a separating funnel, with 1 M aq. KHSO$_4$ (3 times), saturated aq. NaHCO$_3$ (3 times), and saturated aq. NaCl (3 times).

7. Dry the organic layer over anhydrous Na$_2$SO$_4$.

8. Filter solution and evaporate using a rotary evaporator.

9. If necessary,[b] purify the peptide on a column packed with silica-gel eluted with AcOEt/hexane (7:3 v/v) containing 0.1% pyridine.

[a] The peptidyl resin is prepared by standard Fmoc solid phase procedures on 3-(*N*-methoxyamino)propionamido functionalized resin. 3-(*N*-Methoxyamino)propionamidomethyl polystyrene is available from Novabiochem. The N-terminal amino acid residue can be introduced by direct coupling of the appropriate Boc-protected amino acid or by capping the terminal residue with Boc$_2$O.
[b] Purification can lead to epimerization of chiral centre at the aldehyde residue.

---

### 1.4.3  Phenyl ester based linker

Zlatoidsky (50) recently described the synthesis of N-protected α-amino aldehydes via reduction of the corresponding phenyl esters by lithium tri(*tert*-butoxy)aluminium hydride. Since 4-hydroxybenzoic acid is commercially available, we decided to explore this reaction in the solid phase strategy.

The linker was simply obtained by condensation of 4-hydroxybenzoic acid with N-protected amino acid carboxyanhydrides (UNCAs) (56) leading to the corresponding N-protected amino-esters. These compounds were then directly anchored to an amino-functionalized resin. Classical peptide elongation on a solid support could then be performed (*Figure 11*).

N-protected α-amino-aldehydes were synthesized in the solid phase as described by Zlatoidsky (50) by reduction with AlLiH(O*t*Bu)$_3$ of the corresponding N-protected amino phenyl esters linked to the support. In all our attempts, we have obtained a mixture of the aldehyde and the alcohol. In our hands, reduction in a solution of N-protected amino phenyl esters led to the aldehyde as a major compound, but also to the corresponding alcohol as a minor side-product.

In addition, several peptide aldehydes were synthesized by reduction of the peptide-linked-resin moiety (55). Both the presence of the peptide aldehyde and alcohol were observed. This phenomenon can be explained by the over-reduction of the corresponding aldehyde in the presence of excess hydride. Hence, this strategy has to be used carefully for the synthesis of peptide aldehydes.

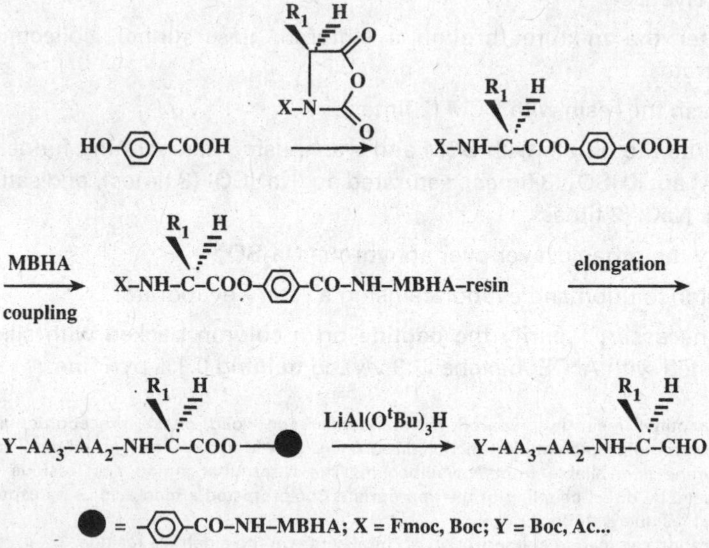

**Figure 11.** Synthesis of peptide aldehydes via the phenyl ester linker.

### 1.4.4 Epimerization of the carbon in position α to the aldehydic function

The stereo-configuration of the C-terminal residue is important for the biological activity of peptide aldehydes. As described earlier (54), the aldehydic signal in $^1$H NMR studies is a very good indicator of the possible epimerization of the C-terminal residue in aldehydic peptides containing three or more residues. It should be noted that no effective method of purification of such peptide aldehydes without racemization is known.

Indeed, we observed epimerization of the α-carbon by $^1$H NMR spectroscopy for the model peptide aldehyde Boc-Phe-Val-Ala-H purified either by flash chromatography on silica gel as described above (52) (0.1 % pyridine as eluant) or by reversed-phase HPLC. These spectra revealed two aldehydic proton peaks (in CDCl$_3$) indicating that some epimerization had occurred during purification. These two signals, corresponding to the two diastereoisomers LLL and LLD (55), could be observed in CDCl$_3$ but not in [$^2$H$_6$]-DMSO. This unique observation could explain why no papers describing the RP–HPLC purification of peptide aldehydes mentioned a loss of optical purity.

### 1.4.5 Solid phase synthesis by ozonolysis

As indicated above, the purity of the crude is very important for the preparation of peptide aldehydes, since no effective purification procedure is possible without epimerization of the C-terminus residue. We found that the cleanest approach for these syntheses was the treatment by ozone of an ethylenic compound linked to a solid support. It produced the ozonide that was treated to yield the corresponding aldehyde. Recently, Sylvain *et al.* (57) demonstrated the stability of the Merrifield resin towards ozonolysis. They reported the use of ozone as a versatile reagent for the generation of aldehydes, carboxylic acids, and alcohols on a solid support. We have used ozonolysis to generate peptide aldehydes on cleavage from the solid support, a strategy that Frechet and Schuerch described for the solid phase synthesis of oligosaccharides (58). The synthesis of peptide aldehydes based on the use of an α,β-unsaturated γ-amino acid as a linker to the solid support is summarized in *Figure 12*. The N-protected α,β-unsaturated γ-amino acid was synthesized by a Wittig reaction between the carboethoxymethylene triphenylphosphorane and the N-protected α-amino aldehyde (49, 53), followed by saponification to yield the corresponding ethylenic compound, which can be anchored to the solid support (59). After removal of the N-protecting group, elongation by classical methods of solid phase peptide synthesis (Boc or Fmoc strategies) was possible.

To validate this methodology, several N-protected peptide aldehydes were synthesized by this method. Three different resins were used to check their compatibility with this strategy: a Merrifield resin with an ester linkage, the

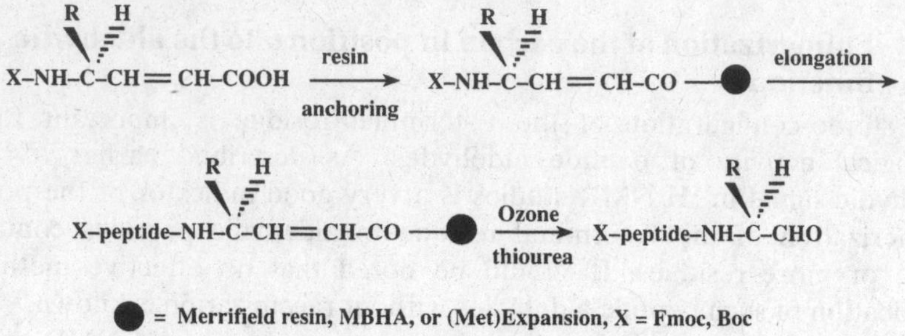

$\bullet$ = Merrifield resin, MBHA, or (Met)Expansion, X = Fmoc, Boc

**Figure 12.** Synthesis of peptide aldehydes by ozonolysis of the linker.

MBHA, and (Met)Expansin (60) resins with an amide linkage. The model tripeptide aldehyde Boc-Phe-Val-Ala-H was synthesized on the three resins. These peptides were checked by RP-HPLC and studied by $^1$H NMR in CDCl$_3$ without purification. The results showed that peptide aldehydes with high purity could be obtained using this strategy with no detectable trace of racemization (within the limit of sensitivity of $^1$H NMR).

An aspartyl-containing peptide aldehyde has also been synthesized using the same approach. Indeed, most of the previously described syntheses of peptide aldehydes are incompatible with the presence of protected Asp or Glu residues. We have shown in this work that benzyl ester derivatives of aspartic acid can be used to obtain aspartyl-containing peptide aldehydes by this methodology *Protocol (10–12)*.

---

**Protocol 10.**   Preparation of Boc-amino acid aldehyde

*Reagents*

- Boc-amino acid *N,N*-methoxymethylamide
- LiAlH$_4$
- 1 M aqueous KHSO$_4$
- Saturated aqueous NaHCO$_3$
- Anhydrous THF
- Ether
- Saturated aqueous NaCl

*Method*

1. Place Boc-amino acid *N,N*-methoxymethylamide (23.3 mmol) in a round-bottomed flask equipped with a magnetic stirrer.

2. Add THF (100 ml) and place in an ice-bath.

3. Add LiAlH$_4$ (1.10 g, 29 mmol) portionwise, allowing evolution of gas to subside between additions.

4. After 40 min, check the reaction by TLC. If starting material is consumed, add 1 M aq. KHSO$_4$ (100 ml).

5. Remove the THF on a rotary evaporator.

6. Redissolve the residue in diethyl ether or ethyl acetate, and wash organic phase in a separating funnel with 1 M aq. KHSO$_4$ (3 times), saturated aq. NaHCO$_3$ (3 times), and saturated aq. NaCl (3 times).

7. Dry the organic layer over anhydrous Na$_2$SO$_4$. Filter solution and evaporate using a rotary evaporator. The residue is used without further purification.

---

**Protocol 11.** Preparation of α,β-unsaturated γ-amino acid linker

*Reagents*

- Boc-amino acid aldehyde
- Anhydrous toluene
- Ethyl acetate/hexane (1:1 v/v)
- 1 M aqueous NaOH
- 1 M aqueous KHSO$_4$
- Saturated aqueous NaCl
- DCM
- DIPEA/DMF ( 1:9 v/v)
- Aminomethyl polystyrene
- Carboethoxymethylene triphenylphosphorane
- Anhydrous Na$_2$SO$_4$
- Silica-gel
- Ethanol
- Saturated aqueous NaHCO$_3$
- TFA/DCM (1:1 v/v)
- DIPEA
- DMF

*Method*

1. Place Boc-amino acid aldehyde (23.2 mmol) and carboethoxy-methylene triphenylphosphorane (8.88 g, 25.5 mmol) in a round-bottomed flask equipped with a magnetic stirrer.

2. Add toluene (100 ml) and stir to dissolve.

3. Heat mixture in an oil bath for 1 h at 80°C.

4. Remove the toluene on a rotary evaporator and dissolve the residue in ether (200 ml).

5. Wash the organic phase with water (3 times) and dry over Na$_2$SO$_4$.

6. Remove solvent on a rotary evaporator.

7. Purify the residue on a column of silica-gel eluted with ethyl acetate/hexane (1:1 v/v).

8. Add to the resulting oil, 1 M aq. NaOH (14 ml) and EtOH (100 ml).

9. Stir for 5 h to effect the saponification of the ester.

10. Remove the ethanol on a rotary evaporator.

11. Wash the aqueous solution in a separating funnel with ether (3 times). Acidify with 1 M aq. KHSO$_4$ and the extract the product from the aqueous solution with ethyl acetate (5 times).

12. Wash organic phase with 1 M aq. KHSO$_4$ (3 times), saturated aq. NaHCO$_3$ (3 times) and saturated aq. NaCl (3 times).

13. Dry the organic layer over anhydrous Na$_2$SO$_4$.

159

**Protocol 11.** *Continued*

**14.** Filter solution and evaporate using a rotary evaporator.

**15.** Couple the resulting linker to aminomethyl polystyrene using standard coupling methods.

**16.** Treat the resultant resin with TFA/DCM mixture for 30 min to remove the Boc group.

**17.** Wash the resin with DCM (80 ml), DIPEA in DMF (80 ml), and DMF (3 × 80 ml).

**18.** Assemble the peptide using standard methods.

---

**Protocol 12.** Typical ozonolysis of the peptide linked via an α,β-unsaturated γ-amino acid linker to the resin

*Reagents*
- Peptidyl resin
- Argon
- Ozone generator
- DCM
- Thiourea (400 mg, 5.25 mmol) dissolved in MeOH/DCM (1:2 v/v)

*Method*

**1.** Place the peptidyl resin (0.3 mmol) in a two-necked round-bottomed flask equipped with a magnetic stirrer and a gas bubbler tube.

**2.** Add DCM (10 ml) to cover resin.

**3.** Cool flask to –80°C.

**4.** Bubble a stream of ozone through the solution.

**5.** Once the reaction mixture turns blue, continue ozone addition for a further 5 min.

**6.** Bubble argon through the solution for 10 min to remove ozone.

**7.** Allow the reaction mixture to warm to room temperature.

**8.** Add thiourea solution and stir for 10 min.

**9.** Isolate the peptide aldehyde according to steps 4–9, *Protocol 9*.

---

### 1.4.6 Summary

We have designed, constructed, and tested three different linkers for the synthesis of aldehydic peptides on a solid support. Two strategies avoided the preliminary synthesis of the C-terminal amino-aldehyde and are based on the reduction by hydrides of the linker to generate the desired aldehydic function. As demonstrated, these syntheses were devoid of epimerization but require a purification step that is always a source of some epimerization of the $C^\alpha$ bearing the aldehyde. These new linkers can be used for further chemical synthesis on a solid support and for the preparation of larger synthetic pseudopeptides as shown by the solid phase preparation of a hexapeptide with a reduced peptide bond Ac-Leu-Leu-LysY[CH$_2$NH]Phe-Asp-Ala-NH$_2$ (55).

The third strategy, the ozonolysis procedure, allowed the preparation of several peptide aldehydes on three different resins (Merrifield resin, MBHA, (Met)Expansin). Furthermore, this methodology could be used in a sequence containing aspartyl or glutamyl residues. However, the presence in the sequence of functional groups sensitive to ozonolysis should be avoided. According to the treatment of the ozonide intermediate, this strategy could also lead to partially protected peptides with a free C-terminal carboxylic acid, which could be useful for peptide synthesis by fragment condensation, or to C-terminal peptide alcohols. On the other hand, this methodology could be applied to organic reactions on a solid support in which aldehydes (or carboxylic acids or alcohols) are the final targets.

More recently, several papers reporting the solid phase synthesis of peptide aldehydes have been published. Dinh and Amstrong (61) outline the use of Weinreb-type amides on solid support for the synthesis of ketones and aldehydes; Ede and Bray (62) report a new linker based on the oxazolidine moiety and use this method with Multipin™ technology; Galeotti *et al.* (63) apply their thiazolidinyl linker to a solid support. The great interest in these aldehydic compounds has also been demonstrated by the fact that the Weinreb amide resin is now commercially available from Bachem and Novabiochem.

# 1.5 Synthesis of C-terminal peptide aldehydes on the Multipin™ system using the oxazolidine linker

## 1.5.1 Introduction

Peptide aldehydes have played a vital role in the early structure–function studies of proteolytic enzyme mechanisms (64). The solid phase syntheses of peptide aldehydes via the oxazolidine linker (62) is one of a number of reported methods, including the semicarbazone (47), the Weinreb amide (55), the ozonolysis (65), the thiazolidine (63), and the backbone amide linker method (66).

The oxazolidine linker satisfies the following criteria for a general linker for the immobilization of aldehyde functionality: (i) it is low cost and easy to construct, (ii) it is chemically stable (long shelf life), (iii) it does not require pre-formation (i.e. the aldehyde couples directly to the solid phase), (iv) aldehyde attachment is generic, and (v) it cleaves under mild conditions which leave no residue upon evaporation. As shown in *Figure 13*, the protected amino acid aldehyde is attached to the solid phase by condensation with a threonine residue 1 to form a support-bound oxazolidine system 2. The oxazolidine 2 is stable to non-aqueous acid and to base, and as the secondary

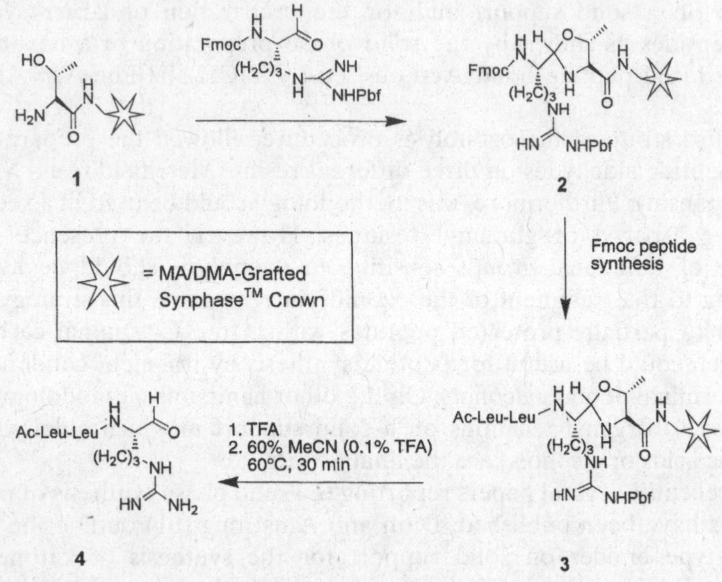

**Figure 13.** Synthesis of leupeptin, Ac-Leu-Leu-Arg-H.

amine is difficult to acylate, it can tolerate subsequent peptide synthesis. Being stable to non-aqueous acid, TFA may be used to remove acid-labile protecting groups prior to release of the aldehyde function (i.e. cleavage). The peptide aldehyde is cleaved using aqueous TFA (0.1%) containing acetonitrile as a co-solvent.

This section describes the derivatization of SynPhase™ crowns with threonine, followed by oxazolidine formation with the arginine aldehyde derivative Fmoc-Arg(Pbf)-H (53) to give **2**, which is used to prepare the peptide aldehyde leupeptin **4** (*Figure 13*). Leupeptin is a naturally occurring microbial peptide aldehyde derived from the culture filtrates of different streptomyces. Acetylation of the peptide aldehyde with acetic anhydride is performed at 4°C for 5 min to avoid acetylation of the oxazolidine secondary amine. Earlier studies with the oxazolidine linker showed that cleavage cannot be obtained if the oxazolidine secondary amine is acetylated (62). Following side-chain deprotection, **4** is cleaved from the crowns using 0.1% TFA in 60% v/v acetonitrile/water. Although a tripeptide is prepared in this example, this linking strategy can be used for generic peptide aldehyde and small molecule synthesis.

### 1.5.2  Peptide synthesis

The protocols presented here describe the synthesis of a single peptide aldehyde. Synthesis using the Multipin™ method can be performed in a variety of vessels depending on the scale of synthesis. Sealable polypropylene tubes and glass vials are suitable for small-scale syntheses, where small numbers of target compounds are prepared in parallel. Larger-volume vessels or round-bottomed flasks are suitable for larger-scale syntheses, whereas the 8 × 12 array format is convenient when larger numbers of target peptide aldehydes are prepared. The Multipin™ method (67) is an effective, low cost, simultaneous multiple peptide synthesis technology. Synthesis can be performed by following the procedures provided in the manuals of commercially available Multipin™ kits. It is also advisable to consult a reference work on peptide synthesis (68, 69; see Chapters 2 and 3) for supplementary information.

---

**Protocol 13.** Preparation of support-bound α-hydroxymethyl amino acid (threonine) aldehyde linker (**1**)

*General equipment and reagents required for Protocols 13–16*

- Solvent-resistant pipettes (e.g. Labsystems and Brand) and tips
- Solvents: DMF, MeOH.
- Deprotection reagent: 20% v/v piperidine in DMF grade
- All reagents should be of analar or higher

*Equipment and reagents*

- SynPhase SP-MD-I-NOF crowns (methacrylic acid/dimethylacrylamide grafted crowns, I-series (loading of 8–10 μmol/crown), Fmoc-protected
- N-(9-fluorenylmethoxycarbonyl-L-threonine (Fmoc-Thr-OH)
- DIC
- HOBt

*Method*

This example describes the preparation of 10 SynPhase crowns coupled with threonine.

1. Place 10 SynPhase crowns in a 20 ml glass vial.
2. Add 20% piperidine/DMF (10 ml) and stand for 20 min.
3. Drain the reaction vial of solution and add DMF (10 ml) to wash the crowns. Stand for 5 min.
4. Repeat step 3 two more times.
5. Dissolve Fmoc-Thr-OH (341 mg, 1.0 mmol) and 1-hydroxybenzotriazole (153 mg, 1.0 mmol) in DMF (10 ml) to give a 0.1 M solution. Add DIC (152 μl, 1.0 mmol). Mix thoroughly and stand for 2 min.
6. Add the activated amino acid solution to the crown and stand for 2 h.
7. Repeat steps 3 and 4.
8. Add 20% piperidine/DMF (10 ml) to the drained crowns and stand for 20 min.
9. Repeat steps 3 and 4.
10. Wash the crowns in methanol (10 ml) twice for 5 min, then air-dry.

---

**Protocol 14.** Attachment of Fmoc-Arg(Pbf)-H to α-hydroxymethyl amino acid (threonine) aldehyde linker crowns to give **2**

*Equipment and reagents*

- SynPhase crowns (SP-MD-I-NOF, 10) derivatized with deprotected threonine as outlined in *Protocol 13*
- Reaction solvent: 1% diisopropylethylamine in methanol
- DPIEA

- $N^\alpha$-Fmoc-$N^\gamma$-2,2,4,6,7-pentamethyldihydrobenzofuran-5-sulphonyl-arginine aldehyde prepared by literature method (53) *see Protocol 10*
- Teflon-capped glass vials (20 ml, Wheaton, Millville, NJ)

*Method*

This example describes the preparation of 10 crowns derivatized with Fmoc-Arg(Pbf)-H.

1. Place threonine-derivatized SynPhase crowns (10) in a 20 ml Teflon-capped glass vial.

2. Dissolve Fmoc-Arg(Pbf)-H (633 mg, 1.0 mmol) in reaction solvent (10 ml) to give a 0.1 M solution. Mix thoroughly.

3. Add the aldehyde solution to the crowns and heat at 60°C (in an oven or heating block) for 2 h.

4. Drain the reaction vial of solution and add methanol (10 ml) to wash the crowns. Stand for 5 min.

5. Repeat step 4 twice and then allow crowns to air-dry.

---

**Protocol 15.** Synthesis of peptide aldehyde Ac-Leu-Leu-Arg-H (**3**)

*Equipment and reagents*

- SynPhase crowns (2) derivatized with Fmoc-Arg(Pbf)-H from *Protocol 14*
- *N*-Fluorenylmethoxycarbonyl-L-leucine (Fmoc-Leu-OH)

- DIC
- DIPEA
- HOBt
- DCM

*Method*

1. Place two SynPhase crowns derivatized with Fmoc-Arg(Pbf)-H (**2**) into a 5 ml glass vial.

2. Add 20% piperidine/DMF (5 ml) and stand for 20 min.

3. Drain the reaction vial of solution and add DMF (5 ml) to wash the crowns. Stand for 5 min.

4. Repeat step 3 two more times.

5. Dissolve Fmoc-Leu-OH (177 mg, 0.5 mmol) and HOBt (80 mg, 0.52 mmol) in DMF (5 ml). Add DIC (78 μl, 0.5 mmol). Mix thoroughly and stand for 2 min.

6. Add the activated amino acid solution (0.1 M solution) to the crowns and stand for 2 h at 20°C.

7. Repeat steps 3 and 4.

8. Couple the second leucine residue by repeating steps 2–6.

9. Remove the Fmoc-protecting group by repeating steps 2–4.

10. To acetylate the N-terminus,[a] dissolve acetic anhydride (150 μl, 2.6 μmol) and DIPEA (20 μl, 0.1 μmol) in DMF (4.8 ml). Mix thoroughly and cool the acetylation mixture in an ice-bath.

11. Add to the crowns and leave for 5 min.

12. Wash the crowns with DMF (2 × 5 ml) in quick succession (10 sec washes).

13. Repeat steps 3 and 4.

14. Wash crowns with DCM (2 × 5 ml, 5 min each) and air-dry (30 min).

---

[a] It is important to acetylate for only 5 min, since longer acetylation times at room temperature result in partial (up to 30%) acetylation of the oxazolidine secondary amine. Earlier results (62) have shown that if the oxazolidine is acetylated, no cleavage of the oxazolidine to yield peptide aldehyde occurs.

---

**Protocol 16.** Side-chain deprotection and cleavage to produce peptide aldehyde (4)

*Equipment and reagents*

- TFA (**Warning:** corrosive and toxic)
- Cleavage solution: acetonitrile/water/TFA (60:40:0.1 v/v)
- DCM
- Teflon-capped glass vials (5 ml, Wheaton, Millville, NJ)

*Method*

1. Side-chain deprotect the peptide aldehyde by immersion of the crown (3) in TFA for 1 h at room temperature.[a] The volume of side-chain deprotection solution used must cover the crown, typically 1 ml for an I-series crown in a polypropylene tube.

2. Remove the crown from the TFA solution and wash with TFA (1 ml, 1 min) and DCM (1 ml, 2 × 1 min).

3. Remove the crown from the DCM and allow to air-dry (2 min).

4. Place the crown in a 5 ml Teflon-capped glass vial.

5. Add cleavage solution (2 ml) and cap the vial.

6. Heat the solution to 60°C for 30 min to cleave the peptide aldehyde.

7. Remove the crown from the cleavage solution and discard.

8. Analyse and lyophilize the solution (supernatant) immediately.

[a] Shorter times (40 min) can be used for other side-chain protecting groups such as *t*-butyl esters and ethers.

### 1.5.3 Peptide aldehyde analysis

In aqueous solutions, peptide aldehydes exist as an equilibrium state between the aldehyde (**A**) and hydrated forms (**B**). In the example presented here (*Figure 14*), leupeptin, the C-terminal arginal, can also cyclize to form the aminol (**C**). The same equilibrium states are seen in the reverse-phase HPLC chromatogram (three broad peaks). The electrospray mass spectrum shows only the $[M + H]^+$ (**A** and **C**) and $[M + H + 18 (H_2O)]^+$ (**B**) peaks.

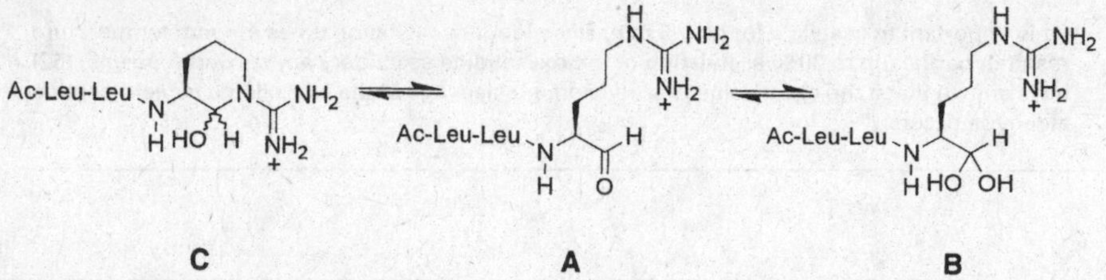

**C**          **A**          **B**

**Figure 14.** Equilibrium state between peptide aldehyde (**A**), hydrated form (**B**), and cyclic aminol (**C**) in aqueous solution.

# 2. N-terminal modifications

Chemical modification of the N-terminus of a peptide is often necessary to accomplish a variety of objectives. First, it can be useful as a device for simplifying the synthesis of difficult sequences; second, it can assist the purification of the synthesized peptide; third, it can provide a useful 'tag' by which to identify the peptide. Finally, peptides bearing a chemically modified N-terminus are not recognized by aminopeptidases and therefore exhibit a longer half-life *in vivo*. Examples of N-terminus modifications are: formylation using 2,4,5-trichlorophenyl formate; acetylation using acetic anhydride; *tert*-butoxycarbonylation using di-*tert*-butyl dicarbonate; and pyroglutamylation using 2,3,4,5,6-pentachlorophenyl pyroglutamate.

## 2.1 Biotinylation

The purpose of biotinylating the N-terminus of a peptide is rooted in the high affinity that biotin exhibits towards avidin. This strong binding has been exploited in several ways, for example as a means of immobilizing molecules onto surfaces for biophysical studies (70), and also to assist in peptide purification whereby a peptide which is biotinylated at the N-terminus is passed through an avidin agarose column (71). Unbound impurities are then eluted before the biotinylated peptide is removed from the column. The procedure outlined in *Protocol 17* gives a general method for attaching biotin to the N-terminus of a peptide. One disadvantage of this is that the biotin cannot subsequently be removed. To this end, reagents that generate a transient biotin 'tag', such as 2-biotinyldimedone, have been developed (72).

---

### Protocol 17.  Biotinylation

*Reagents*
- Biotin
- DMF
- DIC
- HOAt

*Method*

1. Following solid phase peptide assembly (0.05 mmol scale), and removal of the final $N^\alpha$-amine protecting group, expel DMF from the resin bed.

2. Dissolve biotin (122 mg, 0.5 mmol) in DMF (5 ml)[a] and add DIC (78 µl, 0.5 mmol).

3. Stir the solution for 10 min at room temperature.

4. Add HOAt (70 mg, 0.5 mmol) to the solution.

5. Add the mixture to the peptide–resin and stir gently for 24 h at room temperature.

6. Wash the derivatized peptide–resin with DMF (e.g. 10 min at 2.5 ml min$^{-1}$) and perform an amine test (see Chapter 3).

7. If the amine test is positive, repeat steps 1–4. If the test is negative, collect the modified peptide–resin, wash with DCM (15 ml), MeOH (10 ml), and dry *in vacuo*.

8. The material is now readily for cleavage of N-terminal modified peptide.

[a] The excessive volume of solvent used here is necessary because of biotin's low solubility in DMF.

---

## 2.2 Reductive alkylation

Reductive alkylation at the both the N- and the C-terminus has been reported for a wide variety of purposes. As a tool for increasing molecular diversity in combinatorial chemistry, its objective is to increase the peptide half-life *in vivo*. However, choosing a suitable alkyl/aryl group to be attached to the N-terminus can also have a wide-reaching application for improving peptide-coupling reactions. The 2-hydroxy-4-methoxybenzyl (Hmb) group developed by Sheppard and co-workers (73) provides a molecular framework that is ideally set up for an intramolecular acyl transfer. Its incorporation into 'difficult sequences' has been shown to completely suppress aspartimide formation (74). The resultant tertiary amides are not susceptible to hydrogen-bond formation and hence secondary structure formation is suppressed, preventing peptide aggregation during synthesis. The advantages outlined above ultimately result in cleaner and easier to purify peptides, and a typical procedure for any reductive alkylation is outlined in *Protocol 18* (75, 76).

---

**Protocol 18.** Reductive alkylation

*Reagents*
- Glacial acetic acid
- Aldehyde, e.g. 2-hydroxy-4-methoxybenzaldehyde
- Trimethylorthoformate
- DCM
- Sodium borohydride
- DMF
- Methanol
- Hexane

*Method*

1. Following solid phase peptide assembly (0.05 mmol scale), and after removal of the final $N^{\alpha}$-amine protecting group, remove excess DMF from the resin bed.

2. Ensure that the peptide–resin is suspended in 0.75 ml DMF.

3. Add 2-hydroxy-4-methoxybenzaldehyde (39 mg, 0.25 mmol), followed by glacial acetic acid (15 μl, giving 2% v/v in DMF).

4. Gently agitate or stir the reaction mixture for 1 h.

5. Wash the resin briefly with DMF (2 min at 2.5 ml min$^{-1}$).

6. Repeat steps 1 and 2.

7. Add 2-hydroxy-4-methoxybenzaldehyde (39 mg, 0.25 mmol), followed by trimethylorthoformate (0.15 ml).

8. Gently agitate the reaction suspension for 75 min.

9. Wash the resin with DMF (5 min at 2.5 ml min$^{-1}$).

10. Remove any excess DMF, and ensure that the resin is suspended in 1.5 ml DMF.

---

11. Add sodium borohydride (8 mg, 0.20 mmol), immediately followed by 0.20 ml of methanol.
12. Add further amounts of sodium borohydride (12 mg, 0.30 mmol) portionwise over a period of 30 min.
13. Stir the reaction mixture gently for a further 90 min.
14. Wash the resin thoroughly with DMF (20 min at 2.5 ml min$^{-1}$).
15. Collect the modified peptide–resin, and wash with DCM (15 ml), MeOH (10 ml), DCM (20 ml), and hexane (10 ml), and dry *in vacuo*.
16. The sample is now readily for further chemical transformations, or for TFA-mediated cleavage.

# 3. Side-chain modifications

## 3.1 Introduction

The chemistry utilized within Fmoc/tBu solid phase peptide synthesis (SPPS) offers a defined, facile approach to the generation of peptide and peptide-based entities in respectable yields. However, the construction of complex, atypical peptides is limited within the current parameters of this approach. Regiospecific chemical modification of carboxylic acid or amino functional groups within a resin-bound construct requires *en route* a highly selective protection/deprotection strategy that must fit within the existing restraints of differential Fmoc/tBu protection. Selective removal of a semi-permanent protecting group with a unique reagent provides an additional orthogonal level to the synthetic strategy. The versatility afforded by this approach is of considerable interest for the generation of complex modified peptides.

## 3.2 Strategy design in the synthesis of atypical peptides

Recently, there has been a concerted effort to develop protecting groups that utilize a different mechanism of cleavage to achieve selectivity, yet are compatible within the rigid confines of the Fmoc/tBu approach. Ideally, regio-specific removal of semi-permanent protection with a unique reagent will afford the desired additional level of orthogonality to the synthetic strategy. However, these protecting groups must fulfil a number of necessary criteria to be considered suitable for Fmoc SPPS. Introduction of these moieties must proceed in a facile manner with no loss of chiral integrity of the Fmoc-amino acid derivative. The protecting group must also remain chemically inert throughout the synthesis, yet be removed in a facile manner to liberate the appropriate functional group.

### 3.2.1 Branched peptides

As the lysine residue is bifunctional, both the $N^\alpha$- and $N^\epsilon$-amino groups require differential protection for effective peptide elongation. Commonly, an acid-labile semi-permanent protecting group, e.g. Boc, is used in conjunction with the temporary $N^\alpha$-Fmoc protecting group. Hence, $N^\epsilon$-Boc removal is achieved during the universal cleavage/deprotection step, which accompanies liberation of the peptidic material from the solid support. If, however, both amino groups are protected as bis(Fmoc), deprotection would afford a bifunctional template, thus increasing the resin substitution twofold. Each subsequent coupling would proceed in parallel on both the $N^\alpha$- and $N^\epsilon$-amino functionalities, affording a branched peptide. This procedure is suitable for the generation of mono-epitopic branched entities. However, in order to generate a branched peptide composed of two individual sequences, ortho-gonal protecting groups are required for the two amino groups. There exist a number of such orthogonal protecting groups suitable for this purpose, each with specific advantages/limitations over other candidates.

### 3.2.2 Orthogonal amine-protecting groups for the solid phase synthesis of branched peptides

The orthogonal amino approach may also be exploited in the synthesis of di-epitopic peptide constructs. In this approach two separate antigenic sequences can be assembled. Suitably protected lysine derivatives, such as Fmoc-Lys(Ddiv)-OH (77), Ddiv-Lys(Fmoc)-OH, Fmoc-Lys(Alloc)-OH, and Alloc-Lys(Fmoc)-OH, allow the assembly of a single sequence after Fmoc deprotection – chemical structures of these protecting groups are detailed in Chapters 2 and 3. Selective removal of the orthogonal protection thus allows the assembly of another sequence in a similar manner. These di-epitopic constructs are of potential use in inducing cytotoxic T-lymphocyte response as they allow the incorporation of B- and T-cell epitopes within the same macromolecule.

Although the procedure outlined in *Protocol 19* may be applied using Fmoc-Lys(Dde)-OH (78, 79) instead of Fmoc-Lys(Ddiv)-OH, this is not recommended as: (i) the $N$-Dde group is relatively more susceptible to removal on either prolonged or repeated treatment with 20% piperidine in DMF (77, 79); (ii) $N$-Dde, in the presence of free amines, is susceptible to migration via intra- or intermolecular transamination pathways (80).

---

**Protocol 19.** Automated assembly of a tetravalent di-epitopic peptide construct

*Reagents*
- Fmoc-PAL-PEG polystyrene
- Fmoc-Lys(Fmoc)-OH
- Hydrazine monohydrate
- Fmoc-β-Ala-OH
- Fmoc-Lys(Ddiv)-OH

*Method*
1. Pre-swell a sample of Fmoc-PAL-PEG PS (0.5 g, 0.1 mmol) in DMF for 1 h in a reaction column.
2. Treat the resin with 20% v/v piperidine in DMF for 10 min (continuous flow rate 2.5 ml min$^{-1}$).

---

3. Wash the resin with DMF (10 min, 3.0 ml min$^{-1}$).
4. Expel any excess DMF from the resin bed.
5. Prepare a solution of Fmoc-βAla-OH (0.8 mmol), TBTU (0.8 mmol), HOBt (0.4 mmol) in DMF (2.5 ml), and add DIPEA (1.6 mmol). Leave the mixture for 2 min.
6. Add the mixture to the resin bed.
7. Gently agitate the reaction suspension for 2 h.
8. Repeat step 3, and then steps 2–4.
9. Prepare a solution of Fmoc-Lys(Fmoc)-OH (0.4 mmol), TBTU (0.4 mmol), HOBt (0.2 mmol) in DMF (1.5 ml), and add DIPEA (0.8 mmol). Leave the mixture for 2 min.
10. Repeat steps 6–8.[a]
11. Prepare a solution of Fmoc-Lys(Ddiv)-OH (0.8 mmol), TBTU (0.8 mmol), HOBt (0.4 mmol) in DMF (2.5 ml), and add DIPEA (1.6 mmol). Leave the mixture for 2 min.
12. Repeat steps 6–8.[b]
13. Perform stepwise assembly of the first antigenic sequence using standard Fmoc/tBu procedures, using a mixture of Fmoc-amino acid (0.4 mmol), TBTU (0.4 mmol), HOBt (0.2 mmol), and DIPEA (0.8 mmol).
14. Following completion of peptide assembly, repeat step 3, then steps 2–4.
15. Treat the peptide–resin with Boc$_2$O (1.0 mmol) in DMF (1.5 ml) for 2 h.
16. Wash the peptide–resin with DMF (10 min, 3.0 ml min$^{-1}$).
17. Treat the resin, by a continuous flow method, with 2% v/v NH$_2$NH$_2$ monohydrate in DMF for 30 min at 3.0 ml min$^{-1}$.[c]
18. Wash the peptide–resin with DMF (10 min, 3.0 ml min$^{-1}$).
19. Perform stepwise assembly of the first antigenic sequence using standard Fmoc/tBu procedures, using a mixture of Fmoc-amino acid (0.4 mmol), TBTU (0.4 mmol), HOBt (0.2 mmol), and DIPEA (0.8 mmol).[d]
20. Repeat steps 2–3.
21. Collect the peptide–resin construct, wash with DCM (30 ml) and hexane (10 ml), and dry *in vacuo*.
22. The peptide construct can be cleaved from the solid support using standard TFA acidolytic mixtures.

[a] There are now 0.2 mmol of reactive amine sites on the solid support.
[b] This results in 0.1 mmol of reactive amine sites (α-amino groups) on the solid support.
[c] Selective deprotection of *N*-Ddiv is effected, which results in 0.1 mmol of reactive amine sites (ε-amino groups) on the solid support.
[d] Due to possible steric effect, it is recommended that a reaction time of 18 h is applied for the coupling of the initial C-terminal amino acid residue.

### 3.2.3 Solid phase synthesis of cyclic peptides

Like their linear counterparts, cyclic peptides are key biomolecules in understanding structure–activity relationships. Furthermore, cyclic peptides have been shown to display an increased resistance to enzymatic degradation, thus enhancing their desirability as therapeutic targets.

Conventional synthesis of lactam-bridged cyclic peptides have in the past utilized the efficient assembly of the linear sequence by SPPS and upon cleavage from the solid support, cyclization of the semi-protected peptide is then carried out in solution. This method requires a dilute medium to minimize dimerization as a result of intermolecular condensation of the linear sequences (81). As a consequence, the reaction rate and yield are drastically reduced.

In contrast, on-resin cyclization requires the selective removal of both the desired amino and the carboxylic acid protecting groups in the presence of all other semi-permanent protection. Activation of the carboxyl group and subsequent intramolecular aminolysis allows the formation of a lactam-bridged cyclic peptide. Subsequently, cleavage/deprotection of the peptide from the solid support affords the fully deprotected cyclic peptide. Thus, the on-resin approach has distinct advantages over the solution phase technique. This method permits the use of an excess of activation reagents, and the removal of these reagents and by-products is achieved by simple filtration, followed by washing the resin with copious solvent (DMF). Common resin-monitoring techniques can be used to monitor the efficiency of this reaction. In addition, the solid support has a direct pseudo-dilution effect on the cyclization reaction. However, on-resin cyclization requires the use of side-chain protecting groups for the Asp/Glu and Lys/Orn/Dpr residues that are orthogonal to other side-chain protecting groups.

*Orthogonal carboxyl-protecting group*

Protection of the carboxyl of Asp/Glu by 2,4-dimethoxybenzyl (Dmb) esters provides an approach for head-to-tail cyclic peptide formation (82). The Dmb ester can be removed selectively by 1% v/v TFA in DCM in the presence of *t*Bu- and sulphonyl-based side-chain protection within minutes. However, the technique is limited as these conditions are sufficient to remove the side-chain trityl protection of Cys and His residues. Minor incidental loss of *t*Bu during this procedure has also been reported (83).

A truly orthogonal approach avoids these impediments, allowing removal without incidental loss of protection. Of the suitable candidates three carboxyl-protecting groups in particular offer a facile, generic route. Trimethylsilylethyl (OTmse) ester protection (84) has been used successfully for the synthesis of cyclic peptides on TFA-labile resin (85). Selective rapid removal is accomplished over 20 min using tetrabutylammonium fluoride (TBAF) in DMF. These favourably non-acidic conditions eliminate the potential for premature loss of semi-permanent protection during removal. Significantly, these conditions are also sufficient to remove *N*-Fmoc. This allows the simultaneous removal of both the terminal *N*-amino and carboxy protection by a single reaction.

The allyl ester protection offers a similar approach (86–88) to that of OTmse esters. However, its removal requires the use of a complex cleavage mixture; typically Pd(PPh$_3$)–DMSO–THF–0.5 M aqueous HCl–4-methyl-morpholine over 2 h (89). Care must be taken as the deprotection mixture is hazardous and air sensitive. It should be noted that *Protocols 20* and *21* outlined below can be applied for the on-resin deprotection of the *N*-Alloc group.

---

**Protocol 20.** Selective on-resin removal of allyl ester protecting group

*Reagents*
- Tetrakistriphenylphosphine palladium
- 0.5 M aqueous HCl
- 4-Methylmorpholine
- 0.5% Diethyldithiocarbamic acid sodium salt in DMF (w/v)
- DMSO
- THF
- 5% DIPEA in DMF (v/v)
- DCM
- DIPEA

*Method*

1. Dry all glassware overnight at 90°C.

2. Transfer the peptidyl resin into a dried round-bottomed flask and desiccate for 1 h at 40°C.

3. Seal the flask with an appropriate rubber septum.

4. Purge the vessel with a gentle stream of Ar, using a hypodermic needle as the release valve.

5. Weigh Pd(PPh$_3$)$_4$ (1.0 eq to resin substitution) into a similar dry round-bottomed flask.

6. Add a mixture of DMSO–THF–0.5 M aqueous HCl–4-methylmorpholine (2:2:1:0.5 eq, 5 ml per 100 mg of resin), and dissolve the catalyst by Ar agitation.

7. When a solution is obtained, seal the flask with a rubber septum.

8. Transfer the solution obtained in step 7 to the flask containing the resin (prepared in steps 2–4) using a canular or a gas-tight syringe (purged with Ar).

9. Gently swirl the suspension intermittently over a 3 h duration at 25°C.

10. Remove the solution by filtration, and wash with 5% v/v DIPEA in DMF (5 × twice resin bed volume) followed by 0.5% sodium diethyl-dithiocarbamate in DMF (5 × twice resin bed volume).

---

**Protocol 20.** *Continued*

**11.** Wash the resin further with DCM (5 × 10 ml) and diethyl ether (5 × 5 ml).

**12.** Dry resin *in vacuo* for 2 h.

---

**Protocol 21.** Alternative on-resin palladium-mediated allyloxycarbonyl and allyl deprotections[a]

*Reagents*

- Tetrakistriphenylphosphine palladium
- 5,5-Dimethylcyclohexane-1,3-dione
- DCM/THF (1:1 v/v)
- 0.5% DIPEA in DMF (v/v)
- 0.5% Diethyldithiocarbamic acid sodium salt in DMF (w/v)
- DCM
- DIPEA
- DMF
- Diethyl ether
- THF

*Method*

**1.** Place the peptidyl resin (200 mg) in a peptide synthesis vessel.

**2.** Swell the resin as described in Chapter 3, *Protocol 1A.*

**3.** Add tetrakistriphenylphosphine palladium (1 eq) and 5,5-dimethyl-cyclohexane-1,3-dione (10 eq) dissolved in degassed DCM/THF (2 ml).

**4.** Flush with argon, seal, and wrap in silver foil to exclude light.

**5.** Agitate vigorously for 2 h.

**6.** Remove the solution by filtration and wash the resin with DIPEA/DMF (5 times), followed by diethyldithiocarbamic acid sodium salt in DMF (10 × 10 ml) to remove the palladium residues.

**7.** Wash the resin further with DCM (5 × 10 ml) and diethyl ether (5 × 5 ml).

**8.** Dry resin *in vacuo* for 2 h.

[a] Protocol supplied by Professor Mark Bradley, University of Southampton.

---

The 4-[*N*-(1-(4,4-dimethyl-2,6-dioxocyclohexylidene)-3-methylbutyl)amino]-benzyl (ODmab) esters display complete stability to 20% v/v piperidine in DMF and acidolysis (TFA), yet are readily removed with 2% v/v hydrazine monohydrate in DMF within minutes. However, it should be noted that this deprotection condition is sufficient to remove the *N*-Fmoc protecting group. Nevertheless, the ODmab ester has distinct advantages over OTmse and allyl esters, since its removal is facile (90, 91) and avoids extensive post-column manipulations. In a manner similar to the allyl ester technique (92), ODmab ester removal can be automated for convenience. It should, however, be noted that hydrazinolysis of Dde/ODmab-based protecting groups in the presence of allyl/Alloc protecting groups could lead to significant reduction of the allyl moiety. This undesired side-reaction can be prevented by the addition of allyl alcohol as a scavenger in the deprotection reagent (93). Thus, in such circumstances, the recommended hydrazinolysis reagent is: 5% v/v hydrazine monohydrate, 3% v/v allyl alcohol in DMF or 10% v/v hydrazine monohydrate, plus 200 eq allyl alcohol in DMF.

---

**Protocol 22.** Selective on-resin removal of N-Dde(Ddiv)/ODmab-based protection

*Reagents*
- 2% v/v Hydrazine monohydrate in DMF
  (**MUST** be freshly prepared and used within 24 h)
- 10% v/v DIPEA in DMF
- DMF
- DIPEA
- DCM

A. *Procedure for manual removal*

1. Suspend the peptidyl resin in DMF in a sintered glass funnel (twice resin bed volume) and allow to swell for a minimum of 1 h. If the peptidyl resin contains >15 amino acid residues, extend this swelling time to >6 h.

2. Remove DMF by suction or positive pressure.

3. Wash the resin with DMF (3 × twice resin bed volume).

4. Add the solution of 2% v/v hydrazine monohydrate in DMF (twice resin bed volume) to the resin and agitate for 5 min.

5. Remove the deprotection reagent by suction.

6. Repeat step 4, but extend the time to 7 min.

7. Remove the deprotection reagent by suction, and wash the resin with DMF (10 × twice resin bed volume).

8. Wash the partially deprotected peptide–resin with 10% v/v DIPEA in DMF (5 × twice resin bed volume).

9. Wash the resin further with DCM (5 × 10 ml) and diethyl ether (5 × 5 ml).

10. Collect, and dry the resin *in vacuo* for 2 h.

B. *Procedure for continuous-flow removal*

1. Suspend the peptidyl resin in DMF within a reaction column (twice resin bed volume) and allow to swell for a minimum of 1 h. If the peptidyl resin contains >15 amino acid residues, extend this swelling time to >6 h.

**Protocol 22.** *Continued*

2. Prepare a solution of 2% hydrazine monohydrate in DMF. Using a manual function, purge the reagent flow-line for 20 min at 2.5 ml min$^{-1}$.

3. Connect the reaction column (containing the peptide–resin), and treat the resin with 2% hydrazine monohydrate in DMF at 2.5 ml min$^{-1}$ for 10 min.

4. Monitor the progress of the deprotection using a flow-cell spectrophotometric method at 266 nm.

5. Wash the resin with DMF (12 min, 2.5 ml min$^{-1}$), followed by 10% v/v DIPEA in DMF (12 min, 2.5 ml min$^{-1}$).

6. Collect the resin, and wash with DCM (5 × 10 ml) and hexane (2 × 5 ml).

7. Dry the partially deprotected peptide–resin *in vacuo* for 2 h.

### 3.2.4 Activation procedure for on-resin lactam formation

On-resin lactam (amide) formation or cyclization can be achieved via HOBt active esters using a number of reagents. Carbodiimide and phosphonium activation have all been used for ring closure with minimal loss of chiral integrity of the activated amino acid residue. The choice of reagent and auxiliary nucleophile (i.e. HOBt, HOAt) is at the discretion of the operator (see Chapters 2 and 3). However, it is worth noting that successful peptide cyclization is, to an extent, sequence dependent. Nominal ring closure is not uncommon and is due in part to the spatial orientation of the peptide backbone and steric hindrance.

**Protocol 23.** Facile cyclization on-resin: synthesis of cyclic lactam-bridged peptides

*Reagents*

- Peptide-resin that is partially deprotected, i.e. contains a free carboxylic acid and amine functionalities
- DMF
- DIC
- HOAt
- DCM
- Methanol

*Method*

1. Suspend the peptidyl resin in DMF (twice resin bed volume), and allow to swell for 6–18 h.

2. Remove DMF from the resin bed, and add fresh DMF.

3. Prepare a solution of DIC (1.5 equivalent to free carboxylic acid) and HOAt (1.0 eq) in DMF (1 ml per 0.1 mmol of DIC).

4. Add the solution to the resin.

5. Agitate the reaction suspension for 18 h at room temperature, and monitor the progress of the cyclization reaction by resin test (e.g. Kaiser, TNBS, etc.; see Chapter 3, *Protocol 13*).

6. Wash the resin with DMF (5 × twice resin bed volume).

7. Collect the resin in a sinter funnel, and wash with DCM (5 × 20 ml) and methanol (2 × 10 ml).

8. Dry the cyclized peptide–resin *in vacuo* for 2 h.

## 3.2.5 On-resin removal of trityl-protecting groups

The side-chain protecting groups of Cys(Mmt) (94), Lys(Mtt) (95), Ser(Trt) (96), Thr(Trt) (96), and Tyr(2-Clt) can be removed with 1% v/v TFA in DCM. As these conditions leave most standard protecting groups and peptide–resin linkages unaffected, this enables the side-chain functionality of these residues to be selectively unmasked whilst the peptide is attached to the solid phase. This capability allows the modification of individual amino acid residues in the peptide chain and has been exploited in the preparation of phosphopeptides (97) and lysine-branched peptides (95), and in the construction of templates for combinatorial chemistry (98).

With the exception of lysine, the process of trityl group removal from the side-chains of these amino acids is an equilibrium. Consequently, in order to push the equilibrium in the desired direction, a cation scavenger such as triisopropylsilane (TIS) or 1,2-ethanedithiol (EDT) is added to the cleavage mixture, or the reaction is conducted in a continuous-flow manner. With the latter approach, the progress of the reaction can be monitored at 400–500 nm and by following the decrease in optical density with time of the reaction effluent.

**Protocol 24.** Removal of trityl on solid phase

*Reagents*
- Dry acid-free DCM
- TFA/DCM/TIS (1:94:5 v/v)

*Method*

1. Place the peptidyl resin in a sintered glass funnel or manual peptide synthesis vessel.

2. Wash resin with DCM (5 × twice resin bed volume).

3. Remove excess DCM by applying nitrogen pressure.

4. Add TFA mixture (3 times resin bed volume) and allow to stand for 2 min.

**Protocol 24.** *Continued*

5. Remove reagent by applying nitrogen pressure.

6. Repeat step 4 three times.

7. Check completeness of reaction by treating a small sample of resin with TFA/DCM (1:1).[a]

8. Repeat step 4 until a colourless solution is obtained.

9. Wash resin with DCM (5 × twice resin bed volume).

10. Collect the partially deprotected peptide–resin, and dry *in vacuo* for 30 min.

[a] This test is not appropriate if the peptide contains other trityl-protected amino acids.

# References

1. du Vigneaud, V., Ressler, C., Swann, J. M., Roberts, C. W., Katsoyannis, P. G., and Gordon, S. (1953). *J. Am. Chem. Soc.*, **75**, 4879.
2. Mutt, V., Magnusson, S., Jorpes, J. E., and Dahl, E. (1965). *Biochemistry*, **4**, 2358.
3. Matsuo, H., Baba, Y., Nair, R. M. G., and Schally, A. V. (1971). *Biochem. Biophys. Res. Commun.*, **43**, 1334.
4. Selsted, M. E., Novotny, M. J., Morris, W. L., Tang, Y.-Q., Smith, W., and Cullor, J. S. (1992). *J. Biol. Chem.*, **267**, 4292.
5. Mirgorodskaya, O. A., Shevchenko, A. A., Abdalla, K. O. M. A., Chernusevich, I. V., Egorov, T. A., Musaliamov, A. X., Kokryakov, V. N., and Shamova, O. V. (1993). *FEBS Lett.*, **330**, 339.
6. Bradbury, A. F. and Smyth, D. G. (1983). *Biochem. Biophys. Res. Commun.*, **112**, 372.
7. Rink, H. (1987). *Tetrahedron Lett.*, **28**, 3787.
8. Albercio, F., Kneib-Cordonier, N., Biancalana, S., Gera, L., Masada, R. I., Hudson, D., and Barany, G. (1990). *J. Org. Chem.*, **55**, 3730.
9. Sieber, P. (1987). *Tetrahedron Lett.*, **28**, 2107.
10. Chan, W. C., White, P. D., Beythien, J., and Steinauer, R. (1995). *J. Chem. Soc., Chem. Commun.*, 589.
11. Chan, W. C. and Mellor, S. L. (1995). *J. Chem. Soc., Chem. Commun.*, 1474.
12. Garigipati, R. S. (1997). *Tetrahedron Lett.*, **38**, 6807.
13. Brown, D. S., Revill, J. M., and Shute, R. E. (1998). *Tetrahedron Lett.*, **39**, 8533.
14. Brill, W. K.-D., Schmidt, E., and Tommasi, R. A. (1998). *Syn. Lett.*, 906.
15. Atherton E, Clive, D. L. J., and Sheppard, R. C. (1975). *J. Am. Chem. Soc.*, **97**, 6584.
16. Atherton, E., Gait, M. J., Sheppard, R. C., and Williams, B. J. (1979). *Biorg. Chem.*, **8**, 351.
17. Arshady R., Atherton, E., Clive, D. J. L., and Sheppard, R. C. (1981). *J. Chem. Soc., Perkin 1*, 529.
18. Atherton, E., Logan, C. J., and Sheppard, R. C. (1981). *J. Chem. Soc., Perkin 1*, 538.

19. Brown, E., Sheppard, R. C., and Williams, B. J. (1980). *J. Chem. Soc., Chem. Commun.*, 1093.
20. Atherton, E. and Woodhouse, D. P. (1987). Zeneca CRB (unpublished work).
21. Pedroso, E., Grandas, A., de las Heras, X., Eritja, R., and Giralt, E. (1986). *Tetrahedron Lett.*, **27**, 243.
22. Atherton, E. and Sheppard, R. C. (ed.) (1989). *Solid phase peptide synthesis: a practical approach*, p. 70. IRL Press, Oxford.
23. Atherton, E. and Sheppard, R. C. (ed.) (1989). *Solid phase peptide synthesis: a practical approach*, p. 132. IRL Press, Oxford.
24. Atherton, E. and Sheppard, R. C. (ed.) (1989). *Solid phase peptide synthesis: a practical approach*, p. 89. IRL Press, Oxford.
25. Grams, F., Reinemer, P., Powers, J. C., Kleine, T., Pieper, M., Tschesche, H., Huber, H., and Bode, W. (1995). *Eur. J. Biochem.*, **228**, 830.
26. Umezawa, H., Aoyagi, T., Ogawa, K., Obata, T., Iinuma, H., Naganawa, H., Hamada, M., and Takeuchi, T. (1985). *J. Antibiotics*, **38**, 1813.
27. Inaoka, Y., Takahashi, S., and Kinoshita, T. (1986). *J. Antibiotics*, **39**, 1378.
28. Tamaki, K., Ogita, T., Tanazawa, K., and Sugimura, Y. (1993). *Tetrahedron Lett.*, **34**, 683.
29. Cody, D. R., DeWitt, S. H. H., Hodges, J. C., Moos, W. H., Pavia, M. R., Roth, B. D., Schroeder, M. C., and Stankovic, C. J. (1994). US Patent 5,324,483.
30. Prasad, V. V. K., Warnes, P. A., and Lieberman, S. (1983). *J. Steroid Biochem.*, **18**, 257.
31. Richter, L. S. and Desai, M. C. (1997). *Tetrahedron Lett.*, **38**, 321.
32. Floyd, C. D., Lewis, C. N., Patel, S. R., and Whittaker, M. (1996). *Tetrahedron Lett.*, **37**, 8045.
33. Mellor, S. L., McGuire, C., and Chan, W. C. (1997). *Tetrahedron Lett.*, **38**, 3311.
34. McConnell, R. M., York, J. L., Frizzel, D., and Ezell, C. (1993). *J. Med. Chem.*, **36**, 1084.
35. Bajusz, S., Szell, E., Badgy, D., Barabas, E., Horvath, G., Dioszegi, M., Fittler, Z., Szabo, G., Juhasz, A., Tomori, E., and Szilagyi, G. (1990). *J. Med. Chem.*, **33**, 1729.
36. Basak, A., Jean, F., Seidah, N. G., and Lazure, C. (1994). *Int. J. Peptide Protein Res.*, **44**, 253.
37. Chapman, K. T. (1992). *Bioorg. Med. Chem. Lett.*, **2**, 613.
38. Graybill, T. L., Dolle, R. E., Helaszek, C. T., Miller, R. E., and Ator, M. A. (1994). *Int. J. Peptide Protein Res.*, **44**, 173.
39. Sarrubi, E., Seneci, P. F., and Angelastro, M. R. (1992). *FEBS Lett.*, **319**, 253.
40. Fehrentz, J. A., Heitz, A., Castro, B., Cazaubon, C., and Nisato, D. (1984). *FEBS Lett.*, **167**, 273.
41. Kawamura, K., Kondo, S., Maeda, K., and Umezawa, H. (1969). *Chem. Pharm. Bull.*, **17**, 1902.
42. Woo, J. T., Sigeizumi, S., Yamaguchi, K., Sugimoto, K., Kobori, T., Tsuji, T., and Kondo, K. (1995). *BioMed. Chem. Lett.*, **5**, 1501.
43. McConnel, R. M., Barnes, G. E., Hoyng, C. F., and Gunn, J. M. (1990). *J. Med. Chem.*, **33**, 86.
44. Balasubramanian, N., St Laurent, D. R., Federici, M. E., Meanwell, N. A., Wright, J. J., Schumacher, W. A., and Seiler, S. M. (1993). *J. Med. Chem.*, **36**, 300.
45. Schuman, R. T., Rothenberger, R. B., Campbell, C. S., Smith, G. F., Gifford-Moore, D. S., and Gesellchen, P. D. (1993). *J. Med. Chem.*, **36**, 314.

46. Galeotti, N., Plagnes, E., and Jouin, P. (1997). *Tetrahedron Lett.*, **38**, 2459.
47. Murphy, A. M., Dagnino, R., Vallar, P. L., Trippe, A. J., Sherman, S. L., Lumpkin, R. H., Tamura, S. Y., and Webb, T. R. (1992). *J. Am. Chem. Soc.*, **114**, 3156.
48. Dagnino, R. and Webb, T. R. (1994). *Tetrahedron Lett.*, **35**, 2125.
49. Nahm, S. and Weinreb, S. (1981). *Tetrahedron Lett.*, **22**, 3518.
50. Zlatoidsky P. (1994). *Helvetica Chimica Acta*, **77**, 150.
51. Fehrentz, J. A., Paris, M., Heitz, A., Velek, J., Liu, C. F., Winternitz, F., and Martinez J. (1995). *Tetrahedron Lett.*, **36**, 7871.
52. Ho, P. T. and Ngu, K. (1993). *J. Org. Chem.*, **58**, 2313.
53. Fehrentz, J. A. and Castro, B. (1983). *Synthesis*, 676.
54. Fehrentz, J. A., Heitz, A., and Castro, B. (1985). *Int. J. Peptide Protein Res.*, **26**, 236.
55. Fehrentz, J. A., Paris, M., Heitz, A., Velek, J., Winternitz, F., and Martinez, J. (1997). *J. Org. Chem.*, **62**, 6792.
56. Fuller, W. D., Cohen, M. P., Shabankareh, M., and Blair, R. K. (1990). *J. Am. Chem. Soc.*, **112**, 7414.
57. Sylvain, C., Wagner, A., and Mioskowski, C. (1997). *Tetrahedron Lett.*, **38**, 1043.
58. Frechet, J. M. and Schuerch, C. (1971). *J. Am. Chem. Soc.*, **93**, 492.
59. Pothion, C., Paris, M., Heitz, A., Rocheblave, L., Rouch, F., Fehrentz, J. A., and Martinez, J. (1997). *Tetrahedron Lett.*, **38**, 7749.
60. Baleux, F., Daunis, J., and Jacquier, R. (1984). *Tetrahedron Lett.*, **25**, 5893.
61. Dinh, T. Q. and Amstrong, R. W. (1996). *Tetrahedron Lett.*, **37**, 1161.
62. Ede, N. J. and Bray, A. M. (1997). *Tetrahedron Lett.*, **38**, 7119.
63. Galeotti, N., Giraud, M., and Jouin, P. (1997). *Letters in Peptide Science*, **4**, 437.
64. Otto, H.-H. and Schirmeister, T. (1997). *Chem. Rev.*, **97**, 133.
65. Paris, M., Heitz, A., Guerlavais, V., Cristau, M., Fehrentz, J.-A., and Martinez, J. (1998). *Tetrahedron Lett.*, **39**, 7287.
66. Jensen, K. J., Alsina, J., Songster, M. F., Vgner, J., Albericio, F., and Barany, G. (1998). *J. Am. Chem. Soc.*, **120**, 5441.
67. Valerio, R. M., Bray, A. M., and Maeji, N. J. (1994). *Int. J. Peptide Protein. Res.*, **44**, 158.
68. Atherton, E. and Sheppard, R. C. (ed.) (1989). *Solid phase peptide synthesis: a practical approach.* IRL Press, Oxford.
69. Bodanszky, M. and Trost, B. (ed.) (1993). *Principles of peptides synthesis.* Springer-Verlag, NY.
70. Roberts, C. J., Davies, M. C., Tendler, S. J. B., Williams, P. M., Davies, J., Dawkes, A. C., Yearwood, C. D. L., and Edwards, J. C. (1996). *Ultramicroscopy*, **62**, 149.
71. Lobl, T. J., Deibel, M. R., Jr., and Yem, A. W. (1988). *Anal. Biochem.*, **170**, 502.
72. Kellam, B., Chan, W. C., Chhabra, S. R., and Bycroft, B. W. (1997). *Tetrahedron Lett.*, **38**, 5391.
73. Johnson, T., Quibell, M., Owen, D., and Sheppard, R. C. (1993). *J. Chem. Soc., Chem. Commun.*, 369.
74. Quibell, M., Owen, D., Packman, L. C., and Johnson, T. (1994). *J. Chem. Soc., Chem. Commun.*, 2343.
75. Ede, N. J., How Ang, K., James, I. W., and Bray, A. M. (1996). *Tetrahedron Lett.*, **37**, 9097.
76. Chan, W. C. (1996). Unpublished data.

77. Chhabra, S. R., Hothi, B., Evans, D. J., White, P. D., Bycroft, B. W., and Chan, W. C. (1997). *Tetrahedron Lett.*, **39**, 1603.
78. Bycroft, B. W., Chan, W. C., Chhabra, S. R., Teesdale-Spittle, P. H., and Hardy, P. M. (1993). *J. Chem. Soc., Chem. Commun.*, 776.
79. Bycroft, B. W., Chan, W. C., Chhabra, S. R., and Hone, N. D. (1993). *J. Chem. Soc., Chem. Commun.*, 778.
80. Augustyns, K., Kraas, W., and Jung, G. (1998). *J. Peptide Res.*, **51**, 127.
81. Ziegler, K., Eberle, H., and Ohlinger, H. (1933). *Liebigs Ann. Chem.*, **504**, 95.
82. McMurray, J. (1991). *Tetrahedron Lett.*, **32**, 7679.
83. McMurray, J. and Lewis, C. A. (1993). *Tetrahedron Lett.*, **34**, 8059.
84. Sieber, P. (1977). *Helv. Chim. Acta*, **60**, 2711.
85. Marlowe, C. (1993). *Bioorg. Med. Chem. Lett.*, **3**, 437.
86. Trzeciak, A. and Bannwarth, W. (1992). *Tetrahedron Lett.*, **33**, 4557.
87. Kates, S. A., Sole, N. A., Johnson, C. R., Hudson, D., Barany, G., and Albericio, F. (1993). *Tetrahedron Lett.*, **34**, 1549.
88. Alsina, J., Rabanal, F., Giralt, E., and Albericio, F. (1994). *Tetrahedron Lett.*, **35**, 9633.
89. Bloomberg, G. B., Askin, D., Gargaro, A. R., and Tanner, M. J. A. (1993). *Tetrahedron Lett.*, **34**, 4709.
90. Chan, W. C., Bycroft, B. W., Evans, D. J., and White, P. D. (1995). *J. Chem. Soc., Chem. Commun.*, 2209.
91. Chan, W. C., Bycroft, B. W., Evans, D. J., and White, P. D. (1997). Unpublished data.
92. Kates, S. A., Daniels, S. B., and Albericio, F. (1993). *Anal. Biochem.*, **212**, 303.
93. Rohwedder, B., Mutti, Y., Dumy, P., and Mutter, M. (1998). *Tetrahedron Lett.*, **39**, 1175.
94. Barlos, K., Gatos, D., Hatzi, O., Koch, N., and Koutsogianno, S. (1996). *Int. J. Peptide Protein Res.*, **47**, 148.
95. Aletras, A., Barlos, K., Gatos, D., Koutsogianni, S., and Mamos, P. (1995). *Int. J. Peptide Protein Res.*, **45**, 488.
96. Barlos, K., Gatos, D., Koutsogianni, S., Schäfer, W., Stavropoulos, G. and Wenging, Y. (1991). *Tetrahedron Lett.*, **32**, 471.
97. Pullen, N., Brown, N. G., Sharma, R. P., and Akhtar, M. (1993). *Biochemistry*, **32**, 3958.
98. Eichler, J., Lucka, A. W., and Houghten, R. A. (1996). In *Innovations and perspectives in solid phase synthesis and combinatorial libraries* (ed. R. Epton), p. 201. Mayflower Scientific Ltd., Birmingham, UK.

## 7

# Phosphopeptide synthesis

### PETER D. WHITE

## 1. Introduction

Protein phosphorylation mediated by protein kinases is the principal mechanism by which eukaryotic cellular processes are modulated by external physiological stimuli (1). Phosphopeptides are essential tools for the study of this process, serving as model substrates for phosphatases, as antigens for the production of antibodies against phosphorylated proteins, and as reference compounds for determining their physical parameters. The development of methods for the production of phosphopeptides has consequently attracted considerable interest over the last few years, and these endeavours have yielded reliable procedures which have now made their synthesis routine.

There are two strategies used currently for the preparation of phosphopeptides: the building block approach, in which pre-formed protected phospho-amino acids are incorporated during the course of chain assembly, and the global phosphorylation method, which involves post-synthetic phosphorylation of serine, threonine, or tyrosine side-chain hydroxyl groups on the solid support.

## 2. Building block approach

### 2.1 Introduction

The building block procedure is certainly the more straightforward of the two approaches and has now become, owing to the availability of suitably protected phosphoamino acids, the standard method for the routine production of phosphopeptides.

For the side-chain protection of phosphotyrosine in Fmoc/$t$Bu-based solid phase synthesis, methyl (2), benzyl (3, 4), $t$-butyl (5, 6), dialkylamino (7), and silyl (8) groups have been employed. Of these, benzyl is most useful as it is the most convenient to introduce and is rapidly removed during the TFA-mediated acidolysis step. Only the mono-benzyl ester, Fmoc-Tyr(PO(OBzl)-OH)-OH 1 (9, 10), is available commercially; the dibenzyl ester offers no practical benefit as it undergoes mono-debenzylation in the course of the piperidine-mediated Fmoc deprotection reaction (11).

Also available commercially is Fmoc-Tyr($PO_3H_2$)-OH **2** (12). This derivative, despite having no phosphate protection, appears to work well, particularly in the synthesis of small- to medium-sized phosphopeptides; although formation of the pyrophosphate **3** can be a problem in peptides containing adjacent Tyr($PO_3H_2$) residues (13, 14).

Phosphate triesters of serine and threonine are not compatible with Fmoc/*t*Bu chemistry as they undergo β-elimination when treated with piperidine, resulting in the formation of dehydroalanine and dehydoamino-butyric acid, respectively (*Figure 1*). For this reason, it was long believed that the building block approach could not be used for preparation of peptides containing these amino acids. Fortunately, in 1994, Wakamiya *et al.* (15) demonstrated that by using a mono-protected derivative, such as Fmoc-Ser(PO(OBzl)OH)-OH **4**, this side-reaction could be almost totally suppressed. The corresponding threonine derivative has now also been described (9, 16), and a growing body of literature has been accumulated attesting to the utility of these derivatives (17–20).

## 2.2 Practical considerations

The following considerations apply to the synthesis of phosphopeptides using Fmoc-Ser(PO(OBzl)OH)-OH, Fmoc-Thr(PO(OBzl)OH)-OH, Fmoc-Tyr-(PO(OBzl)OH)-OH, and Fmoc-Tyr($PO_3H_2$)-OH.

1. Incorporation of these derivatives into the peptide chain is best effected using either OBt or OAt esters, generated from the corresponding uronium

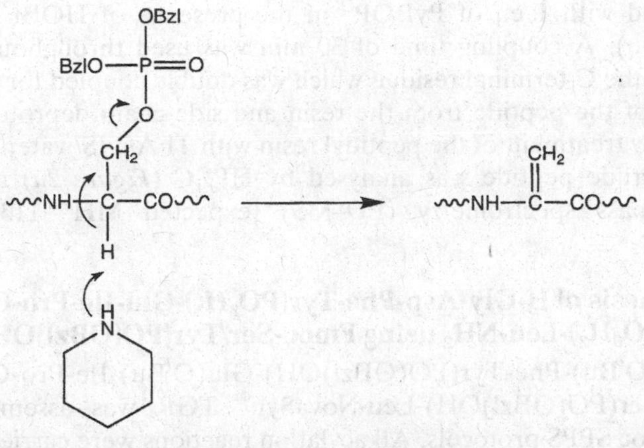

**Figure 1.** Formation of dehydroamino acids by β-elimination.

(aminium) or phosphonium activating reagent as described in Chapter 3, *Protocol 11*. Addition of the residue following the phosphoamino acid can be sluggish, so it is advisable that the completeness of this reaction be checked using either the TNBS or the Kaiser test (Chapter 3, *Protocol 14*). Particular problems can occur when assembling strings of consecutive phosphoamino acids; in these cases, repeat coupling might be necessary.

2. Fmoc group removal can be effected with piperidine in DMF in the normal manner, as described in Chapter 3, *Protocol 7*. To the author's knowledge, the stability of the mono-benzyl phosphoserine and phosphothreonine derivatives to DBU has not been determined.

3. The side-chain benzyl group is normally removed with 95% TFA in 1–3 h, although peptides containing multiple benzyl-protected phosphoamino acid residues may require longer. Trialkylsilane containing cleavage mixtures, such as TFA/TIS/water (95:2.5:2.5), give excellent results (Chapter 3, *Protocol 16*).

4. Introduction of a phosphoamino acid can profoundly affect the ease of assembly of a given sequence. Peptides that can be efficiently assembled using unphosphorylated building blocks are frequently difficult to prepare in the phosphorylated form. In these cases, post-synthetic phosphorylation often proves to be a superior method.

## 2.3 Illustrative examples

### 2.3.1 Synthesis of H-Ala-Pro-Gly-Asp-Ala-Tyr(PO₃H₂)-Gly-Pro-Lys-Leu-Ala-NH₂ using Fmoc-Tyr(PO₃H₂)-OH

H-Ala-Pro-Gly-Asp(O$^t$Bu)-Ala-Tyr(PO₃H₂)-Gly-Pro-Lys(Boc)-Leu-Ala-NovaSyn® KR was assembled using standard Fmoc SPPS protocols. All acylation reactions were carried out using a fivefold excess of Fmoc-amino acid activated with 1 eq of PyBOP® in the presence of HOBt (1 eq) and DIPEA (2 eq). A coupling time of 30 min was used throughout, with the exception of the C-terminal residue which was double coupled for 45 min.

Cleavage of the peptide from the resin and side-chain deprotection were carried out by treatment of the peptidyl resin with TFA/TIS/water (90:5:5) for 2.5 h. The crude peptide was analysed by HPLC (*Figure 2a*) and plasma desorption–mass spectrometry (PD–MS) [expected MH⁺ 1139.2, found 1139.1].

### 2.3.2 Synthesis of H-Gly-Asp-Phe-Tyr(PO₃H₂)-Glu-Ile-Pro-Glu-Glu-Ser(PO₃H₂)-Leu-NH₂ using Fmoc-Ser/Tyr(PO(OBzl)OH)-OH

H-Gly-Asp(OᵗBu)-Phe-Tyr(PO(OBzl)OH)-Glu(OᵗBu)-Ile-Pro-Glu(OᵗBu)-Glu(OᵗBu)-Ser(PO(OBzl)OH)-Leu-NovaSyn® TGR was assembled using standard Fmoc SPPS protocols. All acylation reactions were carried out using a 10-fold excess of Fmoc-amino acid activated with TBTU (1 eq) in the presence of HOBt (1 eq) and DIPEA (2 eq). A coupling time of 1 h was used throughout. Cleavage of the peptide from the resin and side-chain deprotection were effected as above. The crude peptide was analysed by HPLC (*Figure 2b*) and electrospray-mass spectrometry (ES-MS) [expected MH⁻ 1456.2, found 1456.1].

### 2.3.3 Synthesis of H-Tyr-Glu-Ser(PO₃H₂)-Leu-Ser(PO₃H₂)-Ser-Ser-Glu-NH₂ using Fmoc-Ser(PO(OBzl)OH)-OH

H-Tyr(ᵗBu)-Glu(OᵗBu)-Ser(PO(OBzl)OH)-Leu-Ser(PO(OBzl)OH)-Ser(ᵗBu)-Ser(ᵗBu)-Glu(OᵗBu)-NovaSyn® TGR was prepared as described in Section 2.3.2. The crude peptide was analysed by HPLC (*Figure 2c*) and ESI [expected MH⁻ 1058.9, found 1058.0].

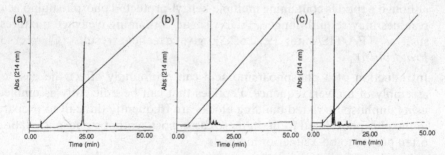

**Figure 2.** HPLC of total crude phosphopeptides. (a) H-Ala-Pro-Gly-Asp-Ala-Tyr(PO₃H₂)-Gly-Pro-Lys-Leu-Ala-NH₂. (b) H-Gly-Asp-Phe-Tyr(PO₃H₂)-Glu-Ile-Pro-Glu-Glu-Ser(PO₃H₂)-Leu-NH₂. (c) H-Tyr-Glu-Ser(PO₃H₂)-Leu-Ser(PO₃H₂)-Ser-Ser-Glu-NH₂. Conditions for (a): column TSK-120T 5 μm; eluant A, 0.1% aq. TFA, eluant B, 90% acetonitrile/10% water/0.1% TFA; gradient 5% B for 5 min then to 100% in 50 min; flow rate 1 ml/min; Conditions for (b) and (c): column Shandon Tris C3; eluant A, 0.06% aq. TFA, eluant B, 90% acetonitrile/10% water/0.06% TFA; gradient 5% B for 5 min then to 100% in 40 min; flow rate 1 ml/min. Optical density was measured at 214 nm.

# 3. Global phosphorylation

## 3.1 Introduction

With the introduction of the mono-benzyl derivatives of serine and threonine, the global phosphorylation methodology has become the technique of second choice, only being used in cases where the building block approach fails to give satisfactory results. It does, however, have the advantage of furnishing both phosphorylated and unphosphorylated peptides in the same synthesis, which can be useful when comparing the biological activity of both forms.

Assembly of the resin-bound peptide can be effected using standard Fmoc protocols. The residue at the phosphorylation site can be incorporated either without side-chain protection or with a side-chain protecting group, such as trityl (21) or *t*-butyldimethylsilyl (22), which can be selectively removed on the solid phase immediately before the phosphorylation reaction. When unprotected serine or threonine is incorporated, coupling of subsequent residues is best achieved using carbodiimide/HOBt activation chemistry: uronium- and phosphonium-type reagents can potentially mediate acylation of unprotected hydroxyl side chains (23). This precaution is not required when the peptide contains unprotected tyrosine residues, as any phenol esters generated will be destroyed during the subsequent Fmoc-deprotection cycle.

The N-terminal residue should be incorporated as an *N*-Boc derivative, or the N-terminal amino group can be capped by treating the resin with di-*t*-butyl-dicarbonate (5 eq) in DMF for 30 min. The former method is preferred when the peptide contains residues with unprotected side-chain hydroxyl functionalities.

Functionalization of the free resin-bound hydroxyl has been carried out using phosphoramidites (3, 4, 24, 25), phosphochloridates (26, 27), and *H*-phosphonates (28). The most widely used approach – with phosphoramidites – is presented in this chapter.

The most useful phosphoramidites described to date are the di-*t*-butyl (29) and dibenzyl derivatives (30, 31), as their respective protecting groups are removed during the course of the normal TFA cleavage reaction. The utility of both these reagents in the global phosphorylation of Tyr, Ser, and Thr residues is now thoroughly established (3, 4, 24, 32–35).

Phosphorylation using phosphoramidites is a two-step procedure (*Figure 3*): the resin-bound hydroxyl group is first converted to a phosphite triester by reaction with the appropriate phosphoramidite in the presence of tetrazole and, in the second step, this phosphorous(III) intermediate is oxidized to the desired phosphate.

For oxidation of the intermediate phosphite ester, *m*-chloroperbenzoic acid (24), aqueous iodine (3) and *t*-butyl hydroperoxide (4, 25, 33) have been used. The use of anhydrous *t*-butyl hydroperoxide gives the best results as this reagent does not appear to cause appreciable oxidation of methionine (4), tryptophan, or cysteine residues (36).

Boc-Gly-Ala-Thr(ᵗBu)-Leu-Asp(OᵗBu)-Gly-Ile-Thr-Ala-⬤

$(BzlO)_2PN^iPr_2$/tetrazole ↓

BzlO—P—OBzl

Boc-Gly-Ala-Thr(ᵗBu)-Leu-Asp(OᵗBu)-Gly-Ile-Thr-Ala-⬤

*t*-BuOOH ↓

$$BzlO\underset{O}{\overset{O}{\|}}OBzl$$

Boc-Gly-Ala-Thr(ᵗBu)-Leu-Asp(OᵗBu)-Gly-Ile-Thr-Ala-⬤

TFA ↓

$$HO\overset{O}{\underset{OH}{\|}}OH$$

H-Gly-Ala-Thr-Leu-Asp-Gly-Ile-Thr-Ala-OH

**Figure 3.** Post-synthetic phosphorylation.

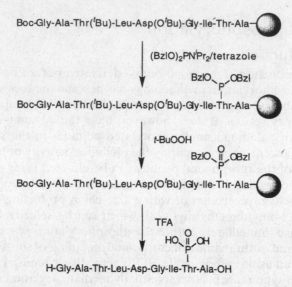

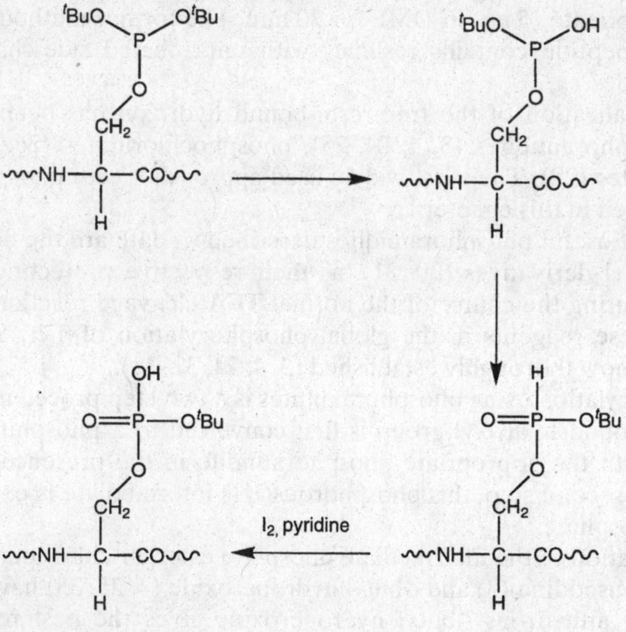

**Figure 4.** *H*-Phosphonate formation.

Benzyl-protected amidites generally give better results than *t*-butyl amidites: in the presence of the weakly acidic tetrazole, di-*t*-butyl phosphite esters have a tendency to lose a *t*-butyl group, leading to formation of the corresponding *H*-phosphonate **5** (12, 37) (*Figure 4*). Fortunately, this by-product can be converted to the desired phosphate by treatment with iodine in pyridine (38).

## 3.2 General protocol for post-synthetic phosphorylation

Phosphoramidites are both moisture- and air-sensitive, hence *Protocol 1* must be carried out using only thoroughly dried reagents and glassware under a dry inert atmosphere of argon or oxygen-free nitrogen. All transfer of reagents should be carried out using gas-tight syringes flushed with argon, or with double-ended needles.

The use of DMF is preferred to acetonitrile in the phosphinylation step, since tetrazole is far more soluble in this solvent, enabling the reaction to be conducted at a much higher concentration. Anhydrous DMF is commercially available stored under nitrogen in bottles fitted with a septum.

Only resublimed DNA-grade tetrazole should be used; this is available from Sigma in septum-sealed, argon-filled vials containing 300 mg of material, sufficient for carrying out a 0.1-mmol-scale reaction. Dibenzyl-*N*,*N*-diisopropylphosphoramidite is available in argon-filled one-shot vials from Novabiochem.

---

**Protocol 1.**   General protocol for phosphorylation

*Equipment and reagents*

- A regulated cylinder of argon, equipped with an outlet line tipped with a narrow-gauge needle
- Dibenzyl-*N*,*N*-diisopropylphosphoramidite (500 mg, 1.44 mmol)
- 3 M *t*-butyl hydroperoxide in isooctane
- 5 ml Gas-tight syringe tipped with a needle

- Anhydrous DMF
- Argon-filled balloon attached to luer of hypodermic needle
- Resublimed DNA-grade tetrazole (300 mg, 4.28 mmol)
- Peptidyl resin.
- 10 ml manual peptide synthesis vessel

*Method*

1. Place the peptidyl resin (0.1 mmol) in the peptide synthesis vessel and dry overnight at 40°C *in vacuo*.

2. Seal the reaction vessel with a rubber septum.

3. Open the tap and flush the vessel with a stream of argon delivered via a needle inserted through the septum. Close tap.

4. Insert needle of argon-filled balloon and the syringe needle into the septum of the DMF bottle and fill the syringe with 5 ml of solvent. Remove syringe and balloon.

---

**Protocol 1.** *Continued*

5. Remove the metal insert from the top of a tetrazole vial. Insert syringe needle and balloon needle into the vial septum. Slowly add DMF to vial, allowing the displaced argon to fill the syringe between additions. Shake vial until contents dissolved.

6. Fill syringe with 4.2 ml of tetrazole solution. Remove syringe and balloon.

7. Remove the metal insert from the top of a vial containing dibenzyl-*N,N*-diisopropylphosphoramidite and insert syringe and balloon needles into septum. Add tetrazole solution and shake vial to mix contents.

8. Fill syringe with amidite/tetrazole solution.

9. Insert syringe and balloon needles into septum of the vessel containing the resin.

10. Add amidite/tetrazole solution. Remove syringe.

11. Gently agitate reaction vessel for 1 h.

12. Open the tap and allow reagents to drain away.

13. Close tap, add fresh DMF using the syringe, agitate, and open tap and allow to drain. Repeat this washing procedure three times. Close tap.

14. Add 3 M *t*-butyl hydroperoxide in isooctane to cover resin. Gently agitate for 30 min.

15. Wash and dry the resin as described in Chapter 3, *Protocol 1C*.

## 3.3 Illustrative examples

### 3.3.1 Synthesis of H-Phe-Phe-Lys-Asn-Ile-Val-Thr-Pro-Arg-Thr(PO$_3$H$_2$)-Pro-Pro-Pro-Ser-Gln-Gly-Lys-NH$_2$ using Fmoc-Thr-OH

Boc-Phe-Phe-Lys(Boc)-Asn-Ile-Val-Thr-Pro-Arg(Pmc)-Thr-Pro-Pro-Pro-Ser('Bu)-Gln-Gly-Lys(Boc)-NovaSyn KR resin was prepared on a NovSyn Crystal® peptide synthesizer. Boc-Phe-OH and Fmoc-Thr-OH were introduced using 5 eq of OBt ester, prepared by *in situ* reaction between the protected amino acid, HOBt, and DIC. All other residues were coupled using a fivefold excess of Fmoc OPfp ester in the presence of 1 eq of HOBt.

Phosphorylation of the resin was carried out as described in *Protocol 1*, except that dibenzyl-*N,N*-diethylphosphoramidite was used.

The phosphorylated peptide was cleaved from the resin using TFA/thioanisole/phenol/ethanedithiol/water (82.5:5:5:2.5:5) for 2 h. The crude peptide was analysed by HPLC (*Figure 5a*) and PD–MS [expected MH$^+$ 1994.5, found 1994.5].

### 3.3.2 Synthesis H-Gly-Asp-Phe-Glu-Glu-Ile-Pro-Glu-Glu-Thr(PO₃H₂)-Leu-NH₂ using Fmoc-Tyr(Trt)-OH

### 3.3.2 Synthesis H-Gly-Asp-Phe-Glu-Glu-Ile-Pro-Glu-Glu-Thr(PO$_3$H$_2$)-Leu-NH$_2$ using Fmoc-Tyr(Trt)-OH

Boc-Gly-Asp(O$^t$Bu)-Phe-Glu(O$^t$Bu)-Glu(O$^t$Bu)-Ile-Pro-Glu(O$^t$Bu)-Glu(O$^t$Bu)-Thr(Trt)-Leu-Rink amide resin was prepared on an ABI 431A® peptide synthesizer. All acylation reactions were carried out for 1 h using a 10-fold excess of Fmoc-amino acids activated with TBTU (1 eq) in the presence of DIPEA (2 eq) and HOBt (1 eq). The N-terminal Gly residue was introduced as the Boc derivative. The Trt group was removed from the side-chain of Thr using *Protocol 8*, Chapter 9.

The resin-bound peptide was then phosphinylated using di-benzyl-*N*,*N*-diisopropylphosphoramidite/tetrazole and oxidized with *t*-butyl hydroperoxide in isooctane as described in *Protocol 1*.

The phosphorylated peptide was cleaved from the resin using TFA/TIS/water (95:2.5:2.5) for 2 h, and analysed by HPLC (*Figure 5b*) and ES–MS [expected MH$^+$ 1381.3, found 1381.3].

## 3.4 Thiophosphorylated peptides

Oxidation of the intermediate phosphite triester can also be effected using elemental sulphur in CS$_2$ (39), tetraethylthiuram disulphide in CH$_3$CN (39), dibenzoyl tetrasulphide (40), and dithiasuccinimide (41), to yield the corresponding resin-bound thiophosphorylated peptide.

For cleavage and side-chain deprotection of thiophosphopeptidyl resins, the use of TFA/thiophenol/water/EDT/TIS (91.5:2.5:2.5:2.5:1 v/v) is recommended.

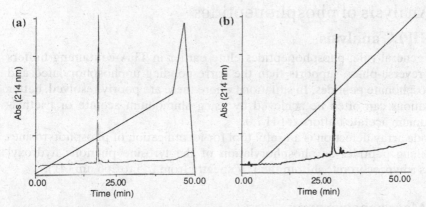

**Figure 5.** HPLC of total crude phosphopeptides. (a) H-Phe-Phe-Lys-Asn-Ile-Val-Thr-Pro-Arg-Thr(PO$_3$H$_2$)-Pro-Pro-Pro-Ser-Gln-Gly-Lys-NH$_2$. Conditions: column Beckman C18; eluant A, 0.1% aq. TFA, eluant B, 0.1% TFA in acetonitrile; gradient 0% to 100% B in 40 min; flow rate 1 ml/min. (b) H-Gly-Asp-Phe-Glu-Glu-Ile-Pro-Glu-Glu-Thr(PO$_3$H$_2$)-Leu-NH$_2$. Conditions: column TSK-120T; eluant A, 0.1% aq. TFA, eluant B, 90% acetonitrile/10% water/0.1% TFA; gradient 5% B for 5 min then to 100% in 50 min; flow rate 1 ml/min. Optical density was measured at 214 nm.

Care should be taken to avoid prolonged exposure of thiophosphoryl peptides to acidic aqueous conditions as this can result in loss of the thiophosphate group (39, 40). Purification of these peptides is best achieved by reverse-phase HPLC using a gradient of $CH_3CN$ in 0.1 M triethylammonium acetate (pH 7.5) (39).

---

**Protocol 2.** Thiophosphorylation

*Equipment and reagents*

- A regulated cylinder of argon, equipped with an outlet line tipped with a narrow-gauge needle
- 5 ml Gas-tight syringe tipped with a needle
- Dry acetonitrile
- Argon-filled balloon attached to luer of hypodermic needle
- 15% tetraethylthiuram in acetonitrile
- 10 ml manual peptide synthesis vessel

*Method*

1. Phosphinylate the resin as described in steps 1–13 of *Protocol 1*.
2. Wash the resin with acetonitrile in the manner described in *Protocol 1*, step 13.
3. Add tetraethylthiuram in acetonitrile (3 ml) using a syringe.
4. Gently agitate for 2 h.
5. Wash and dry the resin as described in Chapter 3, *Protocol 1C*.

---

# 4. Analysis of phosphopeptides

## 4.1 HPLC analysis

As a general rule, phosphopeptides elute earlier in TFA-containing buffers from reverse-phase supports than the corresponding unphosphorylated and *H*-phosphonate peptides. In situations where these are poorly resolved, better separations can often be achieved by using ammonium acetate or triethylammonium acetate buffers at pH 7.0.

Diode array detection is a useful tool for identification of phosphotyrosine-containing peptides as phosphorylation of the tyrosine phenolic hydroxyl causes a hypsochromic shift in the UV spectra from 275 to 266 nm (41).

## 4.2 Mass spectroscopy

Phosphopeptides have a tendency to fragment during electrospray and plasma-desorption mass spectrometry: phosphoserine- and threonine-containing peptides β-eliminate with loss of 98 mass units, corresponding to cleavage of $H_3PO_4$; phosphotyrosine-containing peptides fragment with loss of 174 mass units, due to cleavage of phenylphosphate (32).

In positive ion mode, phosphopeptides form strong sodium and potassium adducts, which often give more intense signals than the parent ion. Peptides containing multiple phosphorylation sites often give better spectra if recorded in negative ion mode.

# References

1. Cohen, P. (1982). *Nature*, **296**, 613.
2. Kitas, E. A., Perich, J. W., Wade, J. D., Johns, R. B., and Tregear, G. W. (1989). *Tetrahedron Lett.*, **30**, 6229.
3. Kitas, E. A., Knorr, R., Trzeciak, A., and Bannwarth, W. (1991). *Helv. Chim. Acta*, **74**, 1314.
4. Andrews, D. M., Kitchin, J., and Seale, P. W. (1991). *Int. J. Peptide Protein Res.*, **38**, 469.
5. Perich, J. W. and Reynolds, E. C. (1991). *Int. J. Peptide Protein Res.*, **37**, 572.
6. Perich, J. W., Ruzzene, M., Pinna, L. A., and Reynolds, E. C. (1994). *Int. J. Peptide Protein Res.*, **43**, 39.
7. Chao, H.-G., Leiting, B., Reiss, P. D., Burkhardt, A. L., Klimas, C. E., Bolen, J. B., and Matsueda, G. R. (1995). *J. Org. Chem.*, **60**, 7710.
8. Chao, H. G., Bernatowicz, M. S., Reiss, P. D., and Matsueda, G. R. (1994). *J. Org. Chem.*, **59**, 6687.
9. White, P. and Beythien, J. (1996). In *Innovations and perspectives in solid phase synthesis and combinatorial libraries: 4th International Symposium* (ed. R. Epton), p. 557. Mayflower Scientific Ltd., Birmingham.
10. Handa, B. K. and Hobbs, C. J. (1998). *J. Pept. Sci.*, **4**, 138.
11. Kitas, E. A., Wade, J. D., Johns, R. B., Perich, J. W., and Tregear, G. W. (1991). *J. Chem. Soc., Chem. Commun.*, 338.
12. Ottinger, E. A., Shekels, L. L., Bernlohr, D. A., and Barany, G. (1993). *Biochemistry*, **32**, 4354.
13. Ottinger, E. A., Xu, Q. H., and Barany, G. (1996). *Pept. Res.*, **9**, 223.
14. Garcia Echeverria, C. (1995). *Lett. Pept. Sci.*, **2**, 93.
15. Wakamiya, T., Saruta, K., Yasuoka, J., and Kusumoto, S. (1994). *Chem. Lett.*, 1099.
16. Vorherr, T. and Bannwarth, W. (1995). *Bioorg. Med. Chem. Lett.*, **5**, 2661.
17. Gururaja, T. L. and Levine, M. J. (1996). *Pept. Res.*, **9**, 283.
18. John, M., Briand, J.-P., and Schnarr, M. (1996). *Pept. Res.*, **9**, 71.
19. Wakamiya, T., Togashi, R., Nishida, T., Saruta, K., Yasuoka, J., Kusumoto, S., Aimoto, S., Kumagaye, K. Y., Nakajima, K., and Nagata, K. (1997). *Bioorg. Med. Chem.*, **5**, 135.
20. Shapiro, G., Buchler, D., Dalvit, C., Fernandez, M. D., Gomezlor, B., Pombovillar, E., Stauss, U., and Swoboda, R. (1996). *Bioorg. Med. Chem. Lett.*, **6**, 409.
21. Pullen, N., Brown, N. G., Sharma, R. P., and Akhtar, M. (1993). *Biochemistry*, **32**, 3958.
22. Shapiro, G., Swoboda. R., and Stauss, U. (1994). *Tetrahedron Lett.*, **35**, 869.
23. Yon, M., White, P., and Groome, N. (1994). In *Innovations and perspectives in solid phase synthesis: 3rd International Symposium* (ed. R. Epton), p. 707. Mayflower Worldwide Ltd., Birmingham.

24. Perich, J. W., Nguyen, D. L., and Reynolds, E. C. (1991). *Tetrahedron Lett.*, **32**, 4033.
25. de Bont, H. B. A., van Boom, J. H., and Liskamp, R. M. J. (1990). *Tetrahedron Lett.*, **31**, 249.
26. Otvos, L., Jr., Elekes, I., and Lee, V. M.-Y. (1989). *Int. J. Peptide Protein Res.*, **34**, 129.
27. Hormozdiari, P. and Gani, D. (1996). *Tetrahedron Lett.*, **37**, 8227.
28. Larsson, E. and Lüning, B. (1994). *Tetrahedron Lett.*, **35**, 2737.
29. Perich, J. W. and Johns, R. B. (1988). *Tetrahedron Lett.*, **29**, 2369.
30. de Bont, H. B. A., Veeneman, G. H., van Boom, J. H., and Liskamp, R. M. J. (1990). *Recl. Trav. Chim. Pays-Bas*, **109**, 27.
31. Bannwarth, W. and Trzeciak, A. (1987). *Helv. Chim. Acta*, **70**, 175.
32. Hoffmann, R., Wachs, W. O., Berger, R. G., Kalbitzer, H. R., Waidelich, D., Bayer, E., Wagnerredeker, W., and Zeppezauer, M. (1995). *Int. J. Peptide Protein Res.*, **45**, 26.
33. Staerkaer, G., Jakobsen, M. H., Olsen, C. E., and Holm, A. (1991). *Tetrahedron Lett.*, **32**, 5389.
34. Perich, J. W. (1996). *Lett. Pept. Sci.*, **3**, 127.
35. Perich, J. W., Meggio, F., and Pinna, L. A. (1996). *Bioorg. Med. Chem.*, **4**, 143.
36. White, P. (1994). Unpublished work.
37. Xu, Q. H., Ottinger, E. A., Sole, N. A., and Barany, G. (1997). *Lett. Pept. Sci.*, **3**, 333.
38. Perich, J. W. (1998). *Lett. Pept. Sci.*, **5**, 49.
39. Kitas, E., Küng, E., and Bannwarth, W. (1994). *Int. J. Peptide Protein Res.*, **43**, 146.
40. Tegge, W. (1994). *Int. J. Peptide Protein Res.*, **43**, 448.
41. Xu, Q., Musier-Forsyth, K., Hammer, R. P., and Barany, G. (1996). In *Peptides: chemistry, structure and biology* (ed. P. T. P. Kaumaya and R. P. Hodges), p. 123. Mayflower Scientific Ltd., Birmingham.

# 8

# Glycopeptide synthesis

JAN KIHLBERG

## 1. Introduction

Most eukaryotic proteins, some bacterial and many viral proteins carry structurally diverse carbohydrates that are covalently attached through $N$- or $O$-glycosidic bonds to the side chains of asparagine, serine, threonine, hydroxylysine, tyrosine, and hydroxyproline (reviewed in ref. 1). In nature, $N$-linked glycoproteins are assembled by post-translational, enzymatic attachment of a common oligosaccharide having the composition $Glc_3Man_9GlcNAc_2$ to the side chain of asparagine (reviewed in ref. 2). This saccharide is then modified enzymatically, thereby giving structural variation to the part remote from the protein. However, $N$-linked glycoproteins have a common penta-saccharide core ($Man\alpha3(Man\alpha6)Man\beta4GlcNAc\beta4GlcNAc$) in which the chitobiose moiety ($GlcNAc\beta4GlcNAc$) is bound to asparagine (**1**) (*Figure 1*). By contrast, $O$-linked glycoproteins are built up by sequential attachment of monosaccharides by different enzymes to hydroxylated amino acids in the protein, and therefore no common core is formed (2). Thus, $N$-acetyl-$\alpha$-D-galactosamine attached to serine (**2**) and threonine (**3**) is found in mucin secreted from epithelial cells. $\beta$-D-Xylosyl serine (**4**) is found in many proteo-glycans, whereas $\beta$-D-galactosyl hydroxylysine (**5**) is common in collagen found in connective tissue. $\alpha$-L-Fucosyl residues linked to serine (**6**) and threonine (**7**) are found in fibrinolytic and coagulation proteins. $N$-Acetyl-$\beta$-D-glucosamine attached to serine (**8**) and threonine (**9**) occurs frequently in glycoproteins located in the nucleus and cytoplasm, whereas glycoproteins produced by yeast have $\alpha$-D-mannosyl residues linked to serine (**10**) and threonine (**11**). Larger structures are usually formed by attachment of additional saccharides to the $O$-linked **2–4** when found in glycoproteins. Structures **5**, **10**, and **11** can also carry additional monosaccharides.

In recent years numerous glycoproteins have been isolated and character-ized, but the roles for the protein-bound carbohydrates have only just begun to be unravelled (1). It is now well established that glycosylation affects both the physiochemical properties and the biological functions of a glycoprotein. For instance, glycosylation has been found to influence uptake, distribution,

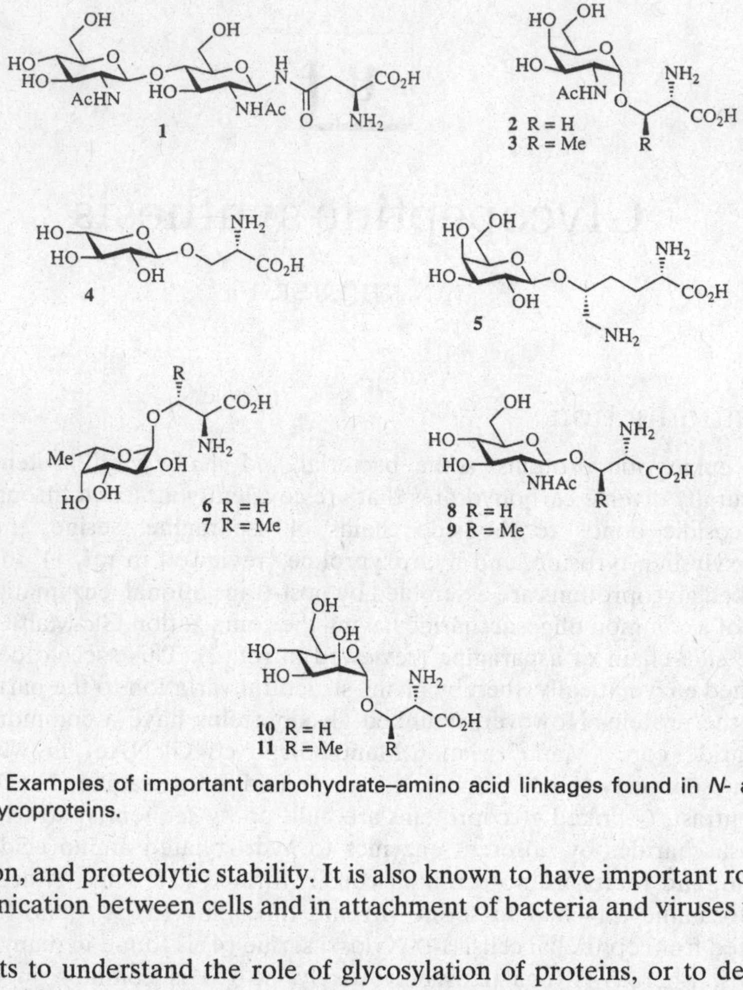

**Figure 1.** Examples of important carbohydrate–amino acid linkages found in *N*- and *O*-linked glycoproteins.

excretion, and proteolytic stability. It is also known to have important roles in communication between cells and in attachment of bacteria and viruses to the host.

Efforts to understand the role of glycosylation of proteins, or to develop glycopeptides as tools in drug discovery and drug design, have led to substantial progress in development of methodology for the synthesis of glycopeptides during the last decades. This chapter attempts to summarize current knowledge on the use of glycosylated amino acids as building blocks in solid phase synthesis of glycopeptides based on the 9-fluorenylmethoxycarbonyl (Fmoc) protective group strategy.

First, some general strategic considerations will be given. Then methods for stereoselective attachment of carbohydrates to amino acids are discussed in general terms, but the reader is referred to other sources to obtain more extensive information concerning assembly of oligosaccharides (reviewed in ref. 3). In solid phase synthesis of both peptides and glycopeptides it is essential to obtain quantitative removal of protective groups without degradation of the product at the end of the synthetic sequence. The additional structural complexity and the chemical properties of glycopeptides make the choice of protective groups more demanding and this will therefore be discussed in some detail together with conditions which allow deprotection under mild conditions. Synthetic procedures that provide glycosylated amino acids corresponding to some of the structures shown in *Figure 1* have been included at the end of the chapter in an attempt to exemplify the most efficient methods available at present. An example of the use of one of these building blocks in solid phase synthesis is also provided.

## 2. Strategic considerations in glycopeptide synthesis

In principal, glycopeptides can be prepared either by attachment of the carbohydrate moiety to the target peptide or by using glycosylated amino acids as building blocks in stepwise synthesis. Condensation between a glycosyl amine and a peptide containing an aspartic or glutamic acid residue has been accomplished in syntheses of *N*-linked glycopeptides (4–6). However, glycosylamines can undergo anomerization (7), resulting in the formation of undesired α-glycosides, which might be more difficult to remove in the case of glycopeptides than for glycosylated amino acids. Direct *O*-glycosylation of serine and threonine residues in peptides has met with only limited success in most cases (8–10), mainly due to the low solubility of peptides under conditions used for glycoside synthesis. Recently, some encouraging progress was made in the glycosylation of serine residues in simple model peptides attached to a solid phase, but it remains to be seen if the approach is compatible with more complex peptides (11). At present the use of glycosylated amino acids as building blocks in stepwise assembly is therefore more reliable and efficient for synthesis of both *N*- and *O*-linked glycopeptides than attempts to attach carbohydrates directly to a peptide.

As for peptides, yields obtained in solution synthesis of glycopeptides are often only modest and isolation of intermediates makes the approach cumbersome when performed on a small scale. Recent efforts have therefore mainly focused on solid phase synthesis, which today constitutes the most reliable and efficient method for preparation of glycopeptides (12–15). Consequently, efficient synthetic routes to glycosylated amino acids are of central importance for the preparation of glycopeptides.

It might be expected that the carbohydrate moiety of a glycosylated amino acid would impose steric hindrance in stepwise glycopeptide synthesis, thereby reducing coupling efficiency and hampering further elongation of the peptide chain. However, an overview of glycopeptides prepared to date reveal that no such problems exist. This conclusion is independent of both the amino acid and the carbohydrate moiety, as revealed by the fact that amino acids which carry oligosaccharides as large as undecasaccharides have been found to couple as well as ordinary, non-glycosylated amino acids (16). Even though any method for activation and coupling used in synthesis of peptides should also be suitable for a glycosylated amino acid, two methods have been used predominantly—the *N,N'*-diisopropylcarbodiimide/1-hydroxybenzotriazole method and the pentafluorophenyl (OPfp) ester method (12, 13).

The preparation of glycosylated amino acids requires considerable effort, so the smallest possible amount should be used in each coupling. A 1–1.5-fold excess of the activated glycoamino acid compared to the capacity (substitution) of the peptide–resin has been shown to be sufficient to obtain complete and fast coupling. Importantly, even an equivalent amount of the glycosylated amino acid has been used without a serious decrease in the overall yield (17, 18). When a minimal amount of glycosylated amino acid is being coupled, monitoring of couplings becomes important. This can be performed visually or in a spectrophotometer using either 3,4-dihydro-3-hydroxy-4-oxo-1,2,3-benzotriazine (DhbtOH) (19) or bromophenol blue (20) as indicator of unreacted amino groups on the solid phase.

## 3. Formation of glycosidic linkages to amino acids

Methods that allow stereoselective formation of glycosidic bonds between carbohydrates and amino acids are crucial in the synthesis of glycopeptides. The equatorial, β-glycosidic bond between *N*-acetyl-D-glucosamine and asparagine (**1**, *Figure 1*) found in *N*-linked glycoproteins is preferably formed by acylation of a glycosyl amine by the side chain carboxyl group of aspartic acid, using standard methods for peptide synthesis (*Figure 2*) (21). The required glycosyl amines are available by reduction of glycosyl azides or by treatment of unprotected carbohydrates with ammonium hydrogen carbonate. For 1,2-*trans-O*-linked glycosylated amino acids (**4, 5, 10**, and **11**, *Figure 1*) the glycosidic bond is most commonly established under the influence of a participating acetyl or benzoyl group at O-2 of the carbohydrate (*cf. Figure 3a*).

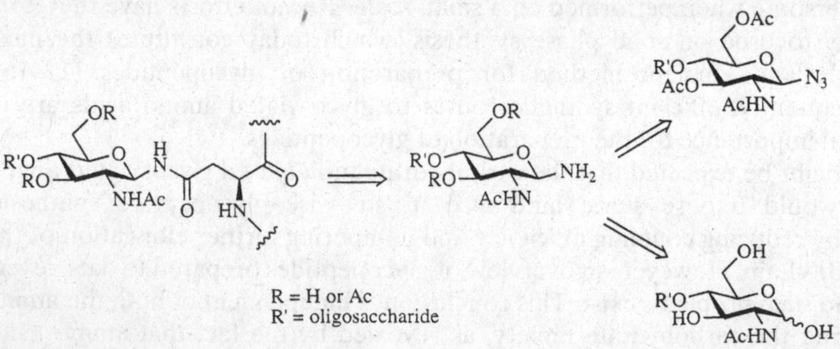

**Figure 2.** *N*-linked glycopeptides can be obtained by acylation of glycosyl amines, prepared either by reduction of glycosyl azides or directly from reducing sugars.

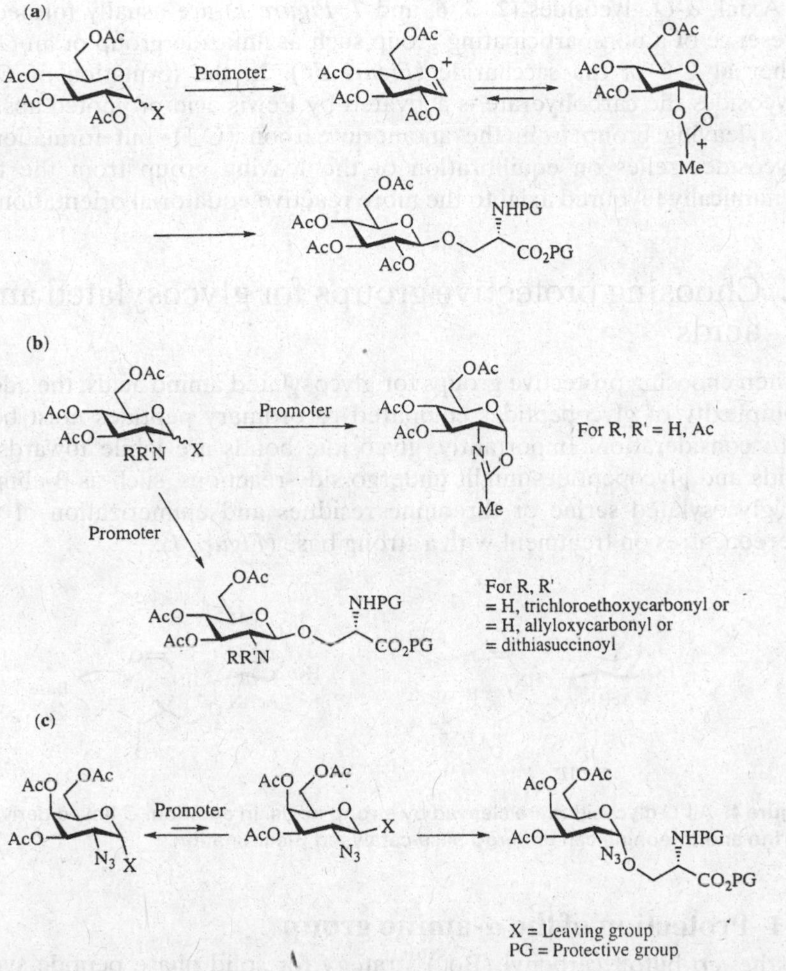

**(a)**

**(b)**

For R, R' = H, Ac

For R, R'
= H, trichloroethoxycarbonyl or
= H, allyloxycarbonyl or
= dithiasuccinoyl

**(c)**

X = Leaving group
PG = Protective group

**Figure 3.** For *O*-linked glycopeptides an equatorial, β-glycosidic bond is usually established using a participating acyl group at O-2 (a). However, the presence of an acetamido group at C-2, as in *N*-acetylglucosamine or *N*-acetylgalactosamine, results in formation of an oxazoline, a reaction which is avoided if, instead, N-2 carries either a trichloroethoxycarbonyl, an allyloxycarbonyl, or a dithiasuccinoyl group (b). A non-participating group, such as the azido group or an *O*-benzyl group, is employed when axial α-glycosides are to be prepared and the substituent at C-2 is equatorial (c).

However, in attempted preparations of 1,2-*trans*-linked *O*-glycosides of *N*-acetyl-D-glucosamine (**8** and **9**, *Figure 1*), the presence of the *N*-acetylamino group at C-2 results in the predominant formation of a 1,2-oxazoline, a problem which can be avoided if the amino group is protected with either a trichloroethoxycarbonyl, an allyloxycarbonyl, or a dithiasuccinoyl group (*Figure 3b*).

199

Axial, α-*O*-glycosides (**2**, **3**, **6**, and **7**, *Figure 1*) are usually formed in the presence of a non-participating group such as an azido group or an *O*-benzyl ether at C-2 of the saccharide (*Figure 3c*). In the formation of *O*-linked glycosides the carbohydrate is activated by Lewis acid-promoted abstraction of a leaving group from the anomeric carbon (C-1), but formation of α-glycosides relies on equilibration of the leaving group from the thermodynamically favoured axial to the more reactive equatorial orientation (3).

# 4. Choosing protective groups for glycosylated amino acids

When choosing protective groups for glycosylated amino acids, the additional complexity of glycopeptides compared to ordinary peptides must be taken into consideration. Importantly, glycosidic bonds are labile towards strong acids and glycopeptides might undergo side-reactions, such as β-elimination of glycosylated serine or threonine residues and epimerization of peptide stereocentres on treatment with a strong base (*Figure 4*).

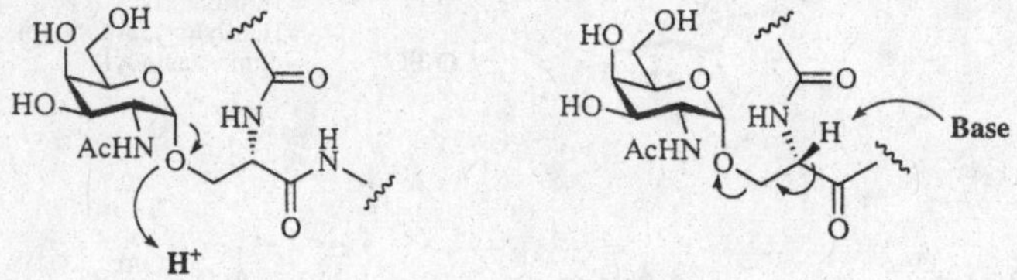

**Figure 4.** All *O*-glycosides are cleaved by strong acids. In addition, *O*-linked derivatives of serine and threonine can undergo base-catalysed β-elimination.

## 4.1 Protection of the α-amino group

In the *tert*-butoxycarbonyl (Boc) strategy for solid phase peptide synthesis, cleavage from the solid phase and side chain deprotection is performed with strong acids, e.g. hydrogen fluoride. Since treatment of a glycopeptide with a strong acid results in partial, or in most cases complete, cleavage of glycosidic bonds, the Boc approach is less suitable for solid phase glycopeptide synthesis. The 9-fluorenylmethoxycarbonyl (Fmoc) group (22), which is removed by a weak base, e.g. piperidine, morpholine, or 1,8-diazabicyclo[5.4.0]undec-7-ene (DBU), provides a more versatile alternative. In the Fmoc strategy cleavage from the solid phase and simultaneous deprotection of amino acid side chains is usually carried out with the moderately strong trifluoroacetic acid (TFA). As discussed in Section 4.3, the glycosidic bonds of most saccharides are stable towards TFA even though it may sometimes be necessary to use acyl protective groups to obtain sufficient stability.

Since glycosides of serine and threonine can undergo β-elimination on treatment with base, morpholine has been advocated for Fmoc removal (23, 24) in synthesis of glycopeptides. It has, however, been shown that the more basic piperidine, which is routinely used for Fmoc cleavage in synthesis of peptides, does not causes β-elimination of *O*-linked glycopeptides. The use of piperidine is in fact preferable to morpholine (25). Numerous synthetic examples have now established that the Fmoc protective group protocol is sufficiently mild to allow efficient and reliable solid phase assembly of glycopeptides.

## 4.2 Protection of the α-carboxyl group

Building blocks for use in solid phase glycopeptide synthesis have been prepared predominantly by glycosylation of Fmoc amino acids having allyl, benzyl, phenacyl, or *tert*-butyl protected carboxyl groups (13, 21). Each of these esters can be removed selectively in the presence of the Fmoc group by palladium-catalysed allyl transfer, hydrogenolysis, zinc reduction, or TFA-treatment, respectively. After deprotection, the liberated carboxyl group is activated, usually by the *N,N'*-diisopropylcarbodiimide/1-hydroxybenzotriazole method.

An attractive alternative takes advantage of the fact that Fmoc amino acid pentafluorophenyl esters, which are in general use as acylating agents in peptide synthesis, are stable enough to survive glycosylation and subsequent purification (13). Protective group manipulations are thus minimized, since the pentafluorophenyl ester plays a dual role, first as a protecting group and then as an acylating reagent. Protective group manipulations can be further decreased by direct *O*-glycosylation of Fmoc amino acids having unprotected carboxyl groups (26). This method has been used to prepare numerous 1,2-*trans*-*O*-glycosylated Fmoc amino acids and, recently, 1,2-*cis*-linked glycosides of fucose (27).

## 4.3 Protection of the carbohydrate hydroxyl groups

Acetyl and benzoyl protective groups have often been used for protection of the carbohydrate hydroxyl groups in glycosylated amino acids. In view of their electron-withdrawing nature, they have been advocated to stabilize *O*-glycosidic bonds during the acid-catalysed cleavage from the solid phase, Several investigations have established, however, that *O*-glycosidic linkages of common monosaccharides, such as glucose, galactose, mannose, *N*-acetyl-glucosamine, and *N*-acetylgalactosamine, survive treatment with TFA for a limited period of time (<2 h) in the absence of acyl protection of the hydroxyl groups (6, 28, 29). However, prolonged treatment with acid, or in some cases addition of nucleophiles such as water, may result in decomposition of the oligosaccharide moiety.

Interestingly, glycopeptides containing the 6-deoxysugar L-fucose, which undergoes acid-catalysed hydrolysis only 5–6 times faster than the corresponding non-deoxygenated monosaccharide galactose, have been shown to require stabilizing acyl protective groups during cleavage with TFA (30, 31). Although acyl protection of carbohydrate hydroxyl groups is not always necessary, it could be considered as a suitable precaution to avoid degradation during cleavage from the solid phase. Furthermore, in many cases the synthetic route to the glycosylated amino acid automatically provides acyl protective groups.

It has been emphasized that side-reactions such as β-elimination and epimerization of peptide stereocentres could be encountered during base-mediated removal of *O*-acyl protective groups from glycopeptides (23). This concern appears to be unwarranted for acetyl groups, since neither β-elimination nor epimerization were encountered when an *O*-linked model glycopeptide was treated under conditions which are in common use for deacetylation of glycopeptides (i.e. hydrazine hydrate in methanol, saturated methanolic ammonia, or dilute sodium methoxide) (32). However, benzoyl groups require more drastic conditions for their removal, which resulted in slow β-elimination and epimerization, suggesting that it might be desirable to avoid benzoyl protection of glycopeptides.

Acid-labile protective groups have recently been introduced as an alternative to acyl protection for glycosylated amino acids. Trimethylsilyl (TMS) protective groups have been employed for 1-aminoalditols and mono-heptasaccharides linked in a non-natural way to asparagine (33, 34). Recently, the more stable *tert*-butyldimethylsilyl (TBDMS), *tert*-butyldiphenylsilyl (TBDPS), *p*-methoxybenzyl, and isopropylidene groups have been used in syntheses of both *N*- and *O*-linked building blocks which correspond to glycosylated amino acids **1**, **3**, and **5** (17, 35–37). When used in glycopeptide synthesis, these protective groups were rapidly and completely removed simultaneously with TFA-mediated cleavage from the solid phase, thereby eliminating the need for separate deprotection of the carbohydrate moiety.

Benzyl ethers are commonly used as protective groups in oligosaccharide synthesis and they have recently begun to find use also in solid phase glyco-peptide synthesis (38, 39). However, benzyl ethers are usually removed by hydrogenolysis, which is incompatible with glycopeptides containing cysteine, methionine, and to some extent tryptophan (40). Moreover, TFA-catalysed cleavage of *O*-benzylated glycopeptides from the solid phase was found to result in the formation of several, partially de-*O*-benzylated products, thereby complicating the final debenzylation and purification. Interestingly, a recent report suggests that this problem may be circumvented by performing the deprotection in two acid-catalysed steps (41).

## 4.4 Suitable linkers and resins

As revealed in the preceding sections, glycopeptides are conveniently synthesized on the solid phase under conditions identical to those employed for synthesis of peptides by the Fmoc/tBu strategy. Consequently, linkers and resins used in Fmoc peptide synthesis can also be used without problems for synthesis of glycopeptides. Glycopeptides having a C-terminal carboxylic acid have therefore usually been prepared attached directly to a 4-alkoxybenzyl alcohol resin (42) or via a 4-hydroxymethylphenoxyacetic acid-type linker (43). The Rink linker, p-[α-(fluoren-9-ylmethoxyformamido)-2,4-dimethoxy-benzyl]phenoxyacetic acid, is suitable for synthesis of C-terminal amides (44). Recently, an allylic linker termed HYCRON, which allowed Pd(0)-catalysed cleavage under almost neutral conditions, was described (45). This linker is suitable for synthesis of glycopeptides which do not withstand the basic and acidic conditions normally employed in the Fmoc-strategy.

# 5. Preparation of glycosylated amino acids

Access to glycosylated amino acids for use as building blocks in solid phase synthesis is the key to success in the preparation of glycopeptides. However, only a few glycosylated amino acids carrying protective groups suitable for direct use in solid phase synthesis are commercially available at present. Thus, $N^\alpha$-(9-fluorenylmethoxycarbonyl)-$N^\gamma$-(2-acetamido-3,4,6-tri-$O$-acetyl-2-deoxy-β-D-glucopyranosyl)-L-asparagine, $N^\alpha$-(9-fluorenylmethoxycarbonyl)-3-$O$-(2-acetamido-3,4,6-tri-$O$-acetyl-2-deoxy-α-D-galactopyranosyl)-L-serine, and $N^\alpha$-(9-fluorenylmethoxycarbonyl)-3-$O$-(2-acetamido-3,4,6-tri-$O$-acetyl-2-deoxy-α-D-galactopyranosyl)-L-threonine can be purchased from Bachem Feinchemikalien AG, Bubendorf, Switzerland. The glycosylated asparagine corresponds to the most important saccharide–amino acid linkage found in $N$-linked glycoproteins, whereas the two $O$-linked building blocks represent the Tn-antigen, a tumour-associated antigen that is extended with further saccharides in a large number of glycoproteins produced by non-malignant cells. In order to allow access to some commonly occurring glycosylated amino acids, which are not commercially available, synthetic procedures are

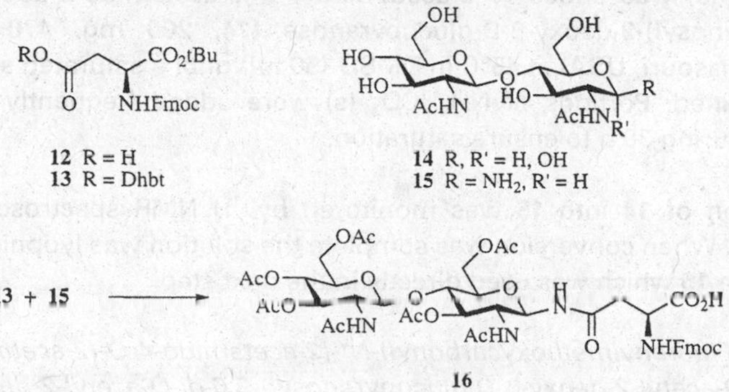

12 R = H
13 R = Dhbt

14 R, R' = H, OH
15 R = NH₂, R' = H

13 + 15 ⟶

16

Scheme 1

given in the following protocols. *Protocol 1* describes attachment of the disaccharide chitobiose to asparagine, but it has also been applied to a large number of other oligosaccharides ranging in size up to undecasaccharides (16). The other four protocols present syntheses of *O*-linked glycosylated amino acids found in the nucleus and cytoplasm (*Protocol 2*), proteoglycans (*Protocol 3*), fibrinolytic and coagulation proteins (*Protocol 4*), and finally in glycoproteins produced by yeast cells (*Protocol 5*).

---

**Protocol 1.** Preparation of $N^\alpha$-(9-fluorenylmethoxycarbonyl)-$N^\gamma$-[2-acetamido-4-*O*-(2-acetamido-3,4,6-tri-*O*-acetyl-2-deoxy-β-D-glucopyranosyl)-3,6-di-*O*-acetyl-2-deoxy-β-D-glucopyranosyl]-L-asparagine (**16**) (16) (Scheme 1)

A. *$N^\alpha$-(9-Fluorenylmethoxycarbonyl)-L-aspartic acid β-3,4-dihydro-4-oxobenzotriazin-3-yl ester α-tert-butyl ester (13)*

Dicyclohexylcarbodiimide (DCC, 376 mg, 1.82 mmol) was added to a solution of $N^\alpha$-(9-fluorenylmethoxycarbonyl)-L-aspartic acid *tert*-butyl ester (**12**, 750 mg, 1.82 mmol) in freshly distilled THF (10 ml) at −35°C. After 10 min DhbtOH (298 mg, 1.82 mmol) was added and the mixture was stirred for a further 2 h before being allowed to attain room temperature. Then the mixture was filtered through Celite and concentrated. The residue was dissolved in THF (7.5 ml), filtered again through Celite and concentrated. Flash column chromatography on SiO$_2$ (heptane/EtOAc, 4:1) gave **13** (890 mg, 88%).

$^1$H NMR (CDCl$_3$) δ 4.65 (ddd, 1H, H-α), 3.32 (m, 2H, H-β), 1.46 (s, 9H, *t*Bu); $^{13}$C NMR (CDCl$_3$) δ 51.2 (C-α), 35.0 (C-β), 28.2 (C*Me*$_3$).

B. *2-Acetamido-4-O-(2-acetamido-2-deoxy-β-D-glucopyranosyl)-2-deoxy-β-D-glucopyranosyl amine (15)*

NH$_4$HCO$_3$ (s) was added to 2-acetamido-4-*O*-(2-acetamido-2-deoxy-β-D-glucopyranosyl)-2-deoxy-β-D-glucopyranose (**14**, 200 mg, 470 μmol, Sigma, Missouri, USA) at 45°C in DMSO (50 ml) until a saturated solution was obtained. Portions of NH$_4$HCO$_3$ (s) were added frequently to the mixture during 30 h to ensure saturation.

Conversion of **14** into **15** was monitored by $^1$H NMR spectroscopy in DMSO-$d_6$. When conversion was complete the solution was lyophilized to give crude **15** which was used directly in the next step.

C. *$N^\alpha$-(9-Fluorenylmethoxycarbonyl)-$N^\gamma$-[2-acetamido-4-O-(2-acetamido-3,4,6-tri-O-acetyl-2-deoxy-β-D-glucopyranosyl)-3,6-di-O-acetyl-2-deoxy-β-D-glucopyranosyl]-L-asparagine (16).*

Diisopropylethylamine (97 µl, 570 µmol) was added to a solution of glycosyl amine **15** (200 mg, ~470 µmol) in DMSO (10 ml) at room temperature. After 10 min, **13** (261 mg, 470 µmol) was added and the coupling was monitored by TLC (SiO$_2$, CHCl$_3$/MeOH/H$_2$O, 10:4:1). It was completed after 30 min and DMSO was removed by lyophilization of the reaction mixture.

The crude product was per-*O*-acetylated by stirring in a mixture of acetic anhydride and pyridine (1:2, 9 ml) for 6 h before being concentrated and co-concentrated with toluene. Purification of the residue by flash column chromatography on SiO$_2$ (CHCl$_3$/MeOH, 30:1 → 10:1) was followed by removal of the *tert*-butyl ester by treatment with TFA (10 ml) for 1 h at room temperature.

Purification of the resulting crude **16** was performed using preparative reversed-phase HPLC on a Delta PAK C18 column (300 Å, 15 µm, 25 × 200 mm, flow rate 10 ml/min) under isocratic conditions (27% CH$_3$CN + 73% H$_2$O containing 0.1% TFA) for 5 min followed by a gradient (→ 90% CH$_3$CN + 10% H$_2$O containing 0.1% TFA) during 60 min. This gave **16** (270 mg, 65%).

$^1$H NMR (DMSO-$d_6$) δ 7.69 (b d, 1H, $J$ = 8.2 Hz, Asn NH), 5.30 (d, 1H, $J$ = 9.3 Hz, GlcNAc H-1), 4.90 (d, 1H, $J$ = 8.3 Hz, GlcNAc H-1′); $^{13}$C NMR (DMSO-$d_6$) δ 100.9 (GlcNAc C-1′), 78.7 (GlcNAc C-1), 50.9 (Asn C-α).

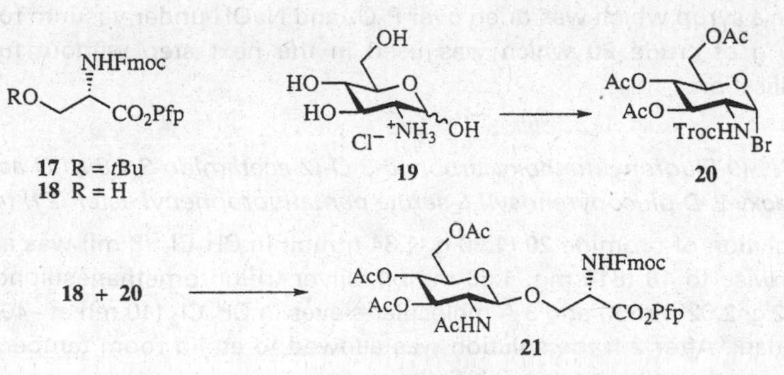

17 R = *t*Bu
18 R = H

19

20

18 + 20

21

(Troc = 2,2,2-trichloroethoxycarbonyl)

Scheme 2

**Protocol 2.** Preparation of $N^\alpha$-(9-fluorenylmethoxycarbonyl)-3-*O*-(2-acetamido-3,4,6-tri-*O*-acetyl-2-deoxy-β-D-glucopyranosyl)-L-serine pentafluorophenyl ester (**21**) (46) (Scheme 2)

A. *$N^\alpha$-(9-Fluorenylmethoxycarbonyl)-L-serine pentafluorophenyl ester (18)*

TFA (20 ml) was added at room temperature to $N^\alpha$-(9-fluorenyl-methoxycarbonyl)-3-*O*-(*tert*-butyl)-L-serine pentafluorophenyl ester (**17**, 1.14 g, 2.06 mmol). The solution was concentrated *in vacuo* after 30 min and the residue was twice co-concentrated with toluene to give **18** (0.96 g, 94%); m.p. 139–140°C (crystallized from diethyl ether).

B. *3,4,6-Tri-O-acetyl-2-deoxy-2-(2,2,2-trichloroethoxycarbonylamino)-α-D-glucopyranosyl bromide (21)*

2,2,2-Trichloroethoxycarbonyl chloride (1.00 g, 4.72 mmol) was added to a solution of 2-amino-2-deoxy-D-glucopyranose hydrochloride (0.68 g, 3.16 mmol) and $NaHCO_3$ (0.68 g, 8.1 mmol) in water (10 ml) at 0°C. After 2 h the mixture was allowed to attain room temperature and was then stirred overnight. After filtration, the colourless precipitate was washed with water and diethyl ether and then recrystallized from EtOH to give 2-deoxy-2-(2,2,2-trichloroethoxycarbonylamino)-D-glucopyranose (0.98 g, 87%); m.p. 183–184°C (dec.); $[\alpha]_D^{19}$ +50° (*c* 0.84, MeOH).

2-Deoxy-2-(2,2,2-trichloroethoxycarbonylamino)-D-glucopyranose (0.70 g, 1.92 mmol) was treated overnight with a mixture of acetic anhydride and pyridine (1:1, 20 ml). The solution was concentrated and co-concentrated twice with toluene. The crude, syrupy tetraacetate was dissolved in dry $CH_2Cl_2$ (200 ml), which was saturated with dry hydrogen bromide at 0°C and then left in a sealed flask at room temperature for 18 h. Concentration gave a syrup which was dried over $P_2O_5$ and NaOH under vacuum to give 1.40 g of crude **20** which was used in the next step without further purification.

C. *$N^\alpha$-(9-Fluorenylmethoxycarbonyl)-3-O-(2-acetamido-3,4,6-tri-O-acetyl-2-deoxy-β-D-glucopyranosyl)-L-serine pentafluorophenyl ester (21) (46)*

A solution of bromide **20** (1.00 g, 1.84 mmol) in $CH_2Cl_2$ (8 ml) was added dropwise to **18** (816 mg, 1.66 mmol), silver trifluoromethanesulphonate (0.52 g, 2.02 mmol) and 3 Å molecular sieves in $CH_2Cl_2$ (10 ml) at –40°C in the dark. After 2 h the solution was allowed to attain room temperature and was then filtered through Celite and concentrated.

Flash column chromatography of the residue on $SiO_2$ (heptane/EtOAc, 3:1) gave $N^\alpha$-(9-fluorenylmethoxycarbonyl)-3-*O*-[2-(2,2,2-trichloroethoxycarbonylamino)-3,4,6-tri-*O*-acetyl-2-deoxy-β-D-glucopyranosyl]-L-serine pentafluorophenyl ester (1.35 g, 85%). This was dissolved in a mixture of THF, acetic anhydride, and AcOH (40 ml, 3:2:1), and zinc dust (3.3 g, 325 mesh) was added with stirring. After 2 h the stirred mixture was filtered through Celite and the filter cake was washed several times with distilled THF. The filtrate was concentrated and the residue was purified by flash column chromatography on dried $SiO_2$ (heptane/EtOAc, 1:2) to give **21** (0.871 g, 75%).

m.p. 207 °C (crystallized from diethyl ether); $[\alpha]_D^{25}$ –10° (*c* 1.0, $CHCl_3$); $^1H$ NMR ($CDCl_3$) δ 4.91 (ddd, 1H, *J* = 2.7, 3.6 and 8.6 Hz, Ser H-α), 4.89 (d, 1H, *J* = 8.3 Hz, GlcNAc H-1), 1.93 (s, 1H, GlcNAc Ac); $^{13}C$ NMR ($CDCl_3$) δ 100.6 (GlcNAc C-1), 54.3 (Ser C-α), 23.2 (GlcNAc, CO*Me*).

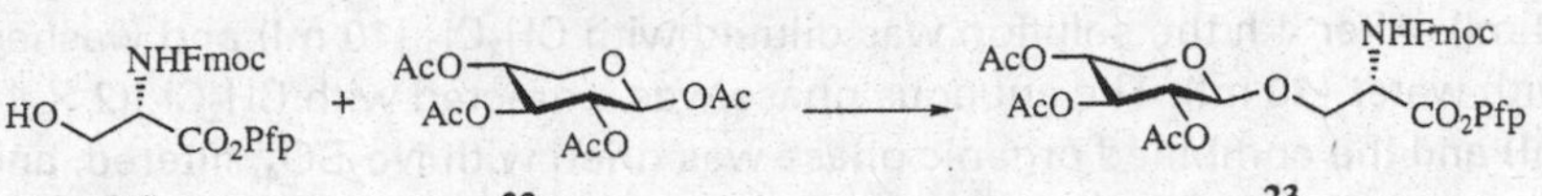

Scheme 3

**Protocol 3.** Preparation of $N^\alpha$-(9-fluorenylmethoxycarbonyl)-3-*O*-(2,3,4-tri-*O*-acetyl-β-D-xylopyranosyl)-L-serine pentafluorophenyl ester (**23**) (47) (Scheme 3)

Boron trifluoride etherate (120 μl, 0.94 mmol) was added to a solution of $N^\alpha$-Fmoc-L-serine pentafluorophenyl ester (**18**, 186 mg, 377 μmol) and 1,2,3,4-tetra-*O*-acetyl-β-D-xylopyranose (**22**, 100 mg, 314 μmol, Sigma, Missouri, USA) in dry $CH_2Cl_2$ (2.0 ml).

After 1.5 h at room temperature the solution was diluted with $CH_2Cl_2$ (8 ml) and washed with saturated aqueous $NaHCO_3$ (10 ml). The aqueous phase was extracted with $CH_2Cl_2$ (2 × 10 ml) and the combined organic solution was dried over sodium sulphate, filtered, and concentrated. The residue was purified twice by flash column chromatography on dried $SiO_2$ (first heptane/EtOAc, 3:2 → 1:1, then toluene/$CH_3CN$, 10:1) to give **23** (139 mg, 60%).

$^1H$ NMR ($CDCl_3$) δ 5.16 (t, 1H, *J* = 7.9 Hz, Xyl H-3), 4.59 (d, 1H, *J* = 6.1 Hz, Xyl H-1), 3.90 (dd, 1H, *J* = 10.4 and 3.4 Hz, Ser H-β); $^{13}C$ NMR ($CDCl_3$) δ 100.5 (Xyl C-1).

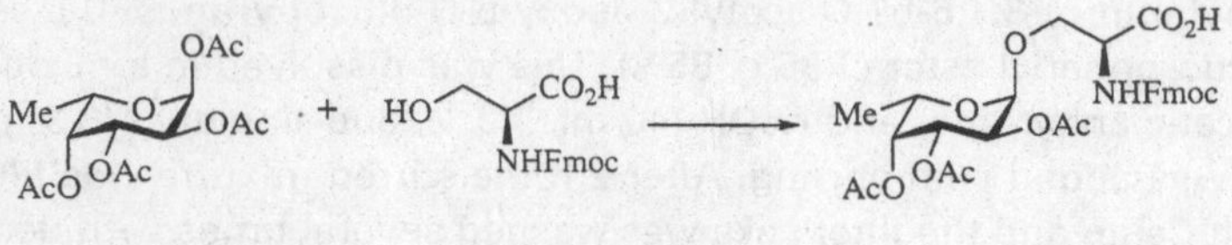

**24**        **25**        **26**

Scheme 4

---

**Protocol 4.** Preparation of $N^{\alpha}$-(9-fluorenylmethoxycarbonyl)-3-$O$-(2,3,4-tri-$O$-acetyl-$\alpha$-L-fucopyranosyl)-L-serine (**26**) (48) (Scheme 4)

Boron trifluoride etherate (110 µl, 0.90 mmol) was added to a solution of 1,2,3,4-tetra-$O$-acetyl-L-fucopyranose (**24**, 50 mg, 150 µmol, Sigma, Missouri, USA) and $N^{\alpha}$-Fmoc-L-serine (**25**, 59 mg, 180 µmol) in dry $CH_3CN$ (3 ml). After 4 h the solution was diluted with $CH_2Cl_2$ (10 ml) and washed with water (10 ml). The aqueous phase was extracted with $CH_2Cl_2$ (2 × 10 ml) and the combined organic phase was dried with $Na_2SO_4$, filtered, and concentrated. The crude product was purified in 30–40 mg portions by preparative reversed-phase HPLC on a Kromasil C8 column (100 Å, 5 µm, 20 × 250 mm, flow rate 10 ml/min) under isocratic conditions ($CH_3CN/H_2O$, 1:1 + 0.1% TFA) to give **26** (39 mg, 44%).

$[\alpha]_D^{20}$ −46° ($c$ 0.58, $CHCl_3$); $^1H$ NMR ($CD_3OD$) δ 4.98–4.88 (m, 2H, Fuc H-1,2), 3.91 (dd, 1H, $J$ = 5.5 and 9.5 Hz, Ser H-β), 3.68 (dd, 1H, $J$ = 3.0 and 9.5 Hz, Ser H-β′), 0.95 (d, 3H, $J$ = 6.5 Hz, Fuc H-6); $^{13}C$ NMR ($CDCl_3$) δ 97.3 (Fuc C-1), 56.8 (Ser C-$\alpha$), 16.1 (Fuc C-6).

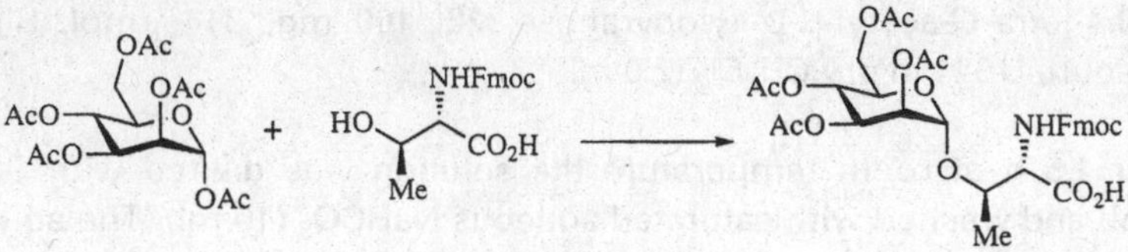

---

**27**        **28**        **29**

Scheme 5

---

**Protocol 5.** Preparation of $N^{\alpha}$-(9-fluorenylmethoxycarbonyl)-3-$O$-(2,3,4,6-tetra-$O$-acetyl-$\alpha$-D-mannopyranosyl)-L-threonine (**29**) (26) (Scheme 5)

Boron trifluoride etherate (154 µl, 1.23 mmol) was added to a solution of 1,2,3,4,6-penta-$O$-acetyl-$\alpha$-D-mannopyranose (**27**, 53.1 mg, 136 µmol,

Sigma, Missouri, USA) and $N^\alpha$-Fmoc-L-threonine (**28**, 56.1 mg, 164 μmol) in dry $CH_2Cl_2$ (2.5 ml) under a nitrogen atmosphere. The resulting solution was stirred at room temperature for 20 h, then diluted with $CH_2Cl_2$ (8 ml), washed with 1 M aqueous HCl (1 ml) and water (1 ml), dried over $Na_2SO_4$, filtered, and concentrated. The residue was purified by preparative reversed-phase HPLC on a Kromasil C8 column (100 Å, 5 μm, 20 × 250 mm, flow rate 10 ml/min) under isocratic conditions ($CH_3CN/H_2O$, 1:1 + 0.1% TFA) to give **29** (38.5 mg, 42%).

$[\alpha]_D^{20}$ +41° ($c$ 0.67, $CHCl_3$); $^1H$ NMR ($CDCl_3$) δ 4.94 (b s, 1H, Man H-1), 4.54 (b d, 1H, $J$ = 9.0 Hz, Thr H-α), 1.32 (d, 3H, $J$ = 6.4 Hz, Thr H-γ); $^{13}C$ NMR ($CDCl_3$) δ 98.7 (Man C-1, $J_{CH}$ = 172 Hz), 17.9 (Thr C-γ).

# 6. Synthesis of a glycopeptide from HIV gp120

This section describes the synthesis of glycopeptide **31**, which corresponds to residues 312–327 of gp120 from the HIV-3B isolate (*Scheme 6*) (25). Since gp120 has been reported to carry 2-acetamido-2-deoxy-α-D-galactopyranosyl moieties, this monosaccharide was attached to a threonine in **31** which constitutes a potential *O*-glycosylation site. In addition, a cysteine residue was incorporated at the C-terminus in order to allow conjugation to carrier proteins in attempts to elicit an immune response to **31**.

Glycopeptide **31** was prepared on the solid phase under standard conditions used in Fmoc synthesis, but some problems were encountered in the synthesis which reflect the more complex structure of glycopeptides compared to ordinary peptides. In a first attempt morpholine was used for Fmoc-removal after incorporation of the *O*-acetylated glycosylated building block **30** to avoid potential β-elimination (23) of the *O*-linked glycan. It was found that use of morpholine led to slow and incomplete removal of the Fmoc group of the following amino acids and after cleavage from the solid phase **31** contained several by-products that were difficult to remove by preparative reversed-phase HPLC.

When the synthesis was repeated using the stronger base piperidine, which is usually employed for synthesis of peptides, fast and complete Fmoc removal was observed and crude **31** of high purity was then obtained. No β-elimination could be detected with piperidine, an observation that has later been confirmed in syntheses of a large number of *O*-linked glycopeptides. It was also found that removal of the *O*-acetyl protective groups from the carbohydrate moiety had to be performed before cleavage from the solid phase in order to avoid cysteine-induced degradation of the glycopeptide under the basic conditions used for de-*O*-acetylation. As pointed out above, the glycosidic bonds of most of the common saccharides are stable during treatment with TFA. This was also the case for **31**, which could be isolated in

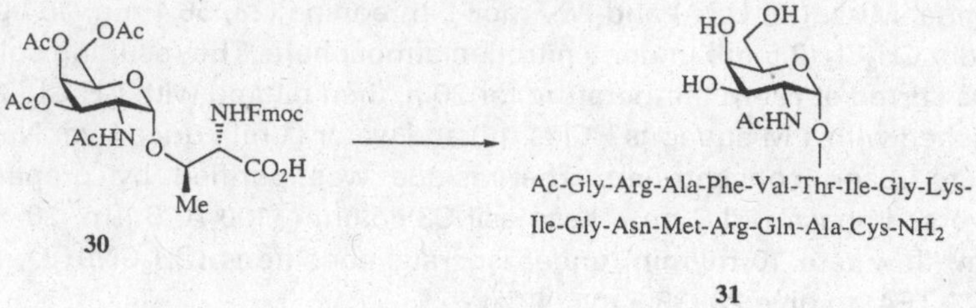

30

31

Scheme 6

an overall yield of 45% after deacetylation with saturated methanolic ammonia and subsequent cleavage from the solid phase with TFA.

---

**Protocol 6.** Preparation of *N*-acetyl-glycyl-L-arginyl-L-alanyl-L-phenylalanyl-L-valyl-*O*-(2-acetamido-2-deoxy-α-D-galactopyranosyl)-L-threonyl-L-isoleucyl-glycyl-L-lysyl-L-isoleucyl-glycyl-L-asparaginyl-L-methionyl-L-arginyl-L-glutaminyl-L-alanyl-L-cysteine amide (**31**) (25) (Scheme 6)

Glycopeptide **31** was synthesized in a custom-made, fully automatic continuous-flow peptide synthesizer constructed essentially as described elsewhere (19). A resin consisting of a cross-linked polystyrene backbone grafted with poly(ethylene glycol) chains (TentaGel™, Rapp Polymer, Germany) and functionalized with the Rink amide linker (*p*-[α-(fluoren-9-ylmethoxyformamido)-2,4-dimethoxybenzyl]phenoxyacetic acid, Novabiochem, Switzerland) was used for the syntheses. DMF was distilled before use.

In the synthesis of glycopeptide **31**, 400 mg (70 μmol) of linker-derivatized resin was used. The $N^\alpha$-Fmoc-amino acids, which had standard side-chain protective groups, and the glycosylated amino acid **30** were coupled to the peptide–resin as 1-benzotriazolyl (HOBt) esters. These were prepared in the synthesizer by reaction of the $N^\alpha$-Fmoc amino acid (0.28 mmol), 1-hydroxybenzotriazole (0.42 mmol), and 1,3-diisopropylcarbodiimide (0.27 mmol) in DMF (1.3 ml). The glycosylated amino acid **30** (0.17 mmol) was activated in the same way. After 45 min bromophenol blue in DMF (0.15 mM, 0.35 ml) was added to the solution of the activated amino acid by the synthesizer, and the solution was recirculated through the column containing the resin. The acylation was monitored (20) using the absorbance of bromophenol blue at 600 nm, and the peptide–resin was automatically washed with DMF after 1 h or when monitoring revealed the coupling to be complete. $N^\alpha$-Fmoc deprotection of the peptide resin was performed by a flow of 20% piperidine in DMF (2 ml/min) through the column for 12.5–17.5 min, and was monitored (49) using the absorbance of the dibenzofulvene–piperidine adduct at 350 nm. After removal of the N-terminal Fmoc group the peptide–resin was again washed automatically with DMF and then acetylated with acetic anhydride (0.8 ml) in the synthesizer.

After completion of the synthesis, the resin was removed from the synthesizer and washed with dichloromethane (5 × 5 ml) and dried under vacuum. Deacetylation of the carbohydrate moiety was performed on the resin (200 mg, 22 µmol) using saturated methanolic ammonia (10 ml). The glycopeptide was then cleaved from the resin and the amino acid side chains were deprotected, by treatment with trifluoroacetic acid/ water/thioanisole/ethanedithiol (87.5:5:5:2.5 v/v, 20 ml) for 2 h, followed by filtration. Acetic acid (20 ml) was added to the filtrate, the solution was concentrated, and acetic acid (20 ml) was added again, followed by concentration. The residue was triturated with diethyl ether (10 ml) which gave a solid, crude glycopeptide that was dissolved in acetic acid/water (1:4, 25 ml) and freeze-dried.

Purification was performed by preparative reversed-phase HPLC on a Kromasil C8 column (100 Å, 5 µm, 20 × 250 mm, flow rate 11 ml/min) under isocratic conditions (21% $CH_3CN$ in $H_2O$ + 0.1% TFA) to give **31** (30 mg, 69% peptide content, 45% overall yield). FABMS, amino acid analysis, and $^1H$ NMR data have been reported (25).

# References

1. Sharon, N. and Lis, H. (1997). In *Glycosciences: status and perspectives* (ed. H.-J. Gabius and S. Gabius), p. 133. Chapman & Hall, Weinheim.
2. Brockhausen, I. and Schachter, H. (1997). In *Glycosciences: status and perspectives* (ed. H.-J. Gabius and S. Gabius), p. 79. Chapman & Hall, Weinheim.
3. Veeneman, G. H. (1998). In *Carbohydrate chemistry* (ed. G.-J. Boons), p. 98. Blackie Academic and Professional, London.
4. Cohen-Anisfeld, S. T. and Lansbury Jr, P. T. (1993). *J. Am. Chem. Soc.*, **115**, 10531.
5. Vetter, D., Tumelty, D., Singh, S. K., and Gallop, M. A. (1995). *Angew. Chem., Int. Ed. Engl.*, **34**, 60.
6. Offer, J., Quibell, M., and Johnson, T. (1996). *J. Chem. Soc., Perkin Trans. 1*, 175.
7. Paul, B. and Korytnyk, W. (1978). *Carbohydr. Res.*, **67**, 457.
8. Hollósi, M., Kollát, E., Laczkó, I., Medzihradszky, K. F., Thurin, J., and Otvös Jr, L. (1991). *Tetrahedron Lett.*, **32**, 1531.
9. Andrews, D. M. and Seale, P. W. (1993). *Int. J. Peptide Protein Res.*, **42**, 165.
10. Paulsen, H., Schleyer, A., Mathieux, N., Meldal, M., and Bock, K. (1997). *J. Chem. Soc., Perkin Trans. 1*, 281.

## Jan Kihlberg

11. Schleyer, A., Meldal, M., Manat, R., Paulsen, H., and Bock, K. (1997). *Angew. Chem., Int. Ed. Engl.*, **36**, 1976.
12. Norberg, T., Lüning, B., and Tejbrant, J. (1994). *Methods Enzymol.*, **247**, 87.
13. Meldal, M. (1994). In *Neoglycoconjugates: preparation and applications* (ed. Y. C. Lee and R. T. Lee), p. 145. Academic Press, San Diego.
14. Kihlberg, J. and Elofsson, M. (1997). *Curr. Med. Chem.*, **4**, 79.
15. Arsequell, G. and Valencia, G. (1997). *Tetrahedron: Asymmetry*, **8**, 2839.
16. Meinjohanns, E., Meldal, M., Paulsen, H., Dwek, R. A., and Bock, K. (1998). *J. Chem. Soc., Perkin Trans. 1*, 549.
17. Elofsson, M., Salvador, L. A., and Kihlberg, J. (1997). *Tetrahedron*, **53**, 369.
18. Jansson, A. M., Jensen, K. J., Meldal, M., Lomako, J., Lomako, W. M., Olsen, C. E., and Bock, K. (1996). *J. Chem. Soc., Perkin Trans. 1*, 1001.
19. Cameron, L. R., Holder, J. L., Meldal, M., and Sheppard, R. C. (1988). *J. Chem. Soc., Perkin Trans. 1*, 2895.
20. Flegel, M. and Sheppard, R. C. (1990). *J. Chem. Soc., Chem. Commun.*, 536.
21. Garg, H. G., von dem Bruch, K., and Kunz, H. (1994). *Adv. Carbohydr. Chem. Biochem.*, **50**, 277.
22. Carpino, L. A. and Han, G. Y. (1972). *J. Org. Chem.*, **37**, 3404.
23. Kunz, H. (1987). *Angew. Chem., Int. Ed. Engl.*, **26**, 294.
24. Kunz, H. and Brill, W. K.-D. (1992). *Trends Glycosci. Glycotechnol.*, **4**, 71.
25. Vuljanic, T., Bergquist, K.-E., Clausen, H., Roy, S., and Kihlberg, J. (1996). *Tetrahedron*, **52**, 7983.
26. Salvador, L. A., Elofsson, M., and Kihlberg, J. (1995). *Tetrahedron*, **51**, 5643.
27. Kihlberg, J., Elofsson, M., and Salvador, L. A. (1997). *Methods Enzymol.*, **289**, 221.
28. Urge, L., Otvos Jr, L., Lang, E., Wroblewski, K., Laczko, I., and Hollosi, M. (1992). *Carbohydr. Res.*, **235**, 83.
29. Urge, L., Jackson, D. C., Gorbics, L., Wroblewski, K., Graczyk, G., and Otvos Jr, L. (1994). *Tetrahedron*, **50**, 2373.
30. Unverzagt, C. and Kunz, H. (1994). *Bioorg. Med. Chem.*, **2**, 1189.
31. Peters, S., Lowary, T. L., Hindsgaul, O., Meldal, M., and Bock, K. (1995). *J. Chem. Soc., Perkin Trans. 1*, 3017.
32. Sjölin, P., Elofsson, M., and Kihlberg, J. (1996). *J. Org. Chem.*, **61**, 560.
33. Christiansen-Brams, I., Meldal, M., and Bock, K. (1993). *Tetrahedron Lett.*, **34**, 3315.
34. Christiansen-Brams, I., Jansson, A. M., Meldal, M., Breddam, K., and Bock, K. (1994). *Bioorg. Med. Chem.*, **2**, 1153.
35. Broddefalk, J., Bäcklund, J., Almqvist, F., Johansson, M., Holmdahl, R., and Kihlberg, J. (1998). *J. Am. Chem. Soc.*, **120**, 7676.
36. Broddefalk, J., Bergquist, K.-E., and Kihlberg, J. (1998). *Tetrahedron*, **54**, 12047.
37. Holm, B., Linse, S., and Kihlberg, J. (1998). *Tetrahedron*, **54**, 11995.
38. Nakahara, Y., Nakahara, Y., and Ogawa, T. (1996). *Carbohydr. Res.*, **292**, 71.
39. Guo, Z.-W., Nakahara, Y., Nakahara, Y., and Ogawa, T. (1997). *Angew. Chem., Int. Ed. Engl.*, **36**, 1464.
40. Bodanszky, M. and Martinez, J. (1981). *Synthesis*, 333.
41. Nakahara, Y., Nakahara, Y., Ito, Y., and Ogawa, T. (1997). *Tetrahedron Lett.*, **38**, 7211.
42. Wang, S.-S. (1973). *J. Am. Chem. Soc.*, **94**, 1328.

43. Sheppard, R. C. and Williams, B. J. (1982). *Int. J. Peptide Protein Res.*, **20**, 451.
44. Rink, H. (1987). *Tetrahedron Lett.*, **28**, 3787.
45. Seitz, O. and Kunz, H. (1997). *J. Org. Chem.*, **62**, 813.
46. Meinjohanns, E., Meldal, M., and Bock, K. (1995). *Tetrahedron Lett.*, **36**, 9205.
47. Sjölin, P. and Kihlberg, J., in preparation.
48. Elofsson, M., Roy, S., Salvador, L. A., and Kihlberg, J. (1996). *Tetrahedron Lett.*, **37**, 7645.
49. Dryland, A. and Sheppard, R. C. (1986). *J. Chem. Soc., Perkin Trans. 1*, 125.

# 9

# Convergent peptide synthesis

KLEOMENIS BARLOS and DIMITRIOS GATOS

## 1. Introduction

Besides the classical step-by-step synthesis, the convergent solid phase peptide synthesis (CSPPS) was developed for the preparation of complex and difficult peptides and small proteins. According to this method, suitably protected peptide fragments spanning the entire peptide sequence and prepared on the solid phase are condensed, either on a solid support or in solution, to the target peptide. Convergent synthesis is reviewed in recent publications (1–3). In this chapter, full experimental details are given for the preparation of complex peptides by applying convergent techniques, using 2-chlorotrityl chloride resin (CLTR) and Fmoc-amino acids.

## 1.1 Strategy in convergent synthesis

In the step-by-step peptide chain elongation the resin-bound C-terminal amino acid is reacted sequentially with suitably protected and activated amino acids. The peptide is thus elongated steadily towards the N-terminal direction. This is advantageous over the opposite direction where the elongation is performed from the N- to the C-terminus, because in the second case the growing peptide is activated at the C-terminal amino acid, which leads to its extensive racemization. This limits considerably the synthetic possibilities of the method. In convergent synthesis, no directional restrictions exist and the chain elongation can be performed with equal possibility to be successful to any direction. *Figure 1* describes schematically the C- to N-terminal synthesis which is the most studied to date. The strategies where the synthesis begins from a central fragment and the peptide chain is extended to both C- and

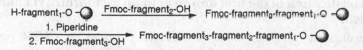

**Figure 1.** CSPPS with C- to N-terminal chain extension.

N-terminal directions and from the N-terminal towards the C-terminal can be considered, at the present time, to be in its infancy.

# 2. Solid phase synthesis of protected peptide fragments

## 2.1 Fragment selection

In general, protected peptide fragments of any length can be used in the condensation reaction, if they are of satisfactory purity and solubility. Usually, fragments of up to 15 amino acids in length are used, because of their simpler purification by RP–HPLC compared with the longer peptides. The solubility of protected peptide acids is independent of their length.

The selection of the correct fragments is very important for the success of convergent synthesis. It is helpful to analyse all available structural information, determined or calculated, for the target peptide. Peptide regions where β-turns are known to occur are readily identified as 'difficult' sequences during their synthesis. The peptide fragments should be selected to contain one part of the β -turn region at the C-terminus of the C-component and the second part at the N-terminus of the N-component. Contrary to other methods, including the step-by-step approach, the CSPPS of rat atriopeptin, using protected fragments selected to cut the peptide β-turn region, give excellent results (4). Pro-residues contained in the peptide chain impede the formation of β-turns. Pro-rich peptides, such as the 115 amino acid residues containing antigen of core mucin (5), can be synthesized very effectively. Similar β-turn interrupting effects can be achieved by the incorporation in the corresponding sequence, if present, of Ser or Thr as their acetonide derivatives or by the use of backbone-amide protection.

Owing to the lack of valuable information in the literature, we performed experiments to determine the suitability of the various amino acids as C- and N-terminal amino acids of the protected fragments. We found that in terms of racemization and condensation rates, Gly, Pro, Lys(Boc), and Glu(*t*Bu) are preferred as the C-terminal amino acid and Gly, Ala, Asp(*t*Bu), Glu(*t*Bu), Lys(Boc), and Ser(*t*Bu) as the N-terminal. In general, the condensation efficiency depends more on the nature of the resin-bound N-component than on that of the C-component.

## 2.2 Synthesis of protected peptide fragments

### 2.2.1 Esterification of the first amino acid on 2-chlorotrityl chloride resin

The esterification of Fmoc-amino acids and peptides to CLTR proceeds rapidly, in high yield, and without racemization (*Protocol 1*). Nevertheless, the amino acid–resin bond is extremely sensitive until the Fmoc-function has been removed. This must be performed immediately after the esterification, or substantial cleavage of the amino acid from the resin occurs.

---

**Protocol 1.** Esterification of amino acids on 2-chlorotrityl chloride resin

*Equipment and reagents*
- Solid phase reactor
- Fmoc-amino acid
- DCM
- DMF
- Hexane
- 2-Chlorotrityl chloride resin
- DIPEA
- MeOH
- Piperidine
- Isopropanol

*Method*
1. In a solid phase reactor, suspend 1 g 2-chlorotrityl chloride resin (1.0–2.0 mmol chloride/g resin) in 8 ml DCM, shake for 5 min and filter the resin.
2. Add to the resin a solution of 2 mmol Fmoc-amino acid and 5 mmol DIPEA dissolved in 8 ml DCM. Shake for 30 min at room temperature and filter the resin.
3. Wash the resin with DMF (2 × 2 min).
4. Add to the resin 10 ml of a mixture of DCM/MeOH/DIPEA (80:15:5), shake for 10 min and filter. Repeat the operation once.
5. Wash the resin with 3 × 10 ml DMF and filter.
6. Add 10 ml 25% piperidine in DMF, shake for 3 min and filter. Repeat for 20 min.
7. Wash the resin with 10 ml DMF (6 × 2 min), 6 ml isopropanol (3 × 5 min), and 4 × 6 ml hexane, filter, and dry by suction in air for 15 min. Dry the resin for 24 h *in vacuo*.
8. Store at 4°C.

---

## 2.2.2 Peptide chain elongation

Starting from the resin-bound amino acid, the first coupling must be performed under mildly basic or neutral conditions. Hence, dehydrating agents such as TBTU/HOBt/DIPEA (1:1:2) or pre-formed benzotriazolyl esters of the Fmoc-amino acids give no side products during the first coupling. *In situ* coupling using Fmoc-amino acid, HOBt and DCC or DIC can cause the formation of 0.5–2.5% tripeptide. The sequence of this tripeptide contains two residues of the resin-bound amino acid. This indicates that the initially formed dipeptide is cleaved from the resin under these conditions and reacts with the resin-bound amino acid to yield the tripeptide. All subsequent couplings can be performed by any coupling procedure used in Fmoc-chemistry, without the formation of side products.

---

**Protocol 2.** Monitoring solid phase synthesis by TLC and RP–HPLC

*Equipment and reagents*
- Eppendorf vial
- Peptide resin ester
- Elution mixture: chloroform/MeOH/AcOH or toluene/AcOH/MeOH
- TFE
- DCM
- DMSO

*Method*
1. Place 1 mg of well-washed peptide resin ester in an Eppendorf vial.
2. Add 5 drops of TFE/DCM (2:8) and shake for 5 min.
3. Apply a spot from the resulting mixture on Kieselgel 60 for TLC analysis. Develop the chromatogram using chloroform/MeOH/AcOH (90:8:2), toluene/AcOH/MeOH (70:15:15), or similar mixtures for the elution. Visualize the chromatogram at 254 nm.
4. Add 10 drops of DMSO to the resin in the vial, and centrifuge. Use 5 μl from the supernatant peptide solution for the HPLC analysis.

---

### 2.2.3 Cleavage of protected fragments from the resin

Protected peptides can be cleaved selectively from the CLTR by various methods, in the presence of side chain protection of the *tert*-butyl type. By following *Protocol 2*, the purity of the resin-bound peptide is generally >97%. The peptide, cleaved by treatment with TFE/DCM (2:8) (*Protocol 3*) and obtained in 70–80% yield, can be subjected to the fragment condensation reactions without further purification. Depending on the individual peptide sequence, His(Trt) can be detritylated by this reagent in 1–4% yield. Fast, selective, and quantitative cleavage occurs using AcOH/TFE/DCM (1:2:7), HFIP/DCM (3:7), and 1% TFA in DCM. AcOH contained in the cleavage mixture is very difficult to remove by drying under vacuum. Reprecipitation of the fragment from TFE/water or DMF/water is necessary for its complete removal. Otherwise, extensive acetylation of the resin-bound N-component occurs during the fragment condensation. HFIP-containing mixtures detritylate to high extent His(Trt) and Ser(Trt) contained in the peptide sequence and must be avoided in such cases.

---

**Protocol 3.** Cleavage of protected peptide fragments from 2-chlorotrityl resin

*Equipment and reagents*
- Round-bottomed flask
- Rotary evaporator
- Peptide-resin
- Ether
- TFE
- DCM
- Hexane

*Method*

1. Place 1 g of peptide-resin in a round-bottomed flask, add 15 ml of the TFE/DCM (2:8) cleavage solution, and stir magnetically for 45 min at room temperature.
2. Filter the resin and wash it twice with 10 ml of the cleavage solution.
3. Concentrate the combined filtrates on a rotary evaporator.
4. Add 20 ml cold ether and 10 ml hexane to the residue.
5. Isolate the resulting solid by centrifugation and wash with ether (4 × 40 ml).
6. Dry the protected peptide for 6 h *in vacuo* and store at –20°C.

## 2.2.4 Purification of protected fragments

The most effective procedure to purify protected peptide fragments is RP–HPLC (*Protocol 4*). Gradient conditions, which start with a high water concentration of the eluant lead in most cases to the peptide precipitation at the beginning of the elution. Therefore, it is advantageous to start the elution with a high MeCN content and to perform the elution under isocratic conditions using RP materials of 40–80 μm. Fragments of medium polarity under these conditions show the best elution properties. Therefore, it is in many cases favourable to use an appropriate protection scheme to obtain fragments of medium polarity. In some cases it is better, for example, to apply unprotected Asn or Gln to increase the polarity of the fragments. Selective removal of a side chain protecting group of a hydroxy amino acid is also possible for achieving medium polarity. In this case the fragment is prepared containing Ser(Trt), Thr(Trt), or Tyr(Clt) at the required position. The protecting groups of these amino acids can be removed easily and simultaneously during the cleavage of the protected peptides from the resin (*Protocol 8*).

---

**Protocol 4.** Purification of protected fragments by RP–HPLC

*Equipment and reagents*

- Lobar column, size C, prepacked with Lichroprep RP-8, 40–63 μm
- Injector, equipped with a 2.5 ml sample loop
- Crude protected peptide
- DMSO
- Metering pump
- UV detector, detection at 300 nm
- Fraction collector
- TFA
- MeCN

*Method*

1. Inject a peptide sample for HPLC analysis using a C8 column of size 4 × 125 mm; elute using a gradient of 50% MeCN in 0.5% TFA–water to 100% MeCN in 0.5% TFA–water for 30 min; flow rate 1 ml/min, detection at 265 nm. Determine the MeCN concentration of the peptide elution.
2. Equilibrate the Lobar column for 60 min with water and MeCN equal to

**Protocol 4.** *Continued*

the determined MeCN concentration of the elution of the peptide in the analytical run plus 10%; flow rate 7 ml/min.

3. Dissolve 400 mg of the crude protected peptide in 2 ml DMSO.
4. Filter through a 45 μm membrane.
5. Inject the clear peptide solution on the Lobar column and elute under isocratic conditions with the mixture used for the equilibration of the column. In most cases the protected peptide will elute in the 90–150 min fractions.
6. Dilute the collected fractions containing the pure peptide with an equal volume of water and lyophilize.
7. Dry the protected peptide for 24 h *in vacuo* and store at –20°C.

# 3. Solid phase fragment condensation

## 3.1 Esterification of the C-terminal fragment on 2-chlorotrityl chloride resin

The loading of the starting resin in the C-terminal fragment (*Protocol 5*) must be chosen so that at the end of the condensation reactions the content in the protected peptide is not higher than 50% of the peptide–resin matrix. In addition, peptide chain aggregation will occur in high loaded resins. Like the other fragments, the resin-bound C-terminal fragment should be of the highest possible purity. This can be achieved by purification and reattachment onto the resin. The size of the C-terminal fragment must be as long as possible. Experiments in our laboratory show that long protected fragments esterify first the more advantageous positions in the outside region of the resin bead. Thus, amino acids and small fragments reach these positions easily and react more readily with the free amino function. By contrast, if small fragments or amino acids are used for the first esterification on the resin, they are able to esterify the more 'difficult' positions deeper inside the resin bead due to their fast movement. This will cause slower rates during the condensation reactions of the longer fragments, which move slower inside the resin beads.

**Protocol 5.** Attachment of the C-terminal fragment onto 2-chlorotrityl resin

*Equipment and reagents*

- Round-bottomed flask
- Eppendorf vial
- Peptide
- DMF
- Piperidine
- Hexane

- 2-Chlorotrityl chloride resin
- DCM
- DIPEA
- MeOH
- Isopropanol

*Method*

1. Place 1 g of 2-chlorotrityl chloride resin in a round-bottomed flask and swell the resin using the minimum amount of DCM for 5 min.

2. Add 4 mmol DIPEA to the resin.

3. In the minimum amount of DMF, dissolve 1.5 M excess over the calculated quantity of the protected fragment that is required to achieve the desired loading. Add the peptide solution to the resin.

4. Immediately transfer 20 μl of the resin suspension into an Eppendorf vial containing 100 μl DMF, and centrifuge. Use 5 μl of the supernatant solution for HPLC analysis and determine the peak area, which corresponds to the initial fragment concentration.

5. Repeat step 4 every 30 min with the remaining peptide in solution.

6. Filter the resin.

7. Wash the resin with DMF (6 × 3 min), DCM/MeOH/DIPEA (80:15:5; 2 × 15 min), and DMF (3 × 50 ml).

8. Treat the resin with 5% piperidine in DMF for 5 min and with 25% piperidine for 20 min.

9. Wash the resin with DMF (3 × 30 ml), DMSO (4 × 30 ml), isopropanol (3 × 30 ml), and hexane (3 × 30 ml), dry by a 15 min suction in air and for 4 h under vacuum. Store the resin at –20°C.

## 3.2 Activation and condensation of protected peptide fragments

The rate of the condensation of protected fragments with the resin-bound N-component increases with the concentration of the fragment. Therefore, fragment solutions of the highest possible concentration should be applied. The solvents and condensing reagents used for the long-lasting fragment condensation play a much more important role for the product purity and the suppression of racemization compared to the fast amino acid coupling. Even treatment of various resin-bound N-components, in the absence of the C-component (N-terminal fragment), with condensing reagents in DMF, DMAC, and *N*-methylpyrrolidone for 10 h at room temperature leads to the formation of several by-products. Similarly, treatment of various C-components with the same solvents leads to the partial removal of the $N^\alpha$-Fmoc group of the fragments. Such by-products are not formed if DMSO is used as solvent.

Modern coupling agents such as BOP, HBTU, TBTU, and HATU are well suited for the racemization-free coupling of amino acids. These reagents were tested in various solvents during the condensation of the resin-bound prothymosin α (ProTα) 95–109 fragment with Fmoc-ProTα(87–94)-OH which contains Glu(*t*Bu) as the C-terminal amino acid (*Figure 2*). The 70–90%

Fmoc-Lys(Boc)-Arg(Pmc)-Ala-Ala-Glu(*t*Bu)-Asp(*t*Bu)-Asp(*t*Bu)-Glu(*t*Bu)-OH

H-Asp(*t*Bu)-Asp(*t*Bu)-Asp(*t*Bu)-Val-Asp(*t*Bu)-Thr(*t*Bu)-Lys(Boc)-
   Lys(Boc)-Gln(Trt)-Lys(Boc)-Thr(*t*Bu)-Asp(*t*Bu)-Glu(*t*Bu)-Asp(*t*Bu)-Asp(*t*Bu)-O –◯

**Figure 2.** Fmoc-ProTα(87–94)-OH and the resin-bound ProTα(95–109) used for the evaluation of racemization suppressing conditions in solid phase fragment condensation.

racemization determined by HPLC was independent of the solvent used. By contrast, in condensations where DCC/HOBt, DIC/HOBt, and DIC/DhbtOH were used as dehydrating agents, racemization in DMSO was <1%. Condensations in solvents of the amide type gave much higher racemization. The use of DCM, even as co-solvent of DMSO, also led to high racemization. DhbtOH as additive gave low racemization but the condensations resulted in the formation of unidentified side-products. The conclusions are:

1. DMSO is the best solvent in terms of solvating properties, purity of products, and suppression of racemization.

2. DCC/HOBt and DIC/HOBt are the best condensing agents.

3. The C-component must be applied in the highest possible concentration.

4. To obtain products of high purity and suppress racemization in long-lasting condensations, it is better to perform a double coupling after 6 h condensation than to leave the condensation to proceed for a long time. Otherwise, the remaining peptide fragment when left in its activated form a long time period undergoes extensive racemization, even if DMSO and racemization suppressing agents are used.

---

**Protocol 6.** Solid phase fragment condensation

*Reagents*

- Protected peptide fragment
- DMSO
- Peptide-resin
- DMF
- Hexane
- HOBt
- DIC
- DIPEA
- Isopropanol
- Piperidine

*Method*

1. Dissolve the protected fragment (3–5-fold excess over the resin bound N-component) and 2.0 eq HOBt in the highest possible concentration in DMSO.

2. Add 2.0 eq DIC.

3. Add to this solution the dry resin without pre-swelling and shake for 6 h.

4. Check the condensation reaction by using the Kaiser test, TLC, and

HPLC. SDS–PAGE (silver staining) can also be used after total deprotection of the peptide.

5. Wash the resin and repeat the condensation if the Kaiser test is positive.

6. Wash the resin with DMSO (6 ×).

7. Add to the resin a 0.5 M solution of a 7-fold excess of acetic anhydride and DIPEA in DMSO, and shake for 1 h at RT.

8. Wash the resin with DMF (6 ×), isopropanol (3 ×), and hexane (3 ×), dry with a 15 min suction in air. Store the resin at −20°C.

9. For the Fmoc-removal, treat the resin with 5% piperidine in DMF for 5 min and with 25% piperidine for 20 min, wash the resin with DMF (4 ×), DMSO (3 ×), isopropanol (4 ×), and hexane (3 ×), and use it, as soon as possible, for the next condensation. If the Fmoc-removal is not complete, repeat the piperidine treatment. In very difficult cases, treat with 20% diethylamine in DMF for 25 min.

With the exception of the two final condensations, after every condensation the remaining unreacted resin-bound amino groups should be acetylated. After the final condensation, acetylation should not be performed because the lipophilicity of the peptide increases and the acetylated smaller peptide can have a similar elution time to the required target peptide. This can make the separation and purification steps more difficult.

# 4. Phase and direction change

## 4.1 Fragment condensation in solution

Notwithstanding the selection of peptide fragments, the protecting groups, and the methods used, some steps of the CSPPS can prove to be very difficult. In such cases, the condensation reactions can be performed in solution. This procedure is often advantageous from an economical point of view, since the C- and N-components are used in equimolar amounts. The required N-component is prepared by the esterification with 2-chlorotrityl chloride in solution of a protected peptide prepared previously either by the step-by-step or fragment condensation on CLTR (*Figure 3*). If a peptide amide is required,

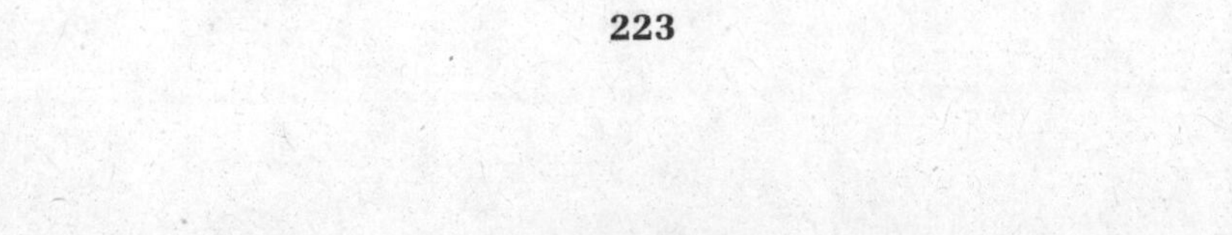

**Figure 3.** Change in the peptide synthesis phase.

a suitable amide linker is used or the fragment is converted to the corresponding amide by treatment with DIC/HOBt•NH$_3$ in DMSO.

---

**Protocol 7.** Synthesis of 2-chlorotrityl esters of protected fragments

*Reagents*
- Protected peptide fragment
- DIPEA
- DCM
- DMSO
- Piperidine
- DMF
- 2-Chlorotrityl chloride
- Ether
- KOH pellets
- THF

*Method*

1. Dissolve the protected fragment in the highest possible concentration in DMF and add 10 eq DIPEA.
2. Prepare a 1 M solution of 5 eq, with respect to the protected fragment, of 2-chlorotrityl chloride in DCM.
3. Add the chloride solution dropwise to the stirred peptide solution and continue stirring for 6–12 h at RT. Check for completion of the esterification by TLC.
4. Remove DCM in the rotary evaporator.
5. Precipitate the protected fragment by the addition of ether to the residual DMF solution.
6. Filter or centrifuge the protected fragment and wash with ether (5 ×).
7. Dry the fragment for 6 h *in vacuo* over potassium hydroxide pellets.
8. Dissolve the protected fragment in 4% piperidine in DMF/THF (50:50) and stir the solution at RT for 2 h. Check for completion of the Fmoc-removal by TLC and HPLC.
9. Remove THF in the rotary evaporator and add ether to the remaining solution until precipitation of the protected peptide occurs.
10. Filter the precipitated peptide, wash with ether (8 ×), and dry under vacuum.

---

The 2-chlorotrityl peptide ester obtained after fragment condensation in solution can be selectively deprotected at the C-terminal carboxy group in the presence of *t*Bu-based protecting groups. This is achieved by a 45-min treatment of the protected peptide with TFE/DCM (30:70), followed by the reattachment of the fragment onto the 2-chlorotrityl chloride resin (*Protocol 5*). The convergent synthesis can then be continued by the sequential condensation of the next fragments on the solid phase. Fragment condensation on a solid support gives, in general, crude peptides of higher purity than that of the corresponding condensation in solution. Therefore, the reattachment of the product on a suitable resin should be performed as soon as possible.

## 4.2 Two-directional synthesis. Attachment of fragments on resins of the trityl-type through an amino acid side chain functional group

An alternative approach to that of changing the phase of the synthesis to overcome difficulties encountered in CSPPS is to begin the synthesis from a central part of the peptide. The starting peptide is synthesized, as usual, on 2-chlorotrityl resin using Fmoc-amino acids. One of the amino acids, specifically His, Cys, Lys, or Ser, contained in the sequence is introduced using His(Mmt), Cys(Mmt), Lys(Mtt), or Ser(Trt), respectively. All other amino acid side-chain protecting groups are of the *t*Bu-type. These Mmt, Mtt, and Trt side-chain protecting groups are extremely acid sensitive and can be removed selectively by treatment with 1% TFA in DCM in the presence of His(Trt), Cys(Trt), or the *t*Bu-type protection (*Protocol 8*). The fragments obtained, selectively deprotected at an amino acid side-chain, can be reattached on a suitable trityl-type resin (*Figure 4*). Instead of 2-chlorotrityl chloride resin, the 4-methoxytrityl chloride resin for the attachment through the His or Cys side chain and the trityl chloride resin for the attachment via the Lys or Ser residues are strongly recommended. Since the ester bonds of peptides to these trityl resins are extremely acid, the esterification reaction is reversible under these conditions; thus, protection of the C-terminal carboxy group in the peptide fragment is not necessary.

---

**Protocol 8.** Selective deprotection of His(Mmt)-, Cys(Mmt)-, Ser(Trt)-, Lys(Mtt)-, Thr(Trt)-, and Tyr(Clt)-containing peptides

*Reagents*
- Crude protected peptide
- DCM/TES (95:5)
- Ether
- TFA
- MeOH

*Method*
1. Add to 250 mg of crude protected peptide (obtained using *Protocol 3*) 40 ml of 1% TFA in DCM/TES (95:5) and stir for 1 h at RT.
2. Check for completion of the deprotection by using TLC and HPLC.
3. Add 5 ml MeOH to the resulting mixture.
4. Concentrate the solution on a rotary evaporator.
5. Precipitate the peptide by addition of ether.
6. Centrifuge the peptide, wash with ether (5 ×), and dry for 12 h *in vacuo*.

---

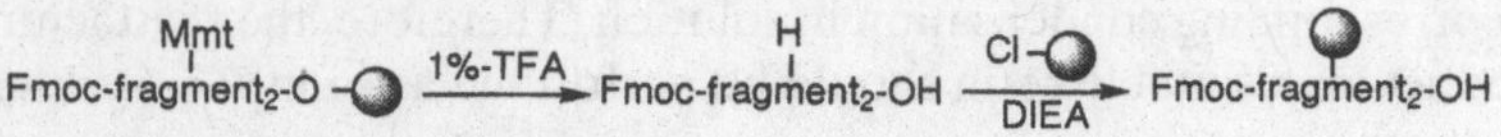

Mmt
|
Fmoc-fragment$_2$-O –⬤ $\xrightarrow{\text{1%-TFA}}$ Fmoc-fragment$_2$-OH $\xrightarrow[\text{DIEA}]{\text{Cl–⬤}}$ Fmoc-fragment$_2$-OH

**Figure 4.** Attachment of protected peptides, through an amino acid side chain functional group, on resins of the trityl type.

In order to extend the peptide chain towards the C-terminus, the resin-bound fragment obtained contains a free carboxy group which can be activated by treating with DIC/HOBt and condensed with excess of a suitably protected N-component (*Figure 5*). In addition, the free carboxy group of side chain attached peptides can be esterified by 2-chlorotrityl chloride by using a very similar procedure to that described in *Protocol 9*, or can be converted by DIC/HOBt•NH$_3$ to the corresponding resin-bound peptide amide. The synthesis is then continued towards the N-terminal of the required peptide. Side chain attached fragments are very useful for the insertion of bifunctional compounds, such as linear and cyclic diamines, into peptide chains.

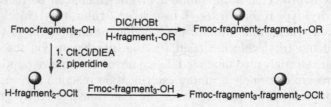

**Figure 5.** Peptide chain extension to both C- and N-terminal directions.

---

**Protocol 9.**  **Attachment of protected fragments through the side chains of His, Cys, Lys, and Ser on resins of the trityl type**

*Equipment and reagents*
- Round-bottomed flask
- Trityl chloride resin
- DMF
- TFE
- Hexane
- DCM
- DIPEA
- MeOH
- Isopropanol

*Method*
1. Place 1 g of 4-methoxyltrityl chloride resin (for attachment through the His or Cys side chain) or trityl chloride resin (for Lys or Ser) in a round-bottomed flask and swell it using the minimum amount of DCM for 5 min.
2. Add 4 mmol DIPEA to the resin.

3. Dissolve in the minimum amount of DMF, a twofold molar excess over the calculated quantity of the protected fragment that is required for achieving the required loading.
4. Add the peptide solution to the resin and stir for 5 h at RT.
6. Filter the resin and wash with DMF (6 × 3 min), DCM/MeOH/DIPEA (80:15:5; 3 × 15 min), and DCM (3 ×).
7. Wash the resin with TFE/DCM (20:80; 3 × 3 min), DMF (3 ×), iPrOH (3 ×), and hexane (3 × ) and dry the resin by suction in air (15 min) and then *in vacuo* for 6 h.

## 5. Deprotection, purification, and purity determination of the synthetic peptides

Several procedures are described in the literature for minimizing side reactions that take place during the peptide deprotection. In the case of convergently synthesized large peptides, a three-step deprotection seems to give the best results. The protected peptide is cleaved from the trityl resin with the $N^\alpha$-Fmoc group intact. Then the peptide is treated with 65% TFA in DCM/EDT (95:5) for 1 h at RT. After evaporation of the solvent and precipitation by the addition of ether, the crude partially deprotected peptide is treated for 4 h with TFA/water/EDT (90:5:5) to remove remaining protecting groups and especially the Pmc or Pbf groups used for the protection of Arg. The mixture is then concentrated *in vacuo*, precipitated, and washed with ether. The $N^\alpha$-Fmoc-peptide, deprotected on the side chains, is then subjected to a first purification step by RP–HPLC. From the partially purified peptide obtained, the Fmoc group is cleaved by treatment with 5% piperidine solution in water for 1–2 h. The mixture is then acidified with 10% acetic acid and extracted with diethyl ether to remove the Fmoc-deprotection products. After lyophilization, the totally deprotected peptide is purified by RP–HPLC using, if possible, isocratic elution conditions.

To establish the correct structure of the synthetic peptide, amino acid analysis, mass spectrometric analysis, and tryptic peptide mapping, including sequencing of the tryptic fragments, are necessary. The peptide purity is best determined by HPLC and CE. Importantly, small peaks observed after a tryptic digestion eluting close to the expected tryptic fragments on RP–HPLC correspond usually to by-products of the synthesis. Their absence is a very important indication of the high purity of the synthetically obtained peptide or protein.

## References

1. Lloyd-Williams, P., Albericio, F., and Giralt, E. (1997). *Chemical approaches to the synthesis of peptides and proteins.* CRC Press, Boca Raton, New York.

2. Lloyd-Williams, P., Albericio, F., and Giralt, E. (1993). *Tetrahedron*, **49**, 11065.
3. Benz, H. (1994). *Synthesis*, 337.
4. Gatos, D., Patrianakou, S, Hatzi, O., and Barlos, K. (1997). *Lett. Pept. Sci.*, **4**, 177.
5. Krambovitis, E., Hatzidakis, G., and Barlos, K. (1998). *J. Biol. Chem.*, **273**, 10874.

# 10

# Methods of preparing peptide–carrier conjugates

JAN W. DRIJFHOUT and PETER HOOGERHOUT

## 1. Introduction

For many applications, peptides should be conjugated to carriers. An important example is the conjugation of peptides to proteins, carbohydrates, or lipids for immunological studies. Further examples are the preparation of peptide affinity media and the conjugation of peptides to suitable coating compounds on surfaces for enzyme-linked immunosorbent assay and plasma resonance.

The most important consideration in designing conjugates is that the peptide part of the conjugate must retain its biological activity. This means that the site of activity of the peptide must not be involved in the conjugation reaction. However, the active part of the peptide is often not precisely known, which complicates the design of a proper conjugation strategy. In addition, if the peptide is a fragment of a larger biologically active protein, the small peptide will frequently not adopt the conformation (for instance, a loop structure) of the corresponding sequence in the native protein. This might also abolish desired biological properties, such as the possibility to induce functional antibodies recognizing the native protein. In that case, appropriate artificial conformational restrictions should be introduced into the peptide – if possible.

From the numerous conjugation methods available (1), only a few are described in this chapter. One example concerns the application of the homobifunctional cross-linker glutaraldehyde. Heterobifunctional cross-linking is illustrated by coupling of thiol-containing peptides or carriers to sulphydryl-reactive carriers or peptides, respectively.

## 2. Homobifunctional cross-linking

A general and easy method of conjugating peptides to proteins is to make use of homobifunctional cross-linkers. An example of a homobifunctional cross-linker is glutaraldehyde (1, 2) (see *Protocol 1*). This bis-aldehyde reacts with amine groups at neutral or basic pH to yield enamines, which can be reduced optionally to amines with sodium cyanoborohydride. The peptide and the protein react selectively via their N-terminus and/or lysine side chains to give not only a very complex mixture of products—peptide–peptide, peptide–protein, and protein–protein conjugates—but also large constructs containing peptide and protein. Typically, several peptide molecules per protein molecule are coupled. In view of the complex reaction pathway, the batch-to-batch reproducibility of the conjugate is difficult to control. If the peptide contains a lysine in the active part, it is recommended to extend the peptide N- or C-terminally with some additional lysines during synthesis in order to provide more amine groups for conjugation.

Numerous homobifunctional cross-linkers are available commercially (e.g. from Pierce), including disuccinimidyl derivatives (3, 4), water-soluble sulphodisuccinimidyl derivatives (5, 6), and diimido esters (7–9). Also available are cross-linkers containing bonds that can be cleaved by reduction (10, 11), hydroxylamine (12, 13), base (14), oxidation (15, 16), etc., which makes it possible to liberate the peptide from the conjugate for analytical purposes.

---

**Protocol 1.** Conjugation of a peptide to bovine serum albumin with glutaraldehyde

*Equipment, buffers, and reagents*

- Disposable gel filtration columns packed with Sephadex G-25 M (PD-10® columns, Pharmacia)
- 0.1 M sodium phosphate buffer (pH 8)
- Bovine serum albumin, BSA (Sigma)
- Phosphate-buffered saline (PBS), i.e. 10 mM sodium phosphate in physiological salt (pH 7.2)
- Glutaraldehyde, 25% w/w solution in water[a]

*Method*

1. Dissolve 3 mg of, for example, a 15-mer peptide and 5 mg BSA in 1.4 ml of 0.1 M sodium phosphate buffer (pH 8).

2. Dilute 5 µl of the stock solution of 25% glutaraldehyde with 500 µl PBS.

3. Add 100 µl of the freshly prepared diluted glutaraldehyde solution to the peptide/BSA solution at $t = 0$, 30, 60, and 90 min and mix well. After the last addition, leave the reaction mixture to stand for a further 2 h.[b]

4. Equilibrate a PD-10® column with at least 25 ml of PBS/water (1:1, v/v). Start the equilibration approximately 30 min before the end of the conjugation reaction.

5. Apply the reaction mixture (1.8 ml) to the PD-10® column and discard the eluent.

---

6. Apply 0.7 ml of PBS/water (1:1) to the column and again discard the eluent.

7. Place a collection tube under the column and apply 3.2 ml of PBS/water (1:1) to the column.

8. Lyophilize the collected column effluent and store the material at −20°C until use.

9. Reconstitute the lyophilized conjugate with 1.6 ml of water. The solution obtained contains about 1 mg/ml of conjugated peptide in PBS.

---

[a] Glutaraldehyde-containing solutions and reaction mixtures should be kept closed to avoid air oxidation.
[b] During the conjugation reaction positive charges of the protein are removed which makes the protein less soluble. Depending on the nature of the peptide to be conjugated, precipitation may occur.

# 3. Heterobifunctional cross-linking

Another approach towards conjugation is the use of cross-linkers containing two different reactive groups. Compared with homobifunctional cross-linkers, heterobifunctional linkers offer the possibility to prepare better-defined conjugates, since peptide–peptide and carrier–carrier conjugation can be prevented. Convenient procedures are based on couplings of thiols to sulphydryl-reactive groups.

## 3.1 Conjugation of thiol-containing peptides to proteins

Cysteine-containing peptides can be conjugated to sulphydryl-reactive proteins, provided that the cysteine is not essential for the biological activity. If the peptide is devoid of Cys, this amino acid can be added N- or C-terminally during synthesis. Depending on the linker attached to the solid phase, C-terminal incorporation is not recommended to avoid possible racemization of Cys (17).

Useful sulphydryl-reactive groups on the protein are maleimide, pyridyl-dithio, and bromoacetyl. Maleimide-activated keyhole limpet haemocyanin, bovine serum albumin, and ovalbumin are available commercially. The degree of functionalization can be determined indirectly (18) by incubation of a small sample of the maleimide-modified protein with an excess of cysteamine, followed by addition of dipyridyldisulphide which re-acts with the remaining cysteamine. The amount of thiopyridone formed can be determined by UV measurement. Conjugation of a Cys-peptide to maleimide-activated proteins proceeds through addition of the thiol to the carbon–carbon double bond of the maleimide group (*Figure 1*, pathway a). The reaction is fast and can therefore be performed at a pH as low as 5, which decreases the rate of undesired (air) oxidation of the Cys-peptide to disulphide dimers. The molar peptide–protein ratio in the conjugate can be estimated by determination of residual maleimide groups—if present—as described above. Finally, it should be noted that the stability of maleimide-conjugates is optimal at about pH 6 (19).

The active ester *N*-succinimidyl 3-(2-pyridyldithio)propionate (SPDP) is the classical reagent to modify proteins with pyridyldithiopropyl (PDP) groups (20). If the protein to be modified contains many free Cys-residues this creates the risk of carrier–carrier conjugation. In this case it is possible to block the free thiols with iodo- or bromoacetic acid (or amide) before reaction with SPDP (or to use a different conjugation method, as described in Section 3.2).

The number of linkers introduced can be determined by incubation of a small sample of the modified protein with an excess of a thiol, for instance dithiothreitol, and analysis of the amount of 2-pyridinethione which is released (see above). Likewise, the formation of a disulphide conjugate from a Cys-peptide and a pyridyldithio-protein (*Figure 1*, pathway b, $R_3$ = H) can be monitored by UV measurement. Since the amount of 2-pyridinethione formed is proportional to the amount of peptide coupled, the peptide–protein ratio is obtained immediately. This is advantageously if the peptide–protein ratio is important for application of the conjugate. The optimal reaction conditions (i.e. the amounts of SPDP or peptide to be used for conjugation) to obtain a desired peptide–protein ratio can be established readily due to the convenient method of analysis. However, the (*in vivo*) lability of the disulphide bond may be disadvantageous in further studies of the conjugate.

*N*-Succinimidyl bromoacetate (BrAc-ONSu) is the reagent of choice for bromoacetylation of proteins (21). The incorporation of bromoacetyl (BrAc) groups can be determined indirectly by measurement of the amount of free primary amino groups with 2,4,6-trinitrobenzenesulphonic acid before and after the modification (22). Cys-peptides couple to BrAc-proteins to give very stable thioether conjugates (*Figure 1*, pathway c). The peptide–protein ratio can be determined by amino acid analysis by quantitation of the amount of *S*-carboxycysteine (21, 23) after hydrolysis of a sample of the conjugate.

Instead of using Cys-containing peptides it is convenient to use *S*-acetylmercaptoacetyl (SAMA) peptides for conjugation purposes. Unlike Cys, SAMA is not sensitive towards (air) oxidation during purification of the peptide. SAMA can be introduced into deprotected peptides in solution by reaction with *N*-succinimidyl *S*-acetylmercaptoacetate (18, 24). Selective introduction of SAMA can be achieved at the end of solid phase synthesis by N-terminal modification (Chapter 6) of the side-chain protected and resin-bound peptide with pentafluorophenyl *S*-acetylmercaptoacetate (SAMA-OPfp) and 1-hydroxybenzotriazole (25). The thiol group of SAMA-peptides can be liberated by reaction with hydroxylamine. This S-deacetylation can be performed in the conjugation reaction mixture (25, 26), i.e. in the presence of the sulphydryl-reactive protein (*Protocol 2*).

If a conjugate is required for immunization experiments, it is convenient to prepare not only the SAMA- or Cys-peptide, but also the corresponding biotinyl-peptide or an N-acetylated MAP-8 (Section 4). The biotinyl-peptide can be bound to (strept)avidin-coated microtitre plates to determine the anti-peptide titre of the sera obtained. The MAP-8 can be coated directly onto regular plates.

---

**Protocol 2.** Conjugation of a SAMA-peptide[a] to bromoacetylated or pyridyldithiopropyl-modified proteins

*Equipment, buffers, and reagents*

- UV/visible spectrophotometer
- PD-10® columns (see *Protocol 1*)
- 0.1 M sodium phosphate buffer, pH 8 (buffer A)
- 0.1 M sodium phosphate buffer, containing 5 mM EDTA, pH 6, deaerated with helium (buffer B)
- PBS (see *Protocol 1*, buffer C)
- Cross-linker: *N*-succinimidyl bromoacetate, BrAc-ONSu (Sigma) or *N*-succinimidyl 3-(2-pyridyldithio)propionate, SPDP (Pierce)

- Carrier protein: bovine serum albumin, BSA (Sigma, A 7030, min 98%) or keyhole limpet haemocyanin, KLH (Sigma, H 7017, lyophilized from stabilizing buffer)[b]
- 2% w/v Sodium dodecylsulphate (SDS) in water
- 2 M Hydroxylamine•HCl in buffer B (140 mg/ml)
- 2-Aminoethanethiol (cysteamine) hydrochloride

*Method*

1. Dissolve the protein (either BSA or KLH) at a concentration of 3.0 mg/ml in buffer A.[c]

2. Dissolve either BrAc-ONSu (18.9 mg/ml) or SPDP (25.0 mg/ml) freshly in *N,N*-dimethylacetamide.

3. Add 50 μl of the solution of the cross-linker (i.e. 4 μmol linker) to 1.75 ml of the protein solution and mix gently. Leave the reaction mixture to stand for 1 h.

4. Apply the reaction mixture to a PD-10® column equilibrated in buffer B (see *Protocol 1*) and discard the eluent. Apply 0.7 ml of buffer B to the column and again discard the eluent.

5. Place a collection tube under the column, apply 3.0 ml of buffer B to the column, and collect the solution of the modified protein.

6. Dissolve 3–4 μmol of the SAMA-peptide in 250 μl of 2% SDS in water.

7. Add 2.0 ml of the solution of the modified protein to the peptide solution and mix gently.

8. *Optional* (only when using SPDP as linker): measure the absorbance at 343 nm ($A_{343}$).

9. Add 25 μl of the solution of hydroxylamine in buffer B to the solution

---

**Protocol 2.** *Continued*

of the SAMA-peptide and the modified protein and mix gently. Leave the reaction mixture to stand for 18–24 h.

10. *Optional* (only when using SPDP as linker): measure $A_{343}$, subtract the absorbance as found in step 8, and calculate the amount of thiopyridone released (and thus the amount of peptide coupled) using $\varepsilon_{343} = 8 \times 10^3$ M$^{-1}$ cm$^{-1}$.[d]

11. Dissolve cysteamine hydrochloride in buffer B (3.0 mg/ml). Add 150 µl of the cysteamine solution to the conjugation reaction mixture, mix gently, and leave the mixture to stand for 4 h.[e]

12. *Optional* (only when using SPDP as linker): measure $A_{343}$ and calculate the amount of thiopyridone released.

13. Apply the reaction mixture to a PD-10® column equilibrated in buffer C and discard the eluent.

14. Place a collection tube under the column, apply 3.5 ml of buffer C to the column, and collect the solution of the conjugate (about 1 mg/ml, total protein).

15. Store the conjugate solution at 2–8°C until use (lyophilization is not recommended).[f]

[a] The protocol can also be used for Cys-peptides. In such a case, step 9 should be omitted.

[b] The protocol has been used extensively (26) with tetanus toxoid (TTd) as the carrier protein. Since TTd is not generally available, the method was tested by conjugation of SAMA-Aha-[Gly-Gly-Ala-Val-Pro]$_3$-NH$_2$ (Aha = 6-aminohexanoyl; the GGAVP-repeat is a sequence from pertactin, a protein from *Bordetella pertussis*) to BSA and KLH.

[c] In the case of KLH, this is a deviation from the supplier's recommendation.

[d] On conjugation of the peptide as described in footnote b to BSA (~ 65 kDa), KLH (~ 400 kDa), and TTd (~ 150 kDa), $A_{343}$ = 2.21, 1.40, and 0.93, respectively. This indicates conjugation of 0.26–0.63 µmol of peptide to approximately 3.5 mg of protein, corresponding to approximate molar peptide–protein ratios of 12, 46, and 11 for the BSA, KLH, and TTd conjugates, respectively.

[e] Remaining linkers on the conjugate are capped in this step.

[f] When conjugates of the peptide as described in footnote b were used in immunogenicity studies in mice, the order of immunogenicity was: KLH and TTd > BSA conjugates and thioethers > disulphide conjugates.

## 3.2 Conjugation of peptides to thiolated carriers

It may be useful to interchange the reactivities of the peptide and the protein (or carrier) as described in Section 3.1. In that case it is convenient to prepare chloroacetyl (27), bromoacetyl (23, 28, 29), or *N*-Boc-*S*-(3-nitro-2-pyridyl-sulphenyl)cysteinyl (30, 31) derivatives by N-terminal acylation of the side-chain protected and resin-bound peptides. (It should be noted that thiol-containing or other reducing scavengers should be avoided during the final deprotection of these peptides.)

If the protein of interest does not contain free Cys, it is sometimes possible

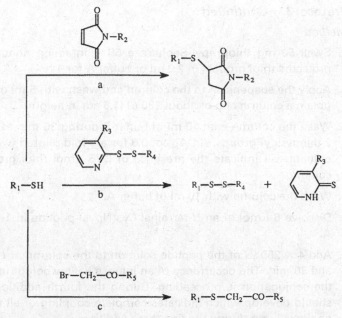

**Figure 1.** Conjugation of thiols to maleimides (pathway a); 2-pyridyldithio derivatives (pathway b, $R_3$ = H); 3-nitro-2-pyridyldithio derivatives (pathway b, $R_3$ = NO$_2$), e.g. a Boc-$S$-(3-nitro-2-pyridylsulphenyl)cysteinyl-peptide; or bromoacetyl derivatives (pathway c).

to generate thiols by disulphide reduction (if unfolding of the protein is allowed). In most cases modification of the protein is required. Direct thiolation of proteins can be achieved by reaction with iminothiolane (32) or ($N$-acyl) homocysteine thiolactone (33). Protected thiol groups can be introduced with $N$-succinimidyl $S$-acetylmercaptoacetate (24), SPDP (20), or cystamine and 1-ethyl-3-(3-dimethylaminopropyl) carbodiimide (34). The degree of thiolation can be determined by colorimetric assay with 5,5'-dithiobis(2-nitrobenzoic acid) (35, 36).

*Protocol 3* provides an example of immobilization of Npys-peptide to thiopropyl-Sepharose (37).

---

**Protocol 3.** Immobilization of an Npys-peptide to thiopropyl-Sepharose

*Equipment, buffers, and reagents*

- Thiopropyl-Sepharose 6B (Pharmacia)
- Polypropylene or glass column (30 × 4 mm) with filter
- 0.1 M sodium phosphate buffer, pH 8 (buffer A)
- 0.1 M sodium phosphate buffer, containing 10 mM DTT, pH 8 (buffer B)
- 3% v/v acetic acid in water (buffer C)

---

---

**Protocol 3.** *Continued*

*Method*

1. Swell 50 mg thiopropyl-Sepharose 6B (containing about 3.75 μmol protected thiol groups) in 2.0 ml of buffer A for 1 h.

2. Apply the suspension to the column and wash with 5 ml of buffer A to obtain a column bed of about 180 μl (1.5 cm in height).

3. Wash the column with 10 ml of buffer B during 30 min to remove the 2-thiopyridyl groups. An $A_{343}$ of 0.6 for a 5-fold diluted (with buffer A) effluent will indicate the presence of 3.75 μmol thiol groups on the column.

4. Wash the column with 10 ml of buffer A.

5. Dissolve 6 μmol of an N-terminal Cys(Npys)-peptide in 1 ml of buffer A.

6. Add $4 \times 250$ μl of the peptide solution to the column at $t = 0, 10, 20$, and 30 min. The occurrence of an intense yellow colour indicates that the conjugation is proceeding. During the fourth addition no colour should develop which indicates complete coupling of all thiol groups on the column during the first three additions.

7. Wash the column with 10 ml of buffer A.

8. *Optional*: wash the column with 10 ml of buffer C in order to acidify the resin before storage.

9. *Optional*: dilute the resin 10 times with Sepharose 4B to increase the volume for convenient column volumes during chromatography.

---

# 4. MAP-core constructs as peptide-carriers

As well as by conjugation to a macromolecule, such as a protein, peptides can be presented as a larger construct in the form of a multiple antigen peptide (MAP). In this type of construct various copies of the peptide are attached to a small core structure (38). The construct, which is fully synthetic, is also known as a dendrimer (39). An MAP might contain one particular peptide (homo-MAP) or two (or even more) distinct peptides (hetero-MAP). MAPs can be prepared by direct sequential synthesis or by fragment condensation of modified peptides and an activated MAP core construct.

HPLC purification of MAPs can be complicated by the fact that these structures might show broad signals when applied to reversed-phase columns. This might be caused by slow molecular motions of the large constructs during chromatography. In addition, mass spectrometric analysis of MAPs is often troublesome.

## 4.1 Sequential synthesis of MAPs

MAPs (40, 41) can be constructed by straightforward peptide synthesis. At a certain stage in the synthesis, acylation is performed with Fmoc-Lys(Fmoc)-OH. After deprotection, two new amino groups are available for coupling of the next amino acid, yielding a double peptide chain. This process can be repeated several times, yielding MAPs with 2–16 peptide copies. To produce MAPs containing multiple copies of two different peptides, coupling can be performed with, for example, Fmoc-Lys(Dde)-OH (42). After synthesis of multiple copies of the first (N-acetylated) peptide, the semi-orthogonal Dde protection is removed. Then synthesis of multiple copies of the second peptide is performed (43).

Especially when MAP-8 or MAP-16 constructs are made, space within the peptide–resin might become limited. It is therefore advisable to use resins with a low loading (e.g. 0.05–0.2 meq/g). The purity of directly prepared MAPs is always lower compared to that of the corresponding peptide, because in the case of a side reaction in one of the branches, the other branches are also part of a failure sequence.

## 4.2 Synthesis of MAPs by fragment condensation

Apart from direct sequential MAP-synthesis, it is also possible to prepare MAPs by fragment condensation (44, 45). The core to which the peptides will be attached and the peptide(s) themselves are synthesized and purified separately. Subsequently, multiple copies of a peptide are conjugated to the core. The core and the peptides should be modified to allow for controlled conjugation. This can be performed in several ways, for example by conjugation of a thiol-containing peptide to a bromoacetylated core, yielding a stable thioether MAP. An example is given of the synthesis of an MAP-8 in which the peptide molecules are attached to the core (*Figure 2*) by disulphide linkages (*Figure 1*, pathway b, $R_3 = NO_2$). This type of MAP can be checked by reduction which liberates the thiopeptide from the core. The core described contains eight glutamic acid residues to increase the solubility and a

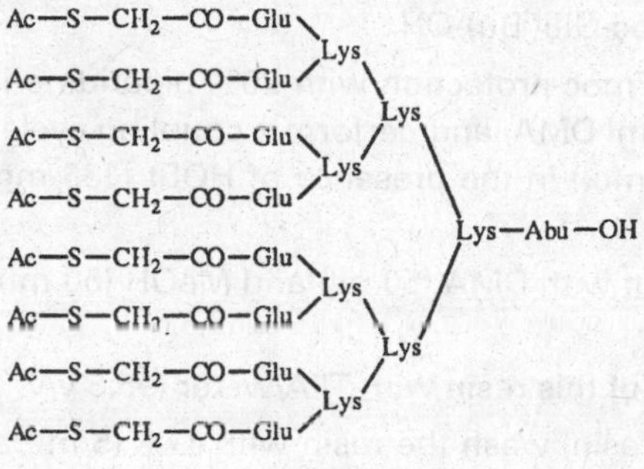

Ac—S—CH$_2$—CO—Glu
Ac—S—CH$_2$—CO—Glu
Lys
Ac—S—CH$_2$—CO—Glu
Lys
Ac—S—CH$_2$—CO—Glu
Lys—Abu—OH
Ac—S—CH$_2$—CO—Glu
Lys
Ac—S—CH$_2$—CO—Glu
Lys
Ac—S—CH$_2$—CO—Glu
Lys
Ac—S—CH$_2$—CO—Glu

**Figure 2.** Structure of the MAP core construct obtained according to *Protocol 4* after step 8.

4-aminobutyric acid residue which allows for determination of peptide/core ratio after amino acid analysis (44).

---

**Protocol 4.** Preparation of an MAP-8 containing disulphide-linked peptide molecules

*Equipment, buffers, and reagents*

- Wang (4-hydroxymethylphenoxymethyl polystyrene) resin, loading 0.092 mmol/g
- Dimethylacetamide (DMA)[a]
- DCM
- Diethyl ether
- Fmoc-4-aminobutyric acid
- DCC
- DMAP
- Piperidine
- Fmoc-Lys(Fmoc)-OH
- Fmoc-Glu('Bu)-OH
- 1-Hydroxybenzotriazole (HOBt)

- Pentafluorophenyl *S*-acetylmercaptoacetate (SAMA-OPfp)
- Trifluoroacetic acid (TFA)
- 0.2 M Sodium phosphate buffer, pH 6.5 (buffer A)
- 0.1 M Sodium phosphate buffer, containing 10 mM DTT, pH 8 (buffer B)
- Hydroxylamine•HCl in water (70 g/l, 1 M)
- 10% v/v Acetic acid in water (buffer C)
- 0.05 M Acetic acid in water (buffer D)
- Sephadex G-25® superfine

*Method*

1. Dissolve Fmoc-4-aminobutyric acid (324 mg, 1 mmol) in dichloromethane (20 ml) and add solid DCC (103 mg, 0.5 mmol) to the stirred solution.

2. Filter after 40 min and evaporate the solvent under vacuum.

3. Immediately dissolve the residue in 3 ml dry DMA,[a] add 500 mg Wang resin (loading 0.092 mmol/g OH) (46), add DMAP (24 mg, 0.2 mmol), and shake over a period of 2 h.

4. Collect and wash the resin with 30 ml of DMA.

5. Remove the Fmoc-protection with 20% v/v piperidine in DMA and perform coupling cycles by applying one of the protocols from Chapter 3 in which the following amino acid derivatives are used: *cycle 1*: 0.2 mmol Fmoc-Lys(Fmoc)-OH; *cycle 2*: 0.3 mmol Fmoc-Lys(Fmoc)-OH; *cycle 3*: 0.4 mmol Fmoc-Lys(Fmoc)-OH; *cycle 4*: 1.0 mmol Fmoc-Glu('Bu)-OH.

6. Remove the Fmoc-protection with 20% piperidine in DMA, wash the resin with 50 ml DMA, and perform a coupling cycle with SAMA-OPfp (300 mg, 1 mmol) in the presence of HOBt (135 mg, 1 mmol) during 2 h.[b]

7. Wash the resin with DMA (50 ml) and MeOH (50 ml) and dry *in vacuo* for 1 h.

8. Treat 350 mg of this resin with TFA/water (95:5 v/v, 10 ml) for 2 h.

9. Filter off the resin, wash the resin with TFA (5 ml) and evaporate the combined filtrates under vacuum to about 3 ml.

10. Precipitate with 30 ml diethyl ether and wash the precipitate twice with 20 ml diethyl ether and dry.

11. Dissolve the crude material (MAP core) in 20 ml buffer C and lyophilize.

12. Synthesize an N-terminal Cys(Npys)-peptide.

13. Prepare a solution of 1 μmol MAP core, obtained in step 11 (containing 8 μmol S-Ac), in 2 ml of buffer A (**solution A**).

14. Prepare a solution of 16 μmol Npys-peptide in 1.6 ml of buffer A (**solution B**).

15. Add **solution B** to **solution A** and then add 100 μl hydroxylamine•HCl solution (100 μmol) and stir in a closed reactor (preferably under nitrogen) for 4 h.

16. Equilibrate a column (150 × 1 cm) of Sephadex G-25® superfine with buffer D, apply the reaction mixture to the column, and elute with buffer D. The yellow band which elutes after the MAP contains 3-nitro-2-thiopyridone (47). The amount of 3-nitro-2-thiopyridone liberated can be measured using $\varepsilon_{329} = 7.3 \times 10^3 \, M^{-1} cm^{-1}$ in buffer D.

17. *Optional*: an aliquot of the MAP-8 can be reduced with buffer B. The HPLC trace of the reaction product should contain signals from the thiopeptide and the deacetylated core. These products should co-elute with the compound obtained from treatment of the Npys-peptide with buffer B and the compound obtained from treatment of the core with hydroxylamine, respectively.

18. *Optional*: the MAP-8 obtained can be analysed by amino acid analysis, from which the peptide content of the MAP can be calculated as the ratio of the amount of amino acids from the peptide (except Glu and Lys) versus the amount of 4-aminobutyric acid in the hydrolysate.

[a] *N,N*-dimethylformamide or *N*-methylpyrrolidinone can also be used
[b] For introduction of *S*-acetylmercapto groups, the use of active esters in the presence of HOBt is preferred since the S-acetyl protection is very labile to base.

# References

1. Hermanson, G. T. (1996). *Bioconjugate techniques*. Academic Press, San Diego.
2. Pfaff, E, Mussgay, M., Böhm, H. O., Schulz, G. E., and Schaller, H. (1982). *EMBO J.*, **1**, 869.
3. Wood, C. L., and O'Dorisio, M. S. (1985). *J. Biol. Chem.*, **260**, 1243.
4. Tsudo, M., Kozak, R. W., Goldman, C. K., and Waldman, T. A. (1987). *Proc. Natl. Acad. Sci. USA*, **84**, 4215.
5. Staros, J. V. (1982). *Biochemistry*, **21**, 3950.
6. D'Souza, S. E., Ginsberg, M. H., Lam, S. C. T., and Plow, E. F. (1988). *J. Biol. Chem.*, **263**, 3943.

7. Hartman, F. C., and Wold, F. (1966). *J. Am. Chem. Soc.*, **88**, 3890.
8. Hartman, F. C., and Wold, F. (1967). *Biochemistry*, **6**, 2439.
9. O'Keeffe, E. T., Mordick, T., and Bell, J. E. (1980). *Biochemistry*, **19**, 4962.
10. Lomant, A. J., and Fairbanks, G. (1976). *J. Mol. Biol.*, **104**, 243.
11. Ioannides, C. G., Itoh, K, Fox, F. E., Pahwa, R., Good, R. A., and Platsoucas, C. D. (1987). *Proc. Natl. Acad. Sci. USA*, **84**, 4244.
12. Moenner, M., Chevallier, B., Badet, J., and Barritault, D. (1986). *Proc. Natl. Acad. Sci. USA'* **83**, 5024.
13. Abdella, P. M., Smith, P. K., and Royer, G. P. (1979). *Biochem. Biophys. Res. Commun.*, **87**, 734.
14. Zarling, D. A., Watson, A., and Bach, F. H. (1980). *J. Immunol.*, **124**, 913.
15. Smith, R. J, Capaldi, R. A., Muchmore, D., and Dahlquist, F. (1978). *Biochemistry*, **17**, 3719.
16. Bragg, P. D., and Hou, C. (1980). *Eur. J. Biochem.*, **106**, 495.
17. Kaiser, T., Nicholson, G. J., Kohlbau, H. J., and Voelter, W. (1996). *Tetrahedron Lett.*, **37**, 1187.
18. Van Bree, J. B. M. M., van Nispen, J. W., Verhoef, J. C., and Breimer, D. D. (1991). *J. Pharmac. Sci.*, **80**, 46.
19. Kitagawa, T., Shimozono, T., Aikawa, T., Yoshida, T., and Nishimura, H. (1981). *Chem. Pharm. Bull.*, **29**, 1130.
20. Carlsson, J., Drevin, H., and Axén (1978). *Biochem. J.*, **173**, 723.
21. Bernatowicz, M. S., and Matsueda, G. R. (1986). *Anal. Biochem.*, **155**, 95.
22. Habeeb, A. F. S. A. (1966). *Anal. Biochem.*, **14**, 328.
23. Kolodny, N., and Robey, F. A. (1990). *Anal. Biochem.*, **187**, 136.
24. Duncan, R. J. S., Weston, P. D., and Wrigglesworth, R. (1983). *Anal. Biochem.*, **132**, 68.
25. Drijfhout, J. W., Bloemhoff, W., Poolman, J. T., and Hoogerhout, P. (1990). *Anal. Biochem.*, **187**, 349.
26. Brugghe, H. F., Timmermans, H. A. M., van Unen, L. M. A., ten Hove, G. J., van de Werken, G., Poolman, J. T., and Hoogerhout, P. (1994). *Int. J. Peptide Protein Res.*, **43**, 166.
27. Lindner, W., and Robey, F. A. (1987). *Int. J. Peptide Protein Res.*, **30**, 794.
28. Robey, F. A., and Fields, F. A. (1989). *Anal. Biochem.*, **177**, 373.
29. Robey, F. A. (1992). In *Current protocols in immunology* (ed. Coico, R.). Vol 2, unit 9.5, John Wiley & Sons, Inc.
30. Drijfhout, J. W., Perdijk, E. W., Weijer, W. J., and Bloemhoff, W. 1988). *Int. J. Peptide Protein Res.*, **32**, 161.
31. Ponsati, B., Giralt, E., and Andreu, D. (1989). *Anal. Biochem.*, **181**, 389.
32. Ghosh, S. S., Kao, P. M., McCue, A. W., and Chapelle, H. L. (1990). *Bioconjugate Chem.*, **1**, 71.
33. Leanza, W. J., Chupak, L. S., Tolman, R. L., and Marburg, S. (1992). *Bioconjugate Chem.*, **3**, 514.
34. Lin, C., Mihal, K. A., and Kruger, R. J. (1990). *Biochim. Biophys. Acta*, **1038**, 382.
35. Ellman, G. L. (1959). *Arch. Biochem. Biophys.*, **82**, 70.
36. Riddles, P. W., Blakeley, R. L., and Zerner, B. (1979). *Anal. Biochem.*, **94**, 75.
37. Welling, G. W., Geurts, T., Van Gorkum, J., Damhof, R. A., Drijfhout, J. W., Bloemhoff, W., and Welling-Wester, S. (1990). *J. Chromatogr.*, **512**, 337.
38. Tam, J. P. (1996). *J. Immunol. Method*, **196**, 17.

39. Spetzler, J. C., and Tam, J. P. (1996). *Pept. Res.*, **9**, 290.
40. Tam, J. P., and Zavala, F. (1989). *J. Immunol. Meth.*, **124**, 53.
41. Tam, J. P., Clavijo, P., Lu, Y. A., Nussenzweig, V., Nussenzweig, R., and Zavala, F. (1990). *J. Exp. Med.*, **171**, 299.
42. Bycroft, B. W., Chan, W. C., Chhabra, S. R. and Hone, N. D. (1993) *J. Chem. Soc., Chem. Commun.*, 778.
43. Chai, S. H., Clavijo, P, Tam, J. P., and Zavala, F. (1992). *J. Immunol.*, **149**, 2385.
44. Drijfhout, J. W., and Bloemhoff, W. (1991). *Int. J. Peptide Protein Res.*, **37**, 27.
45. Lu, Y. A., Clavijo, P., Galantino, M., Shen, Z. Y., Liu, Y., and Tam, J. P. (1991). *Mol. Immunol.*, **28**, 623.
46. Wang, S. S. (1973). *J. Am. Chem. Soc.*, **95**, 1328.
47. Matsueda, R., and Aiba, K. (1978). *Chem. Lett.*, 951.

# Chemoselective and orthogonal ligation techniques

JAMES P. TAM and Y.-A. LU

## 1. Introduction

Peptide synthesis through segment ligation of unprotected peptides (total synthesis) and peptides to proteins (semi-synthesis) in aqueous solution is appealingly simple and efficient because protection and activation steps are not required. In addition, this method offers the potential to access a diverse group of macromolecules such as circular proteins, branched peptides, and protein conjugates which are difficult to obtain through conventional approaches using protecting group strategies. Furthermore, the use of unprotected peptide segments overcomes the problem of solubility encountered in the conventional approach to the synthesis of large peptides or proteins in solution.

Conceptually, ligation can be approached two ways. In the first approach, a non-amide bond is formed between two peptide segments through a pair of mutually reactive functional groups (*Figure 1A*). Typical methods of non-amide ligation include oxime, hydrazone, and thiazolidine as the coupling linkages. This type of reaction is traditionally referred to as chemoselective ligation. Non-amide ligation is characteristically flexible in joining two segments that result in amino-to-amino end (**3a**), carboxyl-to-amine (**3b**) or end-to-side chain (**3c**) structures (*Figure 1*). This flexibility permits synthesis of protein mimetics and branched peptide dendrimers.

In the second approach, an amide bond is formed through a two-step reaction sequence involving four functional moieties, two nucleophiles and two electrophiles in the reaction centre (*Figure 1B*). This reaction is usually used for end-to-end coupling between the $C^{\alpha}$-moiety of one peptide segment and the $N^{\alpha}$-terminus of another peptide segment resulting in a peptide-backbone product. Similar to the non-amide chemoselective ligation, the first step in orthogonal ligation is a capture reaction by a pair of mutually reactive groups. In general, two nucleophiles, a weak-base nucleophile on the side chain and an α-amino, are located at the N-terminus as an N-terminal nucleophile **5** (NTN). The two electrophiles, usually an $O$-glycol-aldehyde or an $S$-ester **4**, are located at the C-terminus of the another peptide segment. The initial non-amide capture of two segments through the side chain NTN with the $O$- or

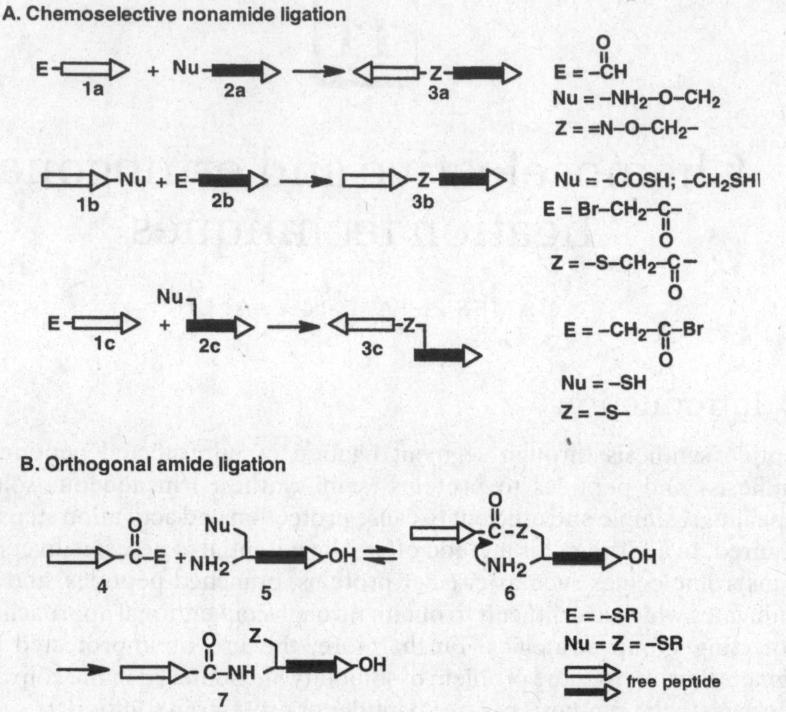

**A. Chemoselective nonamide ligation**

E—▷ + Nu—▶ → ◁—Z—▶    E = –CH (=O)
1a        2a           3a           Nu = –NH₂–O–CH₂
$$E = -\overset{\overset{\displaystyle O}{\|}}{C}H$$
$$Nu = -NH_2-O-CH_2$$
$$Z = =N-O-CH_2-$$

▷—Nu + E—▶ → ▷—Z—▶    Nu = –COSH; –CH₂SH
1b      2b         3b
$$E = Br-CH_2-\overset{\overset{\displaystyle O}{\|}}{C}-$$
$$Z = -S-CH_2-\overset{\overset{\displaystyle O}{\|}}{C}-$$

E—▷ + Nu⌐▶ → ◁—Z⌐▶    
1c    2c        3c
$$E = -CH_2-\overset{\overset{\displaystyle O}{\|}}{C}-Br$$
$$Nu = -SH$$
$$Z = -S-$$

**B. Orthogonal amide ligation**

▷—C(=O)–E + NH₂–Nu⌐▷—OH ⇌ ▷—C(=O)–Z⌐(NH₂)▷—OH
4              5                  6

$$E = -SR$$
$$Nu = Z = -SR$$
▶ free peptide

→ ▷—C(=O)–NH⌐Z▷—OH
7

**Figure 1.** Concepts in orthogonal ligation to form a non-amide linkage between a pair of mutually reactive functional groups, a nucleophile and an electrophile, in the reaction centre (A), and amide ligation with a nucleophile and an electrophile in addition to α-acyl and α-amine moieties (B).

$S$-ester to form a covalent intermediate **6** enables the spontaneous proximity-driven intramolecular acyl transfer to occur. This intramolecular acyl migration achieves orthogonality in amide bond formation **7** between a specific α-amine in the presence of other free α- and ε-amines. Because the amide bond forms through entropic activation, no enthalpic activation reagent is required. Based on the character of this reaction which distinguishes one α-amine from other α- and ε-amines to form an amide bond, we termed this reaction orthogonal ligation. The concept of coupling peptide segments through entropic activation is not new, and has been proposed by Brenner *et al.* (1) and Kemp and colleagues (2, 3) for some time. However, it is only recently that we, as well as others, have been able to achieve peptide ligation through entropic activation (4–6).

# 2. Chemoselective non-amide ligation

Two characteristics of chemoselective ligation are: (i) the use of unprotected peptide segments and (ii) the reaction is performed in aqueous conditions. To

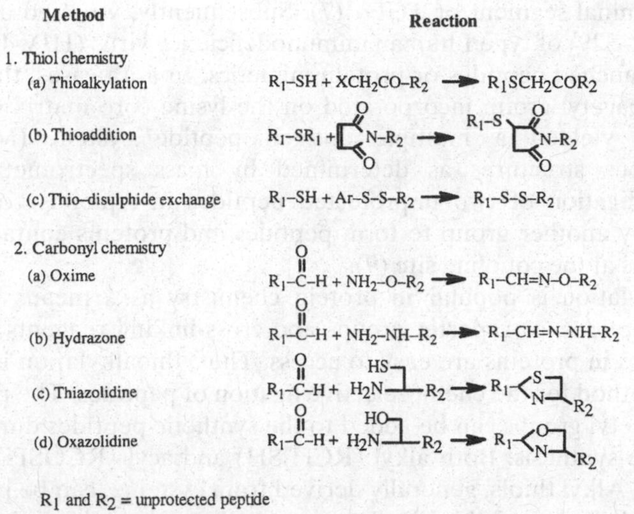

| Method | Reaction |
|---|---|
| **1. Thiol chemistry** | |
| (a) Thioalkylation | $R_1\text{-SH} + XCH_2CO\text{-}R_2 \longrightarrow R_1SCH_2COR_2$ |
| (b) Thioaddition | $R_1\text{-SR}_1 + \underset{O}{\overset{O}{\diagup}}N\text{-}R_2 \longrightarrow R_1\text{-}S\underset{O}{\overset{O}{\diagdown}}N\text{-}R_2$ |
| (c) Thio–disulphide exchange | $R_1\text{-SH} + Ar\text{-}S\text{-}S\text{-}R_2 \longrightarrow R_1\text{-}S\text{-}S\text{-}R_2$ |
| **2. Carbonyl chemistry** | |
| (a) Oxime | $R_1\text{-}\overset{O}{\overset{\|}{C}}\text{-H} + NH_2\text{-}O\text{-}R_2 \longrightarrow R_1\text{-}CH{=}N\text{-}O\text{-}R_2$ |
| (b) Hydrazone | $R_1\text{-}\overset{O}{\overset{\|}{C}}\text{-H} + NH_2\text{-}NH\text{-}R_2 \longrightarrow R_1\text{-}CH{=}N\text{-}NH\text{-}R_2$ |
| (c) Thiazolidine | $R_1\text{-}\overset{O}{\overset{\|}{C}}\text{-H} + \underset{H_2N}{\overset{HS}{\diagup}}R_2 \longrightarrow R_1\text{-}\underset{N}{\overset{S}{\diagup}}R_2$ |
| (d) Oxazolidine | $R_1\text{-}\overset{O}{\overset{\|}{C}}\text{-H} + \underset{H_2N}{\overset{HO}{\diagup}}R_2 \longrightarrow R_1\text{-}\underset{N}{\overset{O}{\diagup}}R_2$ |

$R_1$ and $R_2$ = unprotected peptide

**Figure 2.** Non-amide orthogonal ligation.

achieve these chemistries, a reactive pair consisting of a nucleophile and an electrophile is placed on the peptides during solid phase synthesis. Usually, the nucleophile is a weak base which has either a $pK_a$ significantly lower than the $\alpha$- or $\varepsilon$-amines or a nucleophilicity much stronger than the $\alpha$-amine, so that the ligation can be selective in aqueous buffered solution at pH $\approx$ 7. Applicable weak bases include alkyl thiol, acyl thiol, 1,2-aminothiol (N-terminal cysteine), acylhydrazine, and arylhydrazine (*Figure 2*). Reactive electrophiles include haloacetyl, activated unsymmetrical disulphide, maleimide, or aldehyde. Chemoselectivity is achieved when these mutually reactive groups are brought together in aqueous solution with the weak base as the sole nucleophile to react with the electrophile. Protection of other functional groups on the peptides therefore becomes unnecessary.

## 2.1 Thiol chemistry

Thiol chemistry exploits the extraordinary reactivity of sulphhydryls in alkylation with $\alpha$-halocarbonyls, in addition to conjugated olefins and sulphur–sulphur exchange with disulphides. In practice, in thiol alkylation and addition to conjugated olefins, disulphides have frequently been used for chemoselective ligation in both peptide synthesis and semi-synthesis because of the stability of the products.

### 2.1.1 Thioalkylation

Thiols are stronger nucleophiles than amines or alcohols in aqueous buffered solutions at neutral to mildly basic pH. Thus, our laboratory has exploited thioalkylation for chemoselective segment ligation using unprotected peptides of a C-terminal segment of TGFα (7). Subsequently, we used a segment of gp120-(312–329) of type I human immunodeficiency virus (HIV-I) (8) in 1991 to form branched peptides or protein mimetics. In both cases, thioalkylation of a chloroacetyl group incorporated on the lysine core matrix of a peptide dendrimer yielded a multiple antigen peptide system (MAPs) with unambiguous structures as determined by mass spectrometric analysis. Similarly, ligation of two unprotected peptides in aqueous condition was achieved by another group to form peptides and proteins containing a non-amide bond at the coupling site (9).

Thioalkylation is popular in protein chemistry as a means of attaching ligands, peptides, or reporter groups and cross-linking reagents because the thiol groups in proteins are easy to access. Thus, thioalkylation is also a convenient method for the chemoselective ligation of peptides. The reactive thiol and haloacetyl groups can be added to the synthetic peptides during stepwise solid phase synthesis. Both alkyl ($RCH_2SH$) and acyl ($RCOSH$) thiols have been used. Alkyl thiols, generally derived from cysteine, can be placed at any position, whereas acyl thiol in the form of thiocarboxylic acid is limited at present to the C-terminus. The haloacetyl moiety, whose C–X bond is activated by the acetyl group, can be attached to the N-terminus or the side chain of lysine positioned anywhere in the sequence. The chloroacetyl group is stable to HF cleavage conditions in the absence of thiol scavengers. The rate of thioalkylation to give thioethers is pH dependent and increases as the pH becomes more basic. Because side reactions, such as oxidation of thiol to disulphide as well as hydrolysis of the haloacetyl group, occur significantly at more alkaline pH, thioalkylation is commonly performed at pH 7.5–8. To minimize disulphide formation, Defoort *et al.* (10) have used phosphine as a reducing agent with significant success. We have found that S–S oxidation can be minimized by using a small amount of phosphine prior to the reaction and including EDTA during the reaction (11, 12).

### 2.1.2 Thiol–disulphide exchange

Thiol–disulphide exchanges to form unsymmetrical disulphide bonds have been exploited extensively both in protein and peptide chemistry. King *et al.* (13) have used an aromatic thiol, the 4-dithiopyridyl group, as a cysteinyl-activating group to cross-link proteins via intermolecular disulphide bonds. Similarly, other aromatic thiols such as 2-thiopyridyl and nitropyridyl sulphenyl (Npys) have been used in peptide synthesis to form unsymmetrical disulphides. These and other similar methods can be applied to the ligation of unprotected peptides.

### 2.1.3 Thiol addition

A thioether can also be formed by adding a cysteine thiol to an activated double bond of a maleimido group. This method is popular for cross-linking proteins with reporter groups, and Kitagawa and Aikawa (14) have shown that insulin-containing maleimido groups can be coupled to glycoproteins. The reaction is specific and is usually carried out in aqueous solution with an optimum pH around 7, although lysine and histidine also react slowly with the maleimido group at this pH. Thus, long reaction times should be avoided. Furthermore, hydrolysis of maleimides to non-reactive maleic acid occurs above pH 8. This method is convenient because *N*-alkyl or *N*-aryl maleimide groups are available commercially either as free carboxylic acids or active esters such as *N*-hydroxysuccinimides, which can be incorporated as a pre-formed unit in solid phase synthesis. The maleimido group on lysine and phenylalanine has been shown to be stable to trifluoroacetic acid for 3 h, and is fully compatible with Fmoc chemistry when the maleimido group is added last in the sequence of peptide assembly.

## 2.2 Carbonyl chemistry

Aldehydes and ketones represent a group of electrophiles that are not present in peptides but are exceptionally reactive electrophiles exploited for chemo-selective ligation. In aqueous conditions, particularly at acidic pH, carbonyls react with amines to form unstable imines. To obtain stable products for select-ive exploitation, weak bases are used, particularly those with a strong tendency for a condensation reaction with aldehydes. In general, the desirable weak bases, except cysteine, do not occur naturally in amino acid sequences, and several types of weak bases for ligation to aldehydes have been developed for this purpose. The first type consists of conjugated amines whose basicities are lowered by neighbouring electron-withdrawing groups that also stabilize imino products. These include hydroxylamine to give oxime, and substituted hydrazines, such as acylhydrazines and phenylhydrazines, to give hydrazone. The second type contains the 1,2-disubstituted pattern which forms a cyclic compound in an addition reaction to the imino intermediates. This includes derivatives of 1,2-aminoethanethiol and 1,2-aminoethanol such as those found in N-terminal cysteine and threonine. These 1,2-disubstituted weak bases react with aldehyde to form a proline-like ring such as thiazolidine from cysteine and oxazolidine from threonine.

### 2.2.1 Oxime

Although hydroxylamine–aldehyde chemistry was used by Erlanger *et al.* (15) to form oximes and later by Pochon *et al.* (16) for site-specific conjugation to protein, this weak base was first used for peptide synthesis by Rose (17), and later by Tuchscherer (18) and Shao and Tam (11). The hydroxylamine com-ponent for oxime formation is usually introduced using the commercially available aminooxyacetic acid, $NH_2OCH_2COOH$, which can also be used as a protected unit for incorporation at any position in the peptide sequence during the solid phase synthesis. Aminooxyacetyl-containing peptides are stable to HF and TFA cleavage conditions and can be ligated to aldehydes to give an oxime linkage. Oxime linkages have been used for preparing end-to-side chain cyclized peptides and peptide dendrimers (11, 19). Such intra-molecular cyclization by oxime formation is highly chemoselective and flexible, since the reactive weak base–aldehyde pair can be placed in different configurations, allowing side chain-to-side chain, end-to-end, and end-to-side chain cyclization (11).

Typically, a serine residue (a masked aldehyde precursor) is coupled to a lysyl side chain and the aminooxyacetyl group is used as the weak base at the N-terminus. Under mildly acidic conditions, this weak base is the only re-active nucleophile present. All other functional groups on the amino acid side chains, including the $N^\varepsilon$-amine of lysine, are protonated or form reversible Schiff's bases.

### 2.2.2 Hydrazone

Ligation by hydrazone formation is probably one of the oldest methods in protein and carbohydrate chemistry (20). In general, acyl- and aryl-sub-stituted hydrazines have been used successfully. Offord, Rose and colleagues (21–23) have applied acylhydrazine–aldehyde chemistry for site-specific con-jugation to proteins, protein semi-synthesis, and backbone engineering. To use hydrazide–aldehyde chemistry to ligate a peptide on the N-terminus of a second peptide, Shao and Tam (11) have prepared Boc-monohydrazide succinic acid, which allows facile introduction to the α-amine and the side chain of lysine at any position. Thus, unprotected peptides containing the hydrazide succinyl group are used to ligate to the aldehyde-containing peptide with great efficiency. Similarly, 4-Boc-monohydrazinobenzoic (Hob) acid has been developed as a derivative for analogous modification strategies during peptide synthesis (12). Similar to the hydroxylamine derivative, Hob-peptides are stable to the usual cleavage conditions involving TFA or HF. The progress of the phenylhydrazone reaction can be conveniently monitored with UV at 340 nm (12). The preparation of a peptide dendrimer through hydrazone ligation is described in the following protocols.

---

**Protocol 1.**  Preparation of glyoxylyl tetravalent lysine core peptide

*Reagents*
- H-Ala-OCH$_3$·HCl
- Methyl dimethoxyacetate
- Boc-Lys(Boc)-OH
- *N,N*-Diisopropylethylamine (DIPEA)

- Benzotriazol-1-yl-oxy-tris(dimethylamino) phosphonium hexafluorophosphate (BOP)
- Silica gel, 130–270 mesh, 60 Å

A. *Preparation of Boc-Lys(Boc)-Ala-OCH$_3$*

1. Cool in an ice-bath a solution (DMF 12 ml:DCM 6 ml) of H-Ala-OCH$_3$·HCl (1.4 g, 10 mmol) and Boc-Lys(Boc)-OH (3.46 g, 10 mmol) cooled in ice-bath.

---

## 11: Chemoselective and orthogonal ligation techniques

2. Add BOP (4.42 g, 10 mmol) and DIPEA (2.84 g, 22 mmol).

3. Stir the reaction mixture for 30 min at 0°C and then at room temperature for 20 h.

4. Remove DMF and DCM *in vacuo*.

5. Add ethyl acetate (50 ml) to residue.

6. Extract organic phase with saturated NaCl(aq) (2 × 50 ml), 2% w/v KHSO$_4$(aq) (2 × 50 ml), 5% w/v NaHCO$_3$(aq) (3 × 50 ml), and water (3 × 50 ml).

7. Dry organic phase using anhydrous Na$_2$SO$_4$.

8. Concentrate organic phase to dryness.

9. Recrystallize from ethyl acetate/hexane give dipeptide Boc-Lys(Boc)-Ala-OCH$_3$. Yield 4.37 g

   (91%). MS: calc. MH$^+$ 479.6. C$_{24}$H$_{37}$N$_3$O$_7$ req. C, 60.11; H, 7.78; N, 8.76.

B. *Preparation of Boc-Lys(Boc)-Lys[Boc-Lys(Boc)]-Ala-OCH$_3$*

1. Add 50% v/v TFA in DCM (20 ml) to dipeptide (A) (0.96 g, 2 mmol), and stir at room temperature for 20 min.

2. Remove TFA/DCM *in vacuo*.

3. Wash residue with diethyl ether (4 × 10 ml) and dissolve in DMF (20 ml).

4. Add DIPEA (1.55 g, 12 mmol) and Boc-Lys(Boc)-OH (1.44 g, 4.2 mmol) to above DMF solution cooled to 0°C.

5. Add BOP (1.89 g, 4.2 mmol), stir for 30 min at 0°C then for 24 h at room temperature.

6. Remove DMF and DCM *in vacuo*.

7. Add ethyl acetate (50 ml) to residue.

8. Wash organic phase with saturated NaCl(aq) (2 × 50 ml), 2% w/v KHSO$_4$(aq) (2 × 50 ml), 5% w/v NaHCO$_3$(aq) (3 × 50 ml), and water (3 × 50 ml).

9. Dry organic phase using anhydrous Na$_2$SO$_4$.

10. Concentrate *in vacuo* organic phase to dryness.

11. Recrystallize from ethyl acetate/hexane to yield Boc-Lys(Boc)-Lys[Boc-Lys(Boc)]-Ala-OCH$_3$. Yield 1.65 g (92%). MS: calc. M$^+$ + Na$^+$ 911. C$_{42}$H$_{77}$N$_7$O$_{13}$ req. C, 56.8; H, 8.74; N, 11.04.

C. *Preparation of (CH$_3$O)$_2$CHCOOH*

1. Add 0.5 M NaOH(aq) (10.5 ml, 5.25 mmol) to a stirred solution of methyl dimethoxyacetate (671 mg, 5 mmol) in methanol (5 ml).

2. Stir for 2 h, monitor completion of reaction by TLC.

3. Remove methanol by evaporation *in vacuo*, and dilute to 25 ml with water.

4. Wash aqueous solution with ethyl acetate (3 × 10 ml).

5. Concentrate to 10 ml by evaporation *in vacuo*, lyophilize to dryness.

D. *Branched tetrapeptide $(CH_3O)_2CHCO$-Lys$((CH_3O)_2CHCO)$-Lys$[(CH_3O)_2CHCO$-Lys-$((CH_3O)_2CHCO)]$-Ala-$CH_3$*

1. Add 50% v/v TFA in DCM (20 ml) to (B) (0.89 g, 1 mmol), and stir for 20 min.

2. Remove TFA/DCM, wash residue with ether (3 × 20 ml) and dissolve in DMF (10 ml).

3. Add powder (C) (5 mmol), BOP (1.86 g, 4.2 mmol), and DIPEA (0.65 g, 5 mmol) to above DMF solution. Stir at room temperature for 20 h.

4. Remove DMF by evaporation *in vacuo*, add ethyl acetate (10 ml) to residue, stand at 4°C overnight. The product precipitates.

5. Filter to collect product and dry *in vacuo*.

6. Take the ethyl acetate supernatant from step 4 and evaporated to dryness *in vacuo*.

7. Dissolve dried product in water (3 ml).

8. Add glacial acetic acid (4 drops), filter out precipitate (HOBt).

9. Concentrate filtrate to dryness *in vacuo*.

10. Triturate residue with diethyl ether (4 × 10 ml) to give a white powder.

11. Combine products from steps 4 and 10, and purify on silica gel (40 g) column (elute with $CHCl_3$/EtOAc/MeOH, 60:25:15). Yield 680 mg (76%). MS: calc. $M^+ + Na^+$ 919.0. $C_{38}H_{69}N_7O_{17}$ req. C, 50.94; H, 7.66; N, 10.94.

E. *Preparation of CHO-CO-Lys(CHO-CO)-Lys[CHO-CO-Lys(CHO-CO)]-Ala-OH*

1. Add 0.1 M NaOH(aq) (1.8 ml, 0.18 mmol) to a solution (10 ml water, 2 ml MeOH) of branched tetrapeptide (D) (134.4 mg, 0.15 mmol), and stir for 2 h at room temperature.

2. Remove MeOH.

3. Add 0.2 M HCl(aq) (0.25 ml) to remaining aqueous solution.

4. Lyophilize to give white powder.

5. Dissolve powder in water (1.5 ml).

6. Add concentrated HCl (5 ml) to 500 μl of the solution prepared in step 5 (50 μmol) and stir for 3 h at room temperature.

7. Concentrate to dryness *in vacuo* on water bath at 35 °C.
8. Purify residue by RP–HPLC to afford the glyoxylyl core peptide (22.5 mg, 64%). MS: calc. M$^+$ 698.7.

---

**Protocol 2.** Synthesis of peptide dendrimer through hydrazone ligation

*Reagents*
- Peptide NH₂NHCOCH₂CH₂CO-[VMEYKARR-KRAAIHVMLALA] (synthesized by Fmoc/*t*-Bu chemistry)
- Dimethylsulphoxide (DMSO)
- CHO-CO-Lys(CHO-CO)-Lys[CHO-CO-Lys(CHO-CO)]-Ala-OH (from *Protocol 1*)
- 0.2 M AcONa/AcOH buffer, pH 5.7

*Method*
1. Add DMSO (800 μl) to a mixture of NH₂NHCOCH₂CH₂CO-[peptide] (in 400 μl water, 2 μmol), CHO-CO-Lys(CHO-CO)-Lys[CHO-CO-Lys(CHO-CO)]-Ala-OH (in 40 μl water, 0.2 μmol), and acetate buffer (400 μl).
2. Adjust pH to 5.7.
3. Incubate at room temperature for 5 h.
4. Purify ligation product by RP–HPLC.
5. Adjust the pH of HPLC fractions collected to 5.5.
6. Lyophilize to give the required product (contains sodium trifluoroacetate).

---

## 2.2.3 Thiazolidine

The reaction of aldehydes with N-terminal cysteine to give thiazolidine has long been known, but only recently has this reaction been exploited for chemoselective segment ligation (5, 11, 12, 24, 25). Thiazolidine and oxazolidine are thia- and oxaproline analogues. This method affords a heterocyclic pseudoproline at the ligation site, and may be useful in imparting conformational constraint to the peptides. Unlike thiol chemistry, thiazolidine ring formation requires both a thiol and an amine in a 1,2-substituted relationship, making the reaction highly specific. Thiazolidine ring formation can be performed at pH 2–8, the reaction rate accelerating with increasing pH. Since the thiol group is readily oxidized at high pH, pH 4.5–5.4 generally provides efficient rates without side reactions. Thiazolidine ligation is generally more practical than oxazolidine ligation since this reaction undergoes rapid ring–chain tautomerization. Furthermore, thiazolidine is >104 times more stable than the oxazolidine and prefers ring formation, whereas oxazolidine generally prefers an open chain imine form and must be acylated to give the stable oxazolidine ring.

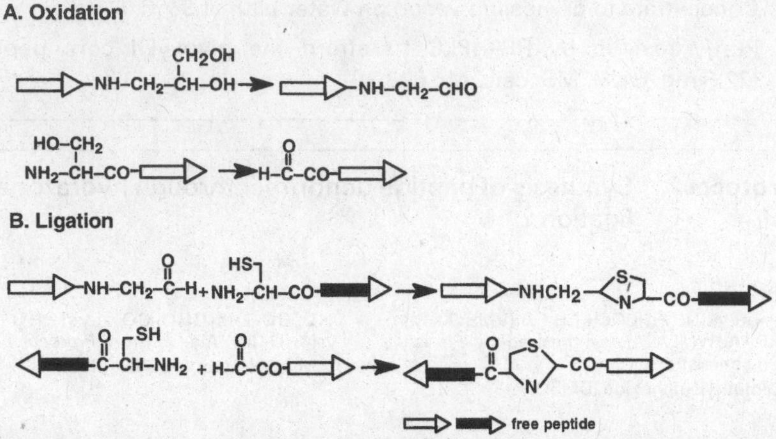

**A. Oxidation**

**B. Ligation**

**Figure 3.** Thiazolidine formation in ligation site by two steps. (A) Oxidation of a C-terminal 1,2-propanediol peptide or an N-terminal serinyl peptide by $NaIO_4$ to form a C-terminal or an N-terminal peptide aldehyde, respectively. (B) An aldehyde ligated with an *N*-cysteinyl peptide to form a thiazolidine peptide with different orientations of the thiazolidine ring.

The peptide aldehyde can be prepared in several ways. Oxidation of an N-terminal serinyl peptide or oxidation of a C-terminal 1,2-propanediol peptide yields the N-terminal or C-terminal peptide aldehyde, respectively (see Section 3.1.2), to give ligation products with different orientations of their thiazolidine ring (*Figure 3*). Typically, use of water-miscible organic solvents, such as DMF, or reactions performed at 37°C significantly accelerate the ligation reaction (11).

# 3. Orthogonal amide ligation

Orthogonal ligation provides an amide bond at the coupling site. Thus, unlike chemoselective ligation, it can produce a peptide with an all-α-amide backbone and can be considered as a strategy for convergent peptide synthesis. To achieve an amide ligation reaction, four functional groups in the reaction centre are required to accommodate a capture reaction between a nucleophile and an electrophile followed by an acyl migration to form an amide bond. Two general methods based on thiol and carbonyl chemistries have been developed to ligate unprotected peptide segments to form the peptide.

## 3.1 Carbonyl chemistry

### 3.1.1 Thiazolidine ligation for pseudoproline

In thiazolidine ligation, the reaction is performed between a glycolaldehyde peptide ester and an $N^{\alpha}$-cysteinyl peptide. The capture step is based on

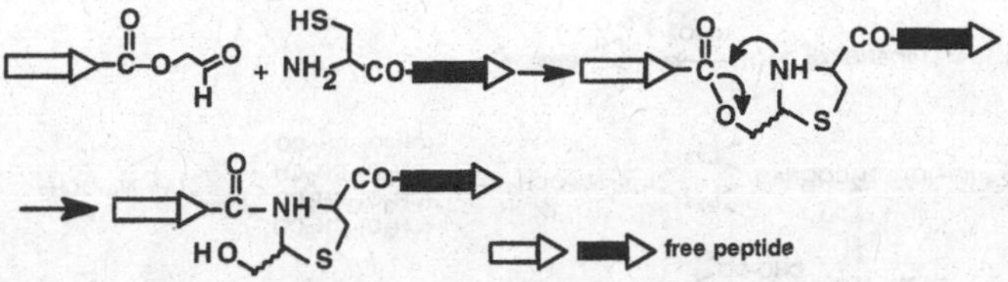

**Figure 4.** Pseudoproline formation at the ligation side by a C-terminal aldehyde peptide reacting with a 1,2-aminothiol peptide at acidic pH. The reactive carboxyl and amino termini are positioned in close proximity for spontaneous peptide bond formation through an entropy-driven intramolecular $O \rightarrow N$ acyl transfer.

thiazolidine (Thz) ligation (see Section 2.2.3) (5, 26, 27), and the orthogonality of this reaction is based on the specific condensation between an aldehyde and the 1,2-aminothiol at acidic pH in the presence of other unprotected and reactive functional groups (*Figure 4*). Consequently, the thiazolidine formation brings the α-amine of one peptide segment close to the α-carboxyl moiety of another peptide segment, thus permitting peptide bond formation through an entropy-driven intramolecular $O \rightarrow N$ acyl transfer. A pseudoproline residue is formed at the ligation site and can be viewed as a proline surrogate (2-hydroxymethyl-3-thiaproline).

### 3.1.2 Preparation of peptide aldehydes

Since peptide aldehydes are involved in both chemoselective and orthogonal ligations, the syntheses of peptide aldehydes involving either Boc or Fmoc chemistry are described in greater detail. Although aldehydes can be obtained by many organic transformations, two methods for preparing N-terminal peptide aldehydes and five methods for C-terminal peptide aldehydes have been developed specifically for peptide ligation (*Figure 5*).

The first method exploits the popular $NaIO_4$ oxidation of N-terminal Ser, Thr, or Cys to give an α-oxoacyl group (28, 29). Free peptides containing these N-terminal amino acids can serve as aldehyde precursors prior to the ligation. Periodate oxidation of a 2-aminoalcohol such as Ser is facile and about 1000 times faster than diol oxidation (21). Hence, oxidation of an N-Ser-containing peptide could be accomplished within a few minutes at neutral pH. Oxidation of Trp or other sensitive groups within the peptide sequence can be minimized by adding a large excess of methionine as scavenger. Oxidation of C-terminal Cys to give the corresponding aldehyde requires a longer reaction time (25). In practice, transformation of a terminal 1,2-aminoalcohol (Ser, Thr) or 1,2-aminothiol (Cys) to an aldehyde is convenient and highly compatible with the overall scheme of peptide synthesis, as both Boc and Fmoc chemistry can be used to generate the peptide aldehyde precursor.

Another method based on the same principle has been developed for the

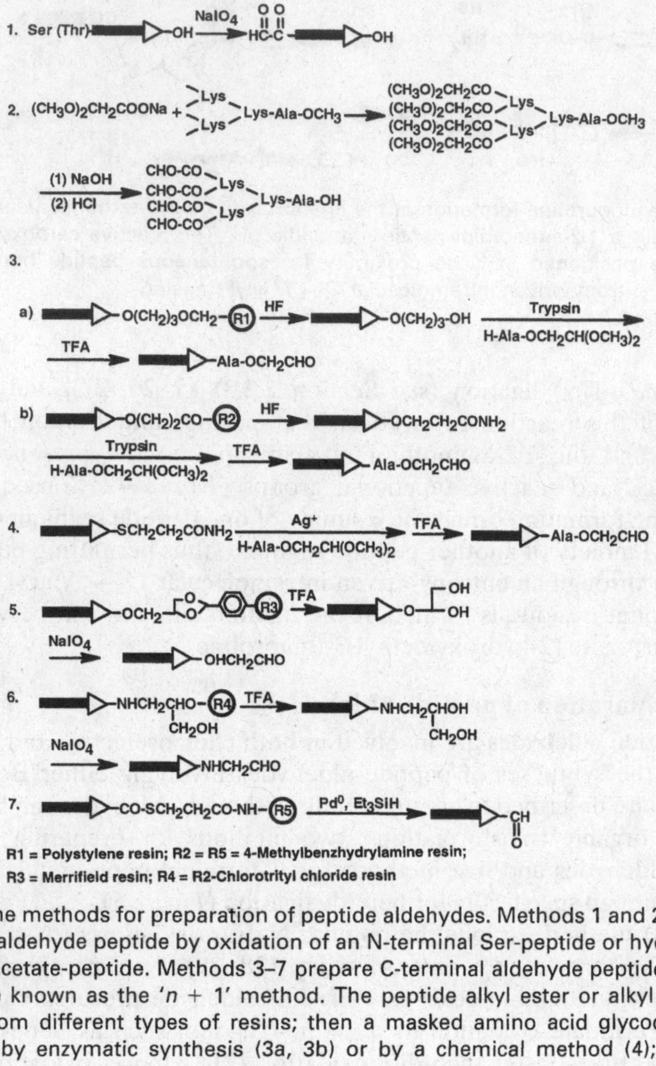

**Figure 5.** The methods for preparation of peptide aldehydes. Methods 1 and 2 prepare an N-terminal aldehyde peptide by oxidation of an N-terminal Ser-peptide or hydrolysis of a dimethoxyacetate-peptide. Methods 3–7 prepare C-terminal aldehyde peptides. Methods 3 and 4 are known as the 'n + 1' method. The peptide alkyl ester or alkyl thioester is obtained from different types of resins; then a masked amino acid glycodiol ester is introduced by enzymatic synthesis (3a, 3b) or by a chemical method (4); finally, the peptide is treated with TFA to give the aldehyde peptide. In methods 5 and 6, aldehyde peptides are obtained from oxidation of a peptide glycol diol ester. In method 7, treatment of an N-protected peptide thioester resin with Pd$^{(0)}$ and Et$_3$SiH gives the cleaved C-terminal peptide aldehyde.

preparation of peptide dendrimers (11). Methyl dimethoxyacetate can be coupled to an amino group such as the branched lysine peptide, and acid hydrolysis will give the N-terminal aldehyde (*Figure 5*, method 2). This method is more suitable for Fmoc than Boc chemistry.

The third and fourth methods developed by Liu and Tam (5, 26) and Liu and co-workers (27) are known as the '$n + 1$' method, in which an amino acid containing a masked aldehyde is added to the C-terminus of a purified peptide segment. This method bypasses problems associated with aldehyde instability during TFA cleavage or $NaIO_4$ oxidation of large peptide segments and allows the synthesis of peptides by Boc chemistry. Both methods involve the solid phase synthesis of a peptide containing either an alkyl ester or alkyl thioester at the C-terminus. The masked amino acid glycolaldehyde ester is then introduced by enzymatic synthesis (*Figure 5*, method 3) or a chemical method (*Figure 5*, method 4). In the enzymatic synthesis, the glycolaldehyde component, amino acid dimethoxyethyl ester prepared from its Z-protected form by hydrogenolysis, is incorporated onto the C-terminus of the peptide segment through the trypsin-catalysed, kinetically controlled aminolysis of the peptide ester bond (30), usually conducted in a solution containing a high content of water-miscible organic solvent such as $DMF/H_2O$ (60:40 v/v).

The enzymatic synthesis is fast and is completed in 15 min. After purification, the peptide obtained, with an acetal-protected aldehyde ester at its C-terminus, is treated with 95% $TFA/H_2O$ for 5 min to release the aldehyde function.

The chemical method via $Ag^+$ ions is more versatile because it overcomes the limitation imposed by substrate specificity of an enzymatic reaction. In this '$n + 1$' coupling, formation of a peptide bond involves a large unprotected peptide and a small functionalized amino acid. An excess of an amino acid containing a masked glycoaldehyde is added to the peptide ester. The selectivity and efficiency of this coupling is driven by the $Ag^+$ ion mediated activation which distinguishes the $C^\alpha$-thioester from other side chain unprotected carboxylic groups and by an overwhelming excess of the small amino acid derivative of the masked glycol. The protected aldehyde can then be unmasked by mild acidic conditions in the presence of the other component bearing the weak base to complete the ligation reaction.

The next three methods are on-resin derivatization based on linkers with a masked aldehyde moiety. The method developed by Botti *et al.* (31) (*Figure 5*, method 5) involves a new resin to yield a C-terminal peptide glyceric ester. After assembling the peptide sequence by Fmoc chemistry on this resin, cleavage gives a glyceric ester diol which is converted to the glycoaldehyde by oxidation with $NaIO_4$ at pH 5. The rate of diol oxidation increases as the pH falls below 5. Thus, at pH 2 the oxidation is complete within a few minutes. At pH 6–7 the rate of oxidation is slow, and requires several hours at pH 7. It should be noted that when the oxidation is performed below pH 5, the Met residues oxidize to Met(O), even in the presence of a large excess of methionine as a scavenger. This side reaction can be avoided at neutral pH. The cyclic acetal resin is prepared as described in the following protocols.

**Protocol 3.** Preparation of the cyclic acetal resin

*Reagents*
- Chloromethyl polystyrene
- Ethylene glycol dimethyl ether
- 1,4-Dioxane
- DMSO
- Glycol
- Hydroxylamine hydrochloride

*Method*

1. Add sodium carbonate (0.2 g) to a suspension of chloromethyl poly-styrene resin (0.5 g, 0.35 mmol) in DMSO (3 ml). Stir for 7 h at 155°C. This gives the benzaldehyde resin.
2. Wash resin with water, methanol, DMF, DCM, and diethyl ether (30 ml each).
3. Dry resin *in vacuo*.
4. FTIR analysis of benzaldehyde resin shows the typical carbonyl absorption at 1700 cm$^{-1}$.
5. *Optional analysis of resin*. Treat the resin (10 mg) with a large excess of hydroxylamine in pyridine for 6 h at 95°C. Wash resin and dry. FTIR analysis shows no carbonyl absorption at 1700 cm$^{-1}$ and microanalysis shows 0.82% nitrogen, indicating a loading of aldehyde of 0.59 mmol/g.
6. Mix and stir the benzaldehyde resin (1 g, 0.59 mmol) with anhydrous ethylene glycol dimethyl ether (10 ml), glycol (1 g), and *p*-toluene-sulphonic acid (10 mg) for 24 h at 90°C.
7. Wash resin with 3% sodium carbonate(aq)/dioxane (1:1 v/v), water, dioxane, DMF, methanol, DCM, and diethyl ether (50 ml each).
8. Dry resin *in vacuo*.

---

**Protocol 4.** Preparation of Fmoc-Gly-OCH$_2$-cyclic acetal resin

*Reagents*
- Fmoc-Gly-OH
- 4-Dimethylaminopyridine (DMAP)
- *N,N'*-Dicyclohexylcarbodiimide (DCC)
- 1,4-Dioxane
- *N*-Hydroxybenzotriazole (HOBt)

*Method*

1. Stir a mixture of Fmoc-Gly-OH (1.66 g, 5.58 mmol), DCC (0.63 g, 3 mmol), and HOBt (44 mg, 0.28 mmol) in DMF (10 ml) for 20 min.
2. Add the solution prepared in step 1 to glycerol-acetal resin (1 g, prepared by *Protocol 3*) in DMF (5 ml) and stir for 18 h.
3. Wash resin with dioxane, DMF, DCM, and diethyl ether (50 ml).

4. Dry resin *in vacuo*.

5. Analyse the resin by microanalysis (typically 0.71% of nitrogen, corresponding to a substitution of 0.5 mmol/g) or by Fmoc-substitution determination (see Chapter 3).

---

Method 6 (*Figure 5*) developed by Zhang *et al.* (32) is based on Fmoc-chemistry synthesis of a C-terminal diol peptide on a diol-derivatized 2-chlorotrityl resin. Oxidation of the free diol peptide gives the peptide aldehyde. A detailed procedure for the preparation of a diol trityl resin and C-terminal peptide aldehyde is given in the following protocols.

---

**Protocol 5.** Preparation of *N*-Fmoc-3-amino-1,2-propanediol derivatized 2-chlorotrityl polystyrene

*Reagents*

• Fluorenylmethyl chloroformate
• Triethylamine
• DMAP
• *t*-Butyl-2,2,2-trichloroacetimidate

• 3-Amino-1,2-propanediol
• 2-Chlorotrityl chloride polystyrene resin
• *N,N*-Dimethylacetamide

*Method*

1. Add fluorenylmethyl chloroformate (5.2 g, 20 mmol) to a solution of 3-amino-1,2-propanediol (1.8 g, 20 mmol) and triethylamine (3.1 ml, 22 mmol) in 100 ml of water/1,4-dioxane (1:3 v/v) at 4°C.

2. Stir for 18 h.

3. Remove solvent by evaporation *in vacuo*.

4. Add ethyl acetate (10 ml) to the residue and crystallize to give 5.6 g (89% yield) of 3-(9-fluorenylmethoxycarbonyl)-amino-1,2-propanediol. m.p. 133–134°C. MS: calc. MH$^+$ 314.

5. Add 3-(9-fluorenylmethoxycarbonyl)-amino-1,2-propanediol (1.25 g, 4 mmol) to 2-chlorotrityl chloride resin (1 g, 1 mmol) and DMAP (0.5 g, 2 mmol) in dry *N,N*-dimethylacetamide (10 ml).

6. Gently agitate the mixture for 24 h.

7. Wash resin with DMF, methanol, and DCM (30 ml each).

8. Dry resin *in vacuo*.

9. *Analysis of resin*: take 5–10 mg resin for determination of resin Fmoc-substitution (0.8 mmol/g) .

10. Add *t*-butyl-2,2,2-trichloroacetimidate (200 μl, 1.07 mmol) in DCM (25 ml) and shake for 30 min to block the free hydroxy group of the derivatized 2-chlorotrityl polystyrene.

---

**Protocol 6.** Preparation of C-terminal peptide aldehyde, AAVALLPAVLLALLA-NHCH$_2$CHO

*Reagents*

- C-terminal 1,2-propanediol peptide: AAVALLPAVLLALLA-NHCH$_2$CH(OH)CH$_2$OH, synthesized by Fmoc chemistry on Fmoc-3-amino-1,2-diol derivatized 2-chlorotrityl resin (prepared by *Protocol 5*)

- Sodium periodate

*Method*

1. Add NaIO$_4$ (20 mM, 0.6 ml in water) solution to the peptide AAVALLPAVLLALLA-NHCH$_2$CH(OH)CH$_2$OH (30 mg, 0.02 mmol) solution in 33% acetic acid(aq) (6 ml).

2. Run RP–HPLC to determine the reaction after 15 min.

3. Purify the reaction mixture by using RP–HPLC.

---

The final method (*Figure 5*, method 7) is an on-resin Pd$^{(0)}$-catalysed reduction of thioesters. Treatment of N-protected peptide thioester resin with Pd$^{(0)}$ in the presence of Et$_3$SiH gives the cleaved C-terminal peptide aldehyde. The reaction is performed typically under N$_2$ in THF or DCM using a 20–30-fold excess of Et$_3$SiH at 4°C to avoid racemization. The catalyst is obtained by *in situ* reduction of Pd(OAc)$_2$. The yield ranges from 80 to 89% (33).

## 3.2 Thiol chemistry

### 3.2.1 Thioester ligation for Cys and Met

Thioester ligation involves two segments, each containing a thioester and a Cys that results in the production of a cysteine residue at the ligation site. In this approach, the first step is formation of a covalent thioester-linked intermediate by nucleophilic attack of the thiol side chain of an $N^\alpha$-cysteine peptide on a thioester segment. An intramolecular $S \rightarrow N$ acyl transfer to form an amide bond then completes the ligation reaction (*Figure 6*).

There are two approaches in thioester ligation. Our laboratory (4) uses a pre-derivatized thioester resin that permits the generation of a thioester directly, following solid phase synthesis. It should be noted that the required peptidyl thioester segment is best prepared using Merrifield Boc-chemistry. The ligation reaction is performed under strongly reductive conditions and an excess of the thiol-component (cysteinyl-peptide) is used. Based on the rates, yields, and product distributions of the thiol–thioester exchange reaction at various pHs ranging from 5.6 to 7.6, an increase of 1.5- to 2-fold product purity with each pH unit is observed. Three groups of by-products are observed: $N^\alpha,S$-diacyl peptides; $N^\alpha$-acyl, disulphide conjugate peptides; and

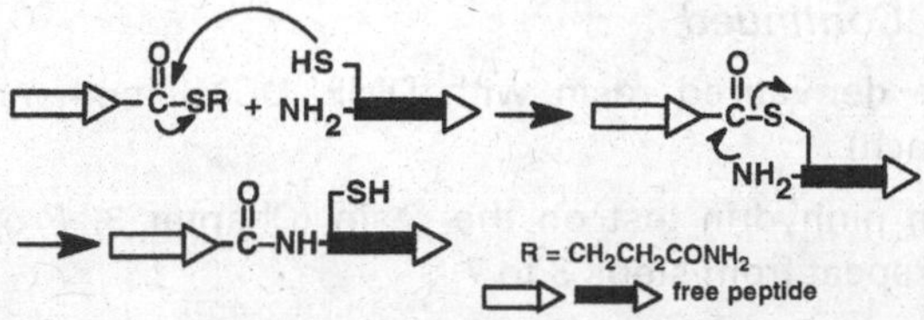

**Figure 6.** Thiol–thioester exchange to form an amide bond. Nucleophilic attack takes place by the side chain thiol functionality in a cysteinyl peptide at a thioester segment to form an intramolecular thioester, which undergoes intramolecular $S \rightarrow N$ acyl transfer to form an amide bond.

the hydrolysed peptidyl thioester. At pH 5.6, the desired ligation reaction is slowest and hydrolysis of the peptidyl thioester greatest (78%). At pH 7.6, hydrolysis occurs at a slow rate, to yield 56% of the desired ligation product.

Collectively, these results support the use of basic pH to inhibit hydrolysis of the peptidyl thioester, and the strong reducing environment of a small thiol and trialkylphosphine to prevent $N^{\alpha}$,$S$-diacyl by-products and to reduce disulphide formation. Using these optimized conditions, peptides ranging from 9 to 88 amino acids with 60–88% yield were prepared. The methods for preparation of the thioester resin and ligation are described below.

The second approach is reported by Kent *et al.* (34) and differs only by the use of a peptidyl thiocarboxylic acid as precursor and the peptidyl thioester is obtained via thioalkylation or disulphide exchange. This method requires an extra step but the ligation mechanism is similar to that described above.

---

**Protocol 7.** Preparation of Boc-Gly-thioester resin

*Reagents*
- 4-Methylbenzhydrylamine polystyrene resin (MBHA resin)
- 3-Mercaptopropionic acid
- N,N'-Diisopropylcarbodiimide (DIC)
- Triphenylphosphine
- Boc-Gly-OH
- HOBt
- Acetyl chloride
- Cysteine methyl ester·hydrochloride
- DIPEA

*Method*
1. Swell MBHA resin (10 g, 5.4 mmol) in DCM (200 ml) for 10 min, in an appropriate SPPS reaction vessel.
2. Filter the resin.
3. Add a solution of 3-mercaptopropionic acid (3.44 g, 32.4 mmol) in 70 ml DMF to resin, and shake for 3 min.
4. Add a solution of HOBt (4.38 g, 32.4 mmol) in 70 ml DMF to reaction vessel, and mix for 3 min.
5. Add a solution of DIC (5.07 ml, 32.4 mmol) in 60 ml DCM, and mix gently for 120 min.

**Protocol 7.** *Continued*

6. Wash the derivatized resin with DMF, DCM, methanol, and DCM (120 ml each).

7. Perform a ninhydrin test on the resin (Chapter 3, *Protocol 13A*). If positive, repeat from steps 3 to 7.

8. Add a solution of triphenylphosphine (0.65 g, 2.5 mmol), cysteine methyl ester·hydrochloride (0.43 g, 2.5 mmol), and DIPEA (0.43 ml, 2.5 mmol) in 200 ml of DCM. Mix for 60 min.

9. Wash resin with DMF, DCM, methanol, and DCM (120 ml each).

10. Add a solution of Boc-Gly-OH (4.73 g, 27 mmol) in 70 ml DMF to the resin.

11. Add a solution of HOBt (4.38 g, 32.4 mmol) in 70 ml DMF to the reaction vessel, and mix for 3 min.

12. Add a solution of DIPCD (5.07 ml, 32.4 mmol) in 60 ml DCM, and mix for 120 min.

13. Wash resin with DMF, DCM, methanol, DCM, and methanol (120 ml each).

14. Dry resin *in vacuo*.

15. Add a solution of acetyl chloride (0.18 ml, 2.5 mmol) in 100 ml DCM followed by a solution of DIPEA (0.43 ml, 2.5 mmol) in 50 ml DCM, and mix for 30 min.

16. Wash resin with DMF, DCM, methanol, and DCM (150 ml each).

17. Dry resin *in vacuo*.

18. Treat 10–15 mg resin with TFA/DCM(50% v/v) and perform a quantitative ninhydrin test (substitution levels typically 0.3–0.4 mmol/g).

---

**Protocol 8.** Thioester ligation

Reaction: TQFCFH-SCH$_2$CH$_2$CONH$_2$ + CKVLTTGLPALISW

*Reagents*

- Sodium phosphate buffer (0.02 M Na$_2$HPO$_4$– 0.1 M citric acid, pH 7.6)
- Tris(2-carboxyethyl)phosphine (TCEP)
- 3-Mercaptopropionic acid

- Peptidyl thioester TQFCFH-SCH$_2$CH$_2$CONH$_2$ (synthesized by Boc-chemistry on thio-ester resin) and cysteinyl peptide CKVLTTG-LPALISW (synthesized by Fmoc-chemistry)

*Method*

1. Add a solution of TCEP (344 µg, 1.2 µmol ) in buffer solution (600 µl) to the cysteinyl peptide (450 µg, 0.3 µmol).

**2.** Add the cysteinyl peptide solution to the peptidyl thioester (94 μg, 0.1 μmol).

**3.** Run analytical RP–HPLC at set time-intervals, e.g. 0 min, 30 min, 1 h, and 3 h.

**4.** Purify the ligated peptide TQFCFHCKVLTTGLPALISW by RP–HPLC.

Thioesters can also be used for intramolecular ligation for the synthesis of end-to-end cyclic peptides (35, 36). The linear peptide precursor should contain both an N-terminal cysteine residue and a C-terminal thioester. Under previously described conditions, intramolecular thiol-exchange produces a thiolactone, which then undergoes S → N migration to yield an end-to-end cyclic peptide.

Thioester ligation to yield a methionine residue can be achieved between a C-terminal thioester peptide and an N-terminal homocysteinyl peptide (37). The ligation reaction results in the formation of a homocysteine residue, which is readily S-methylated with excess methyl *p*-nitrobenzenesulphonate, to yield a methionine residue at the ligation site, or another alkylating reagent to form unusual amino acid residues.

### 3.2.2 Perthioester ligation for Cys

Similar to thioester ligation, perthioester ligation produces Cys at the ligation site (*Figure 7*). Two segments containing, respectively, a thiocarboxylic acid and a cysteinyl derivative are involved. However, in this case, the intramolecular acyl transfer reaction involves an acyl disulphide functionality (38). The method features specific capture of a peptide segment containing the N-terminal thiol-activated Cys residue by the $C^{\alpha}$-thiocarboxylic acid of the second segment. The mixed acyl disulphide perthioester then undergoes a rapid intramolecular $S \rightarrow N$ acyl transfer via a six-membered ring intermediate. Thiolytic reduction of the resulting hydrodisulphide (S–SH) gives the native Cys residue at the ligation site.

This approach has been used for the synthesis of a 32-residue peptide (38) from two purified segments, specifically the acyl segment (15 residues) containing a thiocarboxylic acid and the amino segment containing an $N^{\alpha}$-Cys(Npys). The reaction was carried out at pH 2 in aqueous acetonitrile, and the segment concentration was 10–15 μM. During the course of the ligation reaction, release of a yellow coloration (due to released Npys-H) was observed, indicating the occurrence of the capture reaction by sulphur–sulphur exchange. Finally, the pH was adjusted to 6, and the reaction mixture treated with 1,4-dithiothreitol (DTT) to afford the desired product.

The high efficiency of the first capture step is attributed to the Npys-activated sulphide, and to the super-nucleophilicity and low $pK_a$ of the thiocarboxylic acid compared to a normal alkyl thiol. The acyl transfer was

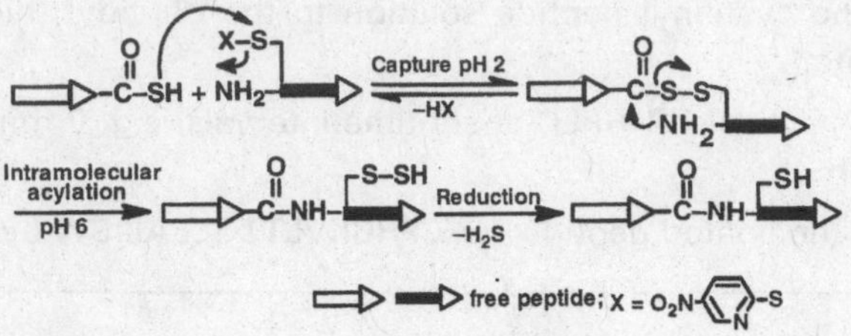

**Figure 7.** General scheme of perthioester ligation. The capture of an activated thiol side chain of an N-terminal Cys residue of the amino segment by the $C^\alpha$-thiocarboxylic acid of the acyl segment results in the formation a mixed acyl disulphide which undergoes a rapid intramolecular $S \rightarrow N$ acyl transfer through a six-membered ring intermediate. Thiolytic reduction of the resulting hydrodisulphide (S–SH) using DTT gives the native Cys residue at the ligation site.

rapid and 90% complete within 5 min, even at pH < 4. The efficiency of this acyl transfer step is attributed to the activated acyl disulphide and to the close proximity of the $C^\alpha$-acyl and $N^\alpha$-amine in the six-membered ring intermediate. This method of orthogonal segment ligation followed by disulphide reduction is a one-pot process, which does not require the isolation of intermediates. Its high efficiency and simplicity make this strategy an attractive approach for the synthesis of large peptides and proteins.

### 3.2.3 Imidazole ligation for His

The imidazole group in the histidine side chain is a well-known weak base and can participate in acyl transfer reactions in a number of enzymatic process. Thus, the NTN of His has also been exploited for orthogonal ligation to give His at the ligation site. In this strategy, one segment contains the $N^\alpha$-histidine and the second acyl segment contains a C-terminal thiocarboxylic acid or thioester. Orthogonal ligation occurs to yield an Xaa–His bond when a suitable thiophilic promoter, such as an aryl disulphide or $Ag^+$ ion, is added to activate the acyl segment. This leads to $N^{im} \rightarrow N^\alpha$ acyl transfer, resulting in a peptide bond formation (39). This method has been used for the ligation of histidine-containing peptides such as human calcitonin and parathyroid hormone. A drawback in this strategy is that the reactive acyl imidazole intermediate is particularly prone to hydrolysis.

## References

1. Brenner, M., Zimmerman, J. P., Wehrmüller, J., Quitt, P., and Photaki, I. (1955). *Experientia*, **11**, 397.
2. Kemp, D. S. (1981). *Biopolymers*, **20**, 1793.
3. Fotouhi, N., Galakatos, N., and Kemp, D.S. (1989). *J. Org. Chem.*, **54**, 2803.

4. Tam, J. P., Lu, Y.-A., Liu, C. F., and Shao, J. (1995). *J. Proc. Natl. Acad. Sci. USA*, **92**, 12485.
5. Liu, C. F. and Tam, J. P. (1994). *Proc. Natl. Acad. Sci. USA*, **91**, 6584.
6. Schnölzer, M. and Kent, S. B. H. (1992). *Science*, **256**, 221.
7. Lu, Y.-A., Clavijo, P., Galantino, M., Shen, Z.-Y., Liu, W., and Tam, J. P. (1991). *Molecular Immunology*, **28**, 623.
8. Defoort, J. P., Nardelli, B., Huang, W., Ho, D. D., and Tam, J. P. (1992). *Proc. Natl. Acad. Sci. USA*, **89**, 3879.
9. Muir, T. W., Williams, M. J., Ginsberg, M. H., and Kent, S. B. H. (1994). *Biochemistry*, **33**, 7701.
10. Defoort, J. P., Nardelli, B., Huang, W., and Tam, J. P. (1992). *Int. J. Peptide Protein Res.*, **40**, 214.
11. Shao, J. and Tam, J. P. (1995). *J. Am. Chem. Soc.*, **117**, 3893.
12. Spetzler, J. C. and Tam, J. P. (1995). *Int. J. Pept. Protein Res.*, **45**, 78.
13. King, T. P., Li, Y., and Kochoumian, L. (1978). *Biochemistry*, **171**, 499.
14. Kitagawa, T. and Aikawa, T. (1976). *J. Biochem. (Tokyo)*, **79**, 233.
15. Erlanger, B. F., Borek, F., Beiser, S. M., and Lieberman, S. J. (1957). *Biol. Chem.*, **288**, 713.
16. Pochon, S., Buchegger, F., Pelegrin, A., Mach, J.-P., Offord, R. E., Ryser, J. E., and Rose, K. (1989). *Int. J. Cancer*, **43**, 1188.
17. Rose, K. (1994). *J. Am. Chem. Soc.*, **116**, 30.
18. Tuchscherer, G. (1993). *Tetrahedron Lett.*, **34**, 8419.
19. Pallin, T. D. and Tam, J. P. (1995). *J. Chem. Soc., Chem. Commun.*, 2021.
20. Bergbreiter, D. E. and Momongan, M. (1991). In *Comprehensive organic synthesis and efficiency in modern organic chemistry* (ed. B. M. Trost and J. Flemming), Vol 2, p. 503, Pergamon, New York.
21. Gaertner, H. F., Rose, K., Cotton, R., Timms, D., Camble, R., and Offord, R. E. (1992). *Bioconjugate Chem.*, **3**, 262.
22. Gaertner, H.F., Offord, R. E., Cotton, R., Timms, D., Camble, R., and Rose, K. (1994). *J. Biol. Chem.*, **269**, 7224.
23. Fisch, I., Kunzi, G., Rose, K., and Offord, R. E. (1992). *Bioconjugate Chem.*, **3**, 147.
24. Rao, C. and Tam, J. P. (1994). *J. Am. Chem. Soc.*, **116**, 6975.
25. Tam, J. P., Rao, C., Liu, C. F., and Shao, J. (1995). *Int. J. Pept. Protein Res.*, **45**, 209.
26. Liu, C. F. and Tam, J. P. (1994). *J. Am. Chem. Soc.*, **116**, 4149.
27. Liu, C. F., Rao, C., and Tam, J. P. (1996). *J. Am. Chem. Soc.*, **118**, 307.
28. Clamp, R. and Hough, L. (1965). *Biochem. J.*, **94**, 17.
29. Dixon, H. B. F. and Fields, R. (1972). In *Methods in enzymology* (ed. C. H. W. Hirs and S. N. Timasheff), Vol. 25, p. 409, Academic Press, New York.
30. Barbas III, C. F., Matos, J. R., West, J. B., and Wong, C. H. (1988). *J. Am. Chem. Soc.*, **110**, 5162.
31. Botti, P., Pallin, T. D., and Tam, J. P. (1996). *J. Am. Chem. Soc.*, **118**, 10018.
32. Zhang, L., Torgerson, T. R., Liu, X. Y., Timmons, S., Colosia, A. D., Hawiger, J., and Tam, J. P. (1998). *Proc. Natl. Acad. Sci. USA*, **95**, 9184.
33. Wyolouoh Cioozynoka, A. and Tam, J. P. (1999). *Frontiers of peptides science: Proceedings of the 15th APS*, (ed. J. P. Tam and P. T. P. Kaumaya), p. 263. Kluwer Academic Publishers, Dordrecht.
34. Dawson, P. E., Muir, T. W., Clark-Lewis, I., and Kent, S. B. H. (1994). *Science*, **266**, 776.

35. Zhang, L. and Tam, J. P. (1997). *J. Am. Chem. Soc.*, **119**, 2363.
36. Tam, J. P. and Lu, Y.-A. (1997). *Tetrahedron Lett.*, **38**, 5599.
37. Tam, J. P. and Yu, Q. (1998). *Biopolymer*, **46**, 319.
38. Liu, C. F., Rao, C., and Tam, J. P. (1996). *Tetrahedron Lett.*, **37**, 933.
39. Zhang, L. and Tam, J. P. (1997). *Tetrahedron Lett.*, **38**, 3.

# 12

# Purification of large peptides using chemoselective tags

PAOLO MASCAGNI

## 1. Introduction

In solid phase peptide synthesis (SPPS), deletion sequences are generated at each addition of amino acid due to non-quantitative coupling reactions. Their concentration increases exponentially with the length of the peptide chain, and after many cycles not only do they represent a large proportion of the crude preparation, but they can also exhibit physicochemical characteristics similar to the target sequence. Thus, these deletion-sequence contaminants present major problems for removal, or even detection.

In general, purification of synthetic peptides by conventional chromatography is based on hydrophobicity differences (using RP–HPLC) and charge differences (using ion-exchange chromatography). For short sequences, the use of one or both techniques is in general sufficient to obtain a product with high purity. However, on increasing the number of amino acid residues, the peptide secondary and progressively tertiary and quaternary structures begin to play an important role and the conformation of the largest peptides can decisively affect their retention behaviour. Furthermore, very closely related impurities such as deletion sequences lacking one or few residues can be chromatographically indistinguishable from the target sequence. Therefore, purification of large synthetic peptides is a complex and time-consuming task that requires the use of several separation techniques with the inevitable dramatic reduction in yields of the final material.

Permanent termination (capping) of unreacted chains using a large excess of an acylating agent after each coupling step prevents the formation of deletion sequences and generates N-truncated peptides. However, even under these more favourable conditions, separation of the target sequence from chromatographically similar N-capped polypeptides requires extensive purification.

If the target sequence could be specifically and transiently labelled so that the resulting product were selectively recognized by a specific stationary phase, then separation from impurities should be facilitated. This chapter deals with such an approach and in particular with the purification of large polypeptides, assembled by solid phase strategy, using lipophilic and biotin-based 9-fluorenylmethoxycarbonyl (Fmoc) chromatographic probes.

# 2. Purification of large polypeptides using Fmoc-based chromatographic probes

## 2.1 The concept of selective and reversible labelling

Assuming that the formation of deletion sequences is prevented by capping unreacted chains (see *Protocol 1*), a reciprocal strategy can be applied that involves functional protection of all polymer-supported peptide chains that are still growing, with a specially chosen affinity reagent or chromatographic probe. In this way, the desired chains are distinguished from all by-products that have been irreversibly terminated either by capping or as a result of chemical side reactions. The chromatographic probe contains an affinity group and a reversible linker (ref. 1 and references cited therein), which are stable under the conditions used to deprotect and cleave the peptide from the resin support. Affinity-labelled peptides are separated from terminated peptides which lack the affinity group by selective binding to a suitable stationary phase. The affinity label is then selectively removed using reagents which do not harm the peptide integrity.

The basic steps of this strategy are shown in *Scheme 1*. It should be emphasized that impurities generated by side reactions occurring either during synthesis or during deprotection and that affect both the target sequence and capped peptides are not removed by this technique.

The first example of purification by selective labelling was described in 1968 for a peptide that was synthesized in solution (2). In 1976, Merrifield and co-workers (3) applied this concept to SPPS and demonstrated that peptides bearing an intentionally introduced Cys–Met N-terminal dipeptide can be purified by absorption onto an organometallic immobilized matrix. The native sequence lacking Cys–Met was recovered through a CNBr-mediated Met–X specific cleavage.

Following these pioneering studies, several chromatographic probes have been proposed as a result of intentional design or modification of other procedures (for a review on the subject see refs 1 and 4).

### 2.1.1 Fmoc-based chromatographic probes

The Fmoc group is ideally suited as a template molecule to develop chromatographic probes, because its urethane link with the amino acid $N^\alpha$ is stable to the harsh, acidic conditions used to cleave peptides from the resin support, but it is labile under mild basic conditions that do not harm the polypeptide chain (5). Furthermore, because the probe is added to the peptide chain at the end of synthesis, Fmoc-probes can aid the purification of peptides synthesized

# 12: Purification of large peptides using chemioselective tags

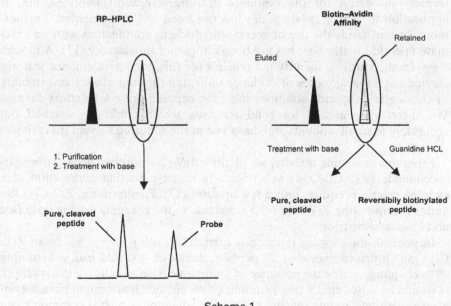

**Scheme 1**

using either of the two most widely used chemistry, i.e. Fmoc-chemistry and *t*-butoxycarbonyl (Boc)-chemistry.

The first Fmoc-based chromatographic probe was described in 1978 by Merrifield (6) and contained negative charges provided by a sulpho group at position 2 of the fluorenyl moiety. The strongly acidic peptides thus derivatized were separated from unlabelled peptides by ion-exchange chromatography.

Another Fmoc-based probe, recently developed, contains benzo-fused analogues of the Fmoc molecule. It can be used to purify peptides by either porous graphitized carbon chromatography (7) or RP–HPLC (8).

In 1989, we proposed the 4-$CO_2$H-Fmoc derivative **1** (9) as a probe for the purification of large synthetic peptides (10). The advantage of **1** over other Fmoc-based molecules is in its carboxylic function which can be derivatized with groups possessing different chromatographic characteristics. For instance, addition of a lipophilic tail to **1** will generate a probe for RP–HPLC, whereas a polyionic tail or biotin will convert **1** in a probe for ion-exchange chromatography and affinity chromatography, respectively (9–14).

The usefulness of lipophilic probes in RP–HPLC purification was demonstrated with the synthesis of several proteins made by either an Fmoc strategy or the alternative Boc chemistry (9–16).

## 3. Capping of unreacted polypeptide chains in SPPS

The classical capping reagent, acetic anhydride, has been shown to be extremely successful for the synthesis of relatively short peptides, but its application to longer polypeptides has not been well documented. Furthermore, in our hands the use of acetic anhydride in conjunction with an automated peptide synthesizer has not been altogether satisfactory (17). Although a systematic study to identify the reasons for this poor performance was not carried out, several pieces of evidence indicated that the quality and stability of the acetic anhydride solutions used for capping are contributory factors. We therefore searched for solid reagents which could be weighed out accurately in small amounts and dissolved in the required solvent just prior to their use.

From this screening activity, we identified *N*-(2-chlorobenzyloxycarbonyl)-succinimide (Z(2-Cl)-OSu) as sufficiently reactive so that completion of a capping reaction requires only a few minutes (17). Furthermore, Z(2-Cl)-OSu and the terminating Z(2-Cl) group are stable to the reagents used in both Boc and Fmoc chemistries.

In general, the capping reaction is carried out using a large excess of Z(2-Cl)-OSu (fourfold excess of all peptide chains or 400-fold excess assuming 99% coupling) and in the presence of an organic base to catalyse the reaction. It should be noted that a small amount of unreacted chains might persist even after repetitive capping reactions. This phenomenon, which is common to all capping reagents, is probably due to inaccessibility to solvent of those peptide chains buried in the bulk of the resin–peptide matrix.

---

**Protocol 1.**   Capping of unreacted polypeptide chains

*Reagents*

- 0.284 g (4-fold excess[a]) Z(2Cl)-OSu in 2 ml of *N*-methylpyrrolidone (NMP)–dichloromethane (DCM) (1:1 v/v) plus 1 ml of 2 M

*N,N*-diisopropylethylamine (DIPEA) in NMP. If not available, NMP can be substituted with dimethylformamide (DMF)

*Method*

1. Prepare a fresh solution of Z(2Cl)-OSu.

2. At the end of the coupling reaction, condition the peptidyl-resin with a mixture of NMP/DCM (1:1, v/v) in preparation for the capping step. Remove the solvent but avoid drying the peptidyl-resin completely as it should be kept moist.

3. Add the Z(2Cl)-OSu solution to the peptidyl-resin and vortex for 5 min.[b] The reaction can be followed by the standard ninhydrin test or other similar colorimetric methods.

4. Remove the reaction solution by filtration and wash the product thoroughly with NMP/DCM (1:1, v/v). The peptidyl-resin is now ready for the next coupling cycle.[c]

[a] The quantities given in the example refer to a concentration of amino groups present on the resin of 0.25 meq.
[b] If other methods of mixing are used, the reaction may take slightly longer.
[c] The capping cycle can be carried out in an automated fashion because it requires the same solvents and reagents used for chain assembly.

---

# 4.  RP–HPLC using lipophilic chromatographic probes

In a comparative study, several hydrophobic probes differing in the type of substitution at position 4 of **1** were prepared (12). Parameters evaluated included stability of the probe during coupling and acid deprotection, RP–HPLC separation between derivatized and underivatized molecules, and time required for probe removal. Probe **2** offered the best compromise between all these parameters; it is easily converted to an activated form (e.g. **2a**) by using the same chemistry as for activation of the Fmoc molecule. If the reaction between **2a** and the peptidyl-resin is carried out immediately after completion of chain assembly, the base-catalysed reaction requires 2–3 h. Longer reaction times might be necessary if derivatization is not carried out on freshly prepared polypeptides. In this case it is advisable to store the peptidyl-resin in an inert solvent (NMP or DMF) at low temperature prior to use. The degree of coupling should be monitored using standard colorimetric methods (e.g. the ninhydrin reaction) and the reaction repeated with a freshly prepared solution of activated probe if coupling is not satisfactory.

The probe and the probe–peptide link are stable to acids and in the presence of the most widely used scavengers. However, both scavengers and acids must be of high quality if side reactions at the Fmoc group and the peptide chain are to be avoided. In general, the solubility profile of large polypeptides is not significantly altered by the presence of lipophilic probe **2**.

The extent of HPLC separation between derivatized and underivatized peptides depends on the length of the polypeptide chain, the type of stationary phase used for separation (e.g. C4–C18), and the elution conditions. When using C18 columns the difference in retention time between the two species varies from about 10 min for peptides of 40–60 residues to about 5 min in the case of larger sequences. A typical elution profile for a 100-residue polypeptide is shown in *Figure 1*. As the length of the chain increases, a progressive reduction in the separation between labelled and unlabelled chains is observed. Unpublished results showed that beyond about 150 residues, the retention times of a probe–peptide adduct and underivatized chains are very similar. However, for sequences up to 120–130 residues, the method is sequence independent and can therefore be standardized.

After semi-preparative RP–HPLC, homogeneous fractions are pooled, lyophilized, and redissolved in an aqueous solvent. In the case of sparingly soluble polypeptides, denaturing solutions (e.g. 6 M guanidinium·HCl) can be used which do not affect the subsequent reaction of probe removal. This is carried out by the addition of an organic base (e.g. NEt₃, see *Protocol 4*) which catalyses the β-elimination typical for the Fmoc protection giving the corresponding dibenzofulvene derivative and free polypeptide (*Scheme 1*). A final semi-preparative RP–HPLC separates the latter from the probe by-product.

The homogeneity and yields of the final preparation depend strongly on the length of the polypeptide chain and on the conditions used for both synthesis and cleavage from the resin support. Under optimized conditions,

**Figure 1.** Purification of rat Cpn10 (101 residues) using Fmoc probe **2**. **(A)** Analytical RP–HPLC (C4 medium) of crude underivatized rat Cpn10. **(B)** Addition of lipophilic probe **2** increases the retention time of the protein (labelled 2) thus facilitating purification from underivatized truncated sequences (labelled 1). **(C)** Purified protein derivatized with **2**. **(D)** Purified protein after treatment with 5% aqueous TEA to remove **2**. **(E)** ESI–MS of purified rat Cpn10. **(F)** Deconvoluted mass spectrum for purified rat Cpn10. The calculated mass of target product is 10770.57 Da (average). The found mass is 10771.0. **(G)** RP–HPLC (C4 medium, gradient TFA–water into 100% TFA–AcCN, 60 min) of purified rat Cpn10. The insert shows the expanded peak. **(H)** CZE of purified rat Cpn10. The concentration of the major peak (protein in its native, heptameric state) is 84%. Separate size-exclusion chromatography experiments showed that the majority of the flanking peaks correspond to protein with correct sequence but having an aggregation state different from the major peak.

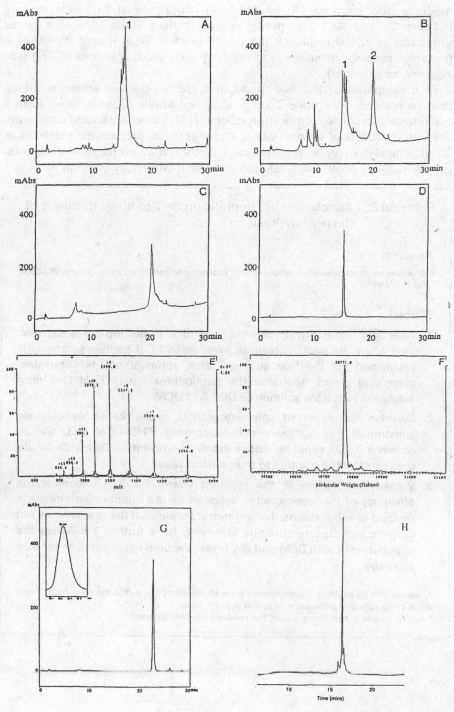

medium-sized peptides (40–60 residues) are obtained in yields ranging between 20 and 40% and purity is greater than about 95%. For larger sequences (e.g. 100 residues), the purity is 85–95% as judged by chromatography, mass spectrometry, and capillary zone electrophoresis (CZE) (see *Figure 1* for example).

In a comparative study we found that yields of a 100 amino acid-long protein purified with probe **2** are 5–15% and about 10-times those from a conventional multistep purification scheme (13). Under optimized conditions, the entire process of derivatization, cleavage, semi-preparative purification, fraction analysis, probe removal, and further semi-preparative chromatography requires about one week using standard laboratory equipment.

---

**Protocol 2.**   Attachment of lipophilic probe 2 to the N-terminus of the peptidyl-resin

*Reagents*

- 4-Dodecylaminocarbonyl-fluoren-9-ylmethyl succinimidyl carbonate **2a** (enquire with author for availability)

*Method*

1. After chain assembly, remove the N-terminal protecting group as usual, then allow the peptidyl-resin to swell in DCM. If peptide synthesis is performed via the Boc strategy, after removal of the N-terminal protecting group, neutralize the peptidyl-resin with brief (1–2 min) washings with a 5% solution of DIPEA in DCM.

2. Dissolve the activated chromatographic probe **2a** (4 eq of resin substitution[a]) in sufficient trifluoroethanol (TFE)–DCM (1:3, v/v) to obtain a 0.2 M solution. Add a catalytic amount of DIPEA (5–10 μl), then transfer the solution to the peptidyl-resin.

3. Vortex the mixture in the reaction vessel for 2 h.[b] Monitor the efficiency of the incorporation reaction by the quantitative ninhydrin method or other standard colorimetric methods. If the reaction has not gone to completion, continue vortexing for a further 2 h. Wash the peptidyl-resin with DCM and dry under vacuum in preparation for acid cleavage.

[a] Assume that all coupling reactions have gone to completion and that the resin substitution value at the *n*th cycle is the same as that at the first cycle.
[b] If other means of agitation are used, the reaction may take longer.

---

## Protocol 3.  Semi-preparative RP–HPLC

### Equipment and reagents

- HPLC apparatus with UV detector
- RP–HPLC semi-preparative column: stationary phase, C18, dimensions, 10 × 250 mm
- Solvent A: $H_2O$–2% AcCN–0.1% TFA
- Solvent B: AcCN–0.08% TFA

### Method

1. Connect the semi-preparative C18 column to the HPLC apparatus and condition it with solvent A.

2. Dissolve crude derivatized polypeptide (50–70 mg) in sufficient solvent A to obtain a clear solution. Submit an aliquot to analytical RP–HPLC and apply the bulk of it to the top of the C18 column.

3. Equilibrate with solvent A for 15 min and then apply the gradient.[a] If monitoring is at 214 nm (peptide bond) two large envelopes will be detected.[b] Monitoring at 300 nm will show only the second, more retained envelope, which contains the probe-derivatized polypeptide.

4. Collect 2–3 ml fractions throughout the second envelope[c] and submit them to analytical RP–HPLC.

5. Combine homogeneous fractions and lyophilize.

[a] Elution conditions: gradient 0–32% B in 91 min, 32–50% B in 90 min, 50–80% B in 30 min; flow rate 2.5 ml/min.
[b] Under these conditions most large polypeptide impurities elute as a broad envelope after about 90–100 min and are followed by another envelope containing the polypeptide–probe adduct.
[c] If separation of the two envelopes is not seen because of high peptide loading, fractions should be collected throughout the combined envelopes.

## Protocol 4.  Probe removal

### Equipment and reagents

- HPLC instrument and C18 RP semi-preparative column (10 × 250 mm)
- Triethylamine (TEA)

### Method

1. Dissolve purified probe–polypeptide adduct in either $H_2O$ or a suitable buffer[a] to a final concentration of approximately 1 mg/ml.

2. While stirring, add small aliquots of TEA to a final concentration of 5% v/v. Stir the solution for 30 min[b] at RT.

3. Quench the reaction by adding acid (e.g. TFA) to pH 2–4.

**Protocol 4.** *Continued*

**4.** Apply the solution to a C18 column and repeat the same semi-preparative purification as described in *Protocol 3.*[c]

[a] In the case of poorly soluble polypeptides, denaturing solutions such as 6 M guanidinium·HCl can be used.
[b] Progression of the reaction can be monitored by HPLC using a short RP column which allows for rapid analysis.
[c] Owing to its highly hydrophobic character the probe released from the polypeptide chain is highly retained on the column.

# 5. Affinity chromatography using biotinylated chromatographic probes

The concept of selective labelling can be applied to biotinylated peptides and their purification by avidin/agarose affinity chromatography. This approach is as effective as that based on lipophilic probes and the choice of method depends on the subsequent applications of the purified polypeptide. Thus, for the biologist the presence of a biotinylated group on a protein at a specific location offers a number of interesting possibilities for bioassays, antibody generation, fluorescent microscopy, cytochemistry/immunochemistry, and indirect affinity chromatography.

The first attempts at purifying chemically synthesized proteins using avidin-based affinity chromatography involved covalent attachment of biotin to the N-terminal residue of the peptidyl-resin (18, 19). The method, although effective, gives peptides which are permanently derivatized with biotin. A reversible linker was described in 1993 (20). The probe, 2-[N-biotinyl amino-ethylsulphonyl]ethyl p-nitrophenyl carbonate, was used to purify peptides of different lengths, i.e. magainin-2, h-GRF, and transducin γ-subunit, although it was less effective for the longer sequences (20).

Our first Fmoc-based probe contained biotin attached directly to the fluorenyl ring. This, however, gave unsatisfactory binding to avidin/agarose presumably because of steric hindrance caused by the bulky Fmoc group. Subsequent probes therefore contained a spacer between the 4-CO group and biotin. In particular, by using molecule **3** highly homogeneous synthetic rat Cpn10 (101 residues) was obtained in 10% yield (14). Furthermore, the same avidin–biotinylated protein complex was used for the isolation of the natural ligand (e.g. tetradecameric Cpn60, MW ≈ 840 000) of rat Cpn10 from a crude cell lysate (21).

The procedure for coupling **3a** to the peptidyl-resin is similar to that for probe **2**. After acid cleavage, a solution of the crude mixture is applied directly to an avidin-based affinity column and elution with buffers removes all underivatized chains. The peptide-containing column can be treated with an organic base to release free peptide (*Protocol 5*) or used directly for binding studies or affinity purification.

**Protocol 5.** Purification of peptides using biotinylated-probe **3**

*Equipment and reagents*
- 4-[5-(Biotinylamino)pentylaminocarbonyl] fluoren-9-ylmethyl succinimidyl carbonate **3b** (enquire with author for availability)
- Immobilized monomeric avidin gel (Pierce)
- Disposable polypropylene columns (2.5 × 1 cm)
- Sodium acetate buffer (0.2 M, pH 4)

*Method*

1. At the end of polypeptide chain assembly, take enough peptidyl-resin so as to have about 3 μmol protected peptide[a] and remove the $N^\alpha$-protection from the terminal residue as usual. Wash the peptidyl-resin with DCM. If chain assembly is performed with Boc chemistry, neutralize (2 min) the freshly deprotected peptidyl-resin with 5% DIPEA in DCM and then wash with DCM.

2. Dissolve 20.7 mg (30 μmol) of 4-[5-(biotinylamino)pentylamino-carbonyl]fluoren-9-ylmethyl succinimidyl carbonate (**3b**) in 200 ml DCM/NMP (3:1, v/v) and add a catalytic amount (5–10 μl) of DIPEA.

3. Transfer the solution to the reaction vessel containing the N-deprotected peptidyl-resin and vortex at room temperature overnight.

4. Wash the peptidyl-resin with DCM-NMP (3:1, v/v) first, then with DCM only and dry.

5. Cleave the peptide from the resin as usual and lyophilize the crude peptide from the appropriate solvent.

6. Load immobilized avidin into the disposable column and equilibrate with the sodium acetate buffer for 1 h.

7. Dissolve crude lyophilized peptide in sufficient sodium acetate buffer to obtain a clear solution. Incubate the peptide solution with the immobilized avidin for 30 min with occasional mixing. Remove an aliquot of the supernatant and determine by RP–HPLC whether all the biotinylated-polypeptide has been bound. If underivatized peptide is still present in the solution continue the incubation for a further 30 min.

8. Wash the resin with sodium acetate buffer and then water. To remove the peptide from the biotinylated probe–avidin complex add a 5% solution of TEA in water to the resin and leave for 15–30 min with occasional mixing.

9. Collect the solution and neutralize it with 10% AcOH.

10. Desalt the purified protein either by RP–HPLC or dialysis.

[a] Work out the number of μmoles of peptide by using the resin substitution value and by assuming that all coupling reactions have gone to completion. In this way the resin substitution value at the *n*th cycle is the same as that at the first cycle.

# References

1. Barany, G. and Merrifield, F. B. (1980). In *Peptides, analysis, synthesis and biology* (ed. E. Gross and J. Meienhofer), Vol. 2, p. 163. Academic Press, London.
2. Camble, R., Garner, R., and Young G. T. (1968). *Nature*, **217**, 247.
3. Krieger, D. E., Erickson, B. W., and Merrifield, R. B. (1976). *Proc. Natl. Acad. Sci. USA*, **73**, 3160.
4. Mascagni, P., Ball, H. L., and Bertolini, G. (1997). *Anal. Chim. Acta*, **352**, 375.
5. Carpino, L. A. and Han, G. Y. (1973). *J. Am. Chem. Soc.*, **92**, 5748.
6. Merrifield, R. B. and Bach, A. E. (1978). *J. Org. Chem.,* **43**, 25.
7. Ramage, R. and Raphy, G. (1992). *Tetrahedron Lett.*, **33**, 385.
8. Brown, A. R., Irving, S. L., Ramage, R., and Raphy, G. (1995). *Tetrahedron*, **51**, 11815.
9. Mutter, M. and Bellof, D. (1984). *Helv. Chim. Acta*, **67**, 2009.
10. Ball, H. L., Grecian, C., Kent, S. B. H., and Mascagni, P. (1990). In *Peptides, chemistry, structure and biology* (ed. J. E. Rivier and G. R. Marshall), p. 435. Escom, Leiden.
11. Ball, H. L. and Mascagni, P. (1992). *Int. J. Peptide Protein Res*, **40**, 370.
12. Ball, H. L., Bertolini, G., Levi, S., and Mascagni, P. (1994). *J. Chromatogr.*, **686**, 73.
13. Ball, H. L. and Mascagni, P. (1996). *Int. J. Peptide Protein Res.*, **48**, 31.
14. Ball, H. L., Bertolini, G., and Mascagni, P. (1995). *J. Pept. Sci.,* **1**, 288.
15. Fossati, G., Lucietto, P., Giuliani, P., Coates, A. R. M., Harding, S., Colfen, H., Legname, G., Chan, E., Zaliani, A., and Mascagni, P. (1995). *J. Biol. Chem.*, **270**, 26159.
16. Ball, H. L., Chan, A. W. E., Gibbons, W. A., Coates, A. R. M., and Mascagni, P. (1996). *J. Pept. Sci.*, **3**, 168.
17. Ball, H. L. and Mascagni, P. (1995). *Lett. Pept. Sci.*, **2**, 49.
18. Lobl, T. J., Deibel Jr, M. R., and Yem, A. W. (1988). *Anal. Biochem.*, **170**, 502.
19. Tomasselli, A. G., Bannow, C. A., Deibel Jr, M. R., Hui, J. O., Zurcher-Neely, H. A., Reardon, I. M., Smith, C. W., and Heinrikson, R. (1992). *J. Biol. Chem.*, **267**, 10232.
20. Funakoshi, S., Fukuda, H., and Fujii, N. (1993). *J. Chromatogr.*, **638**, 21.
21. Ball, H. L. and Mascagni, P. (1996). *J. Pept. Sci.*, **3**, 252.

# 13

# Instrumentation for automated solid phase peptide synthesis

LINDA E. CAMMISH and STEVEN A. KATES

## 1. Introduction

The concept of solid phase peptide synthesis introduced by Merrifield in 1963 (1) involves elongating a peptide chain on a polymeric support via a two-step repetitive process: removal of the $N^\alpha$-protecting group and coupling of the next incoming amino acid. A second feature of the solid phase technique is that reagents are added in large excesses which can be removed by simple filtration and washing. Since these operations occur in a single reaction vessel, the entire process is amenable to automation. Essential requirements for a fully automatic synthesizer include a set of solvent and reagent reservoirs, as well as a suitable reaction vessel to contain the solid support and enable mixing with solvents and reagents. Additionally, a system is required for selection of specific solvents and reagents with accurate measurement for delivery to and removal from the reaction vessel, and a programmer to facilitate these automatic operations is necessary.

The current commercially available instruments offer a variety of features in terms of their scale (15 mg to 5 kg of resin), chemical compatibility with 9-fluorenylmethyloxycarbonyl/*tert*-butyl (Fmoc/tBu) and *tert*-butyloxycarbonyl/benzyl (Boc/Bzl)-based methods, software (reaction monitoring and feedback control), and flexibility (additional washing and multiple activation strategies). In addition, certain instruments are better suited for the synthesis of more complex peptides such as cyclic, phosphorylated, and glycosylated sequences while others possess the ability to assemble a large number of peptide sequences. The selection of an instrument is dependent on the requirements and demands of an individual laboratory. This chapter will describe the features of the currently available systems.

As the field of solid phase synthesis evolved, manufacturers designed systems based on the synergy between chemistry and engineering. A key component to an instrument is the handling of amino acids and their subsequent activation to couple to a polymeric support. The goal of an automated system is to duplicate conditions that provide stability to reactive species that might decompose. Standard protocols for automated synthesis incorporate carbodiimide, phosphonium, and aminium/uronium reagents, preformed active esters, and acid fluorides. For further details on coupling methods, see Chapter 3. A second issue related to coupling chemistry is the time required to dissolve an amino acid and store this solution.

From an automation perspective, preformed active esters and acid fluorides are the simplest reagents to incorporate into a system (for reviews see refs 2 and 3). Preferably, solid derivatives are placed in tubes and dissolved at point of use in *N,N*-dimethylformamide (DMF) or *N*-methylpyrrolidone (NMP). For pentafluorophenyl esters, residues are dissolved typically in a solution of 1-hydroxybenzotriazole (HOBt) in DMF at a concentration specific to an individual instrument. Preformed solutions of activated amino acid esters or fluorides (4) are not stable for long periods of time and are not recommended.

*N,N'*-dicyclohexylcarbodiimide (DCC) (5) and *N,N'*-diisopropylcarbodiimide (DIC) (6) in conjunction with an $N^\alpha$-protected amino acid are widely used coupling methods for Boc- and Fmoc-based methods, respectively, with the latter forming a more soluble urea by-product in DMF. To prevent many undesired side reactions and to improve coupling efficiencies with carbodiimide chemistry, HOBt (7) and, more recently, 1-hydroxy-7-azabenzotriazole (HOAt) (8) were introduced into the reaction mixture. Typical protocols for this activation strategy include: (i) a preformed solution of DCC or DIC adding to an $N^\alpha$-protected amino acid and HOBt or HOAt (referred to as HOXt), followed by dissolution and activation; (ii) a preformed solution of DCC or DIC and a preformed solution of HOXt adding to an $N^\alpha$-protected amino acid, followed by dissolution and activation; (iii) a preformed solution of DCC or DIC adding to a solution of an $N^\alpha$-protected amino acid for on-line dissolution and activation. Prior to delivering an activated amino acid to a peptidyl-resin, preactivation might occur in a vial that contained the solid amino acid residue or the activated solution can be transferred to a designated vessel.

Phosphonium benzotriazolyl *N*-oxytrisdimethylaminophosphonium hexafluorophosphate (BOP) (9), benzotriazol-1-yloxytris(pyrrolidino)phosphonium hexafluorophosphate (PyBOP) (10), and 7-azabenzotriazol-1-yloxytris(pyrrolidino)phosphonium hexafluorophosphate (PyAOP) (11), and aminium salts of *N*-[(1*H*-benzotriazol-1-yl)(dimethylamino)methylene]-*N*-methylmethanaminium hexafluorophosphate *N*-oxide (HBTU) (12), *N*-[(1*H*-benzotriazol-1-yl)(dimethylamino)methylene]-*N*-methylmethanaminium tetrafluoroborate (TBTU) (13), and *N*-[(dimethylamino)-1*H*-1,2,3-triazolo[4,5-*b*]pyridin-1-ylmethylene]-*N*-methylmethan aminium hexafluorophosphate *N*-oxide (HATU) (14) are used in conjunction with $N^\alpha$-protected amino acids and a tertiary base. Recently, it was reported that the use of HOXt in HXTU- and PyXOP-mediated couplings did not significantly increase the yield and purity of a sequence (15). Since there are at least three components required for an efficient acylation, protocols are different than those used with carbodiimide and active ester/fluoride methods to accommodate the additional reagents. An $N^\alpha$-protected amino acid, aminium/phosphonium salt, and HOXt (if desired) are placed in a tube and treated with a solution of base in DMF or NMP followed by activation. Alternatively, separate solutions of activator and base are added to the $N^\alpha$-protected amino acid, followed by activation. In some instances separate solutions of activator and base are added to an aliquot of a predissolved $N^\alpha$-protected amino acid solution, followed by activation. For optimal results, the tertiary base should be stored as a solution in a vessel separate from the aminium/phosphonium salt and amino acid.

Since both expense and time are associated with performing a stepwise synthesis, monitoring this process is a critical step to automation. The repetitive nature of chain elongation, in conjunction with exposing a primary amine, provides an opportunity to analyse the progression of the synthesis. Qualitative monitoring can be accomplished by removing an aliquot of the peptidyl resin from the reaction vessel and performing a ninhydrin (16) test for free amino and imino groups. Other methods for monitoring amines include picric acid (17), 4,4'-dimethoxytritylchloride (18), and bromophenol blue (19). Quinoline yellow has been incorporated as a dye to monitor acylation reactions in continuous-flow instrumentation using counterion distribution monitoring techniques (20). With Fmoc chemistry, on removal of the $N^\alpha$-amino Fmoc protecting group during piperidine treatment, the by-products, dibenzofulvene–piperidine adduct, and piperidine carbamate salt can be monitored by ultraviolet (UV) (21–23) and conductivity (24–26) detectors, respectively. These monitoring techniques should only be used as a guide to assist in the determination of the quality of the peptide, difficulty of synthesis, and performance of the instrument. In addition, since these techniques are not quantitative they cannot be used to calculate coupling efficiencies. Amino acid analysis (27) and Edman degradation (28–30) are the appropriate methods to determine the sequence and composition of a peptide.

## 2. Batchwise peptide synthesis

In automated batch peptide synthesis instruments, the reaction vessel is designed to enable the addition and removal of solvents and reagents through a filter via application of gas pressure or vacuum. Agitation of the solid support is achieved by shaking, vortexing, inverting the reaction vessel, or bubbling an inert gas from the bottom of the reaction vessel. Solvent delivery and the flow rate are generally accomplished by application of an external gas pressure. In this instance, to ensure accurate solvent flow rates, a volume calibration is required so that the external gas pressure is set to facilitate a specific flow rate. In some systems, solvent flow is achieved by motor driven syringe delivery. A typical wash protocol for a batch instrument includes the delivery of a solvent to the reaction chamber, agitation for a designated time, solvent removal, and repetition of this process several times to ensure efficiency of the specific washing step.

Batch instruments are generally compatible with both Fmoc/tBu and Boc/Bzl methods as well as polystyrene, polyethylene glycol–polystyrene graft (PEG–PS), and polyethylene oxide–polystyrene (PEO–PS) supports (for recent reviews see refs 31 and 32). For amino acid activation, protocols have been developed to include carbodiimide, aminium and phosphonium salts, and active esters (pentafluorophenyl esters and acid halides). For batch instruments designed to monitor the Fmoc function, UV or conductivity detectors are used.

## 2.1 PE Biosystems Model 433A peptide synthesis system

PE Applied Biosystems, a Division of the Perkin-Elmer Corporation, introduced the Model 433A peptide synthesis system (*Figure 1*) in 1993. The instrument design was based on the Model 431A platform (introduced in 1989) with additional monitoring and programming capabilities. The benchtop batch reactor instrument can perform both Boc/Bzl or Fmoc/tBu chemistries with reaction mixing achieved by a vigorous patented vortexing procedure. The automated cycles for tBoc amino acids (HOBt ester in DMF or NMP, preformed symmetrical anhydrides in DMF) and Fmoc amino acids (HOBt ester in either DMF or NMP) have been described in detail in the literature (33). The tBoc amino acid preformed symmetrical anhydride cycles include a solvent exchange so that activation takes place in dichloromethane (DCM) and coupling in DMF. The system also uses fully automated HBTU with HOBt *in situ* cycles (Fmoc chemistry in NMP) called *FastMoc™* (34).

The system incorporates a vertically positioned single reaction vessel manufactured from an inert fluorocarbon polymer that generally operates at the 0.1 or 0.25 mmol scale with 10 ml and 40 ml reaction vessels, respectively.

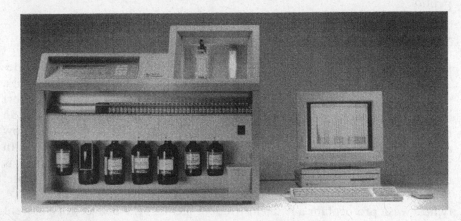

**Figure 1.** PE Biosystems Model 433A peptide synthesis system.

A range of reaction vessels are available to enable synthesis scales ranging from 0.005 to 1.0 mmol (3 ml reaction vessel for 0.005, 0.010, and 0.025 mmol; 10 ml reaction vessel for <0.1 mmol; 40 ml reaction vessel for <0.25 mmol; 55 ml reaction vessel for <1.0 mmol). Different sizes of reaction vessels are required to enable resin swelling and efficiency of washing steps. Mixing in the reaction vessel occurs by vortex agitation that allows for efficient resin–fluid interaction.

Amino acid derivatives are dry packed in cartridges provided by the manufacturer. Pre-weighed cartridges are available commercially. The cartridges are placed in the desired sequence in a guideway that can hold a maximum of 50 cartridges in a single loading. The cartridges pass through a cartridge guide and bar-code reader prior to amino acid dissolution to enable correct positioning and verification of the amino acid sequence, respectively.

Amino acid derivatives are dissolved via a pressure-driven needle-based assembly that punctures the amino acid cartridge septum and delivers solvents for dissolution and gas for mixing. The amino acid solution is then transferred via the needle-based assembly to the preactivation vessel where conversion of the amino acid residue to its corresponding activated derivative takes place. Since the activation process is generally carried out in a preactivation vessel separated from the reaction vessel, the formation of an activated Fmoc or *t*Boc amino acid species occurs immediately prior to the coupling reaction. Subsequent to completion of the activation of an amino acid residue, the solution is transferred to the reaction vessel via gas pressure delivery. Any precipitated by-product is filtered during this step and left behind in the preactivation vessel where it is dissolved and washed out to waste with a solution of methanol (MeOH) and DCM. Alternatively, according to the activation method used, the solutions of the amino acid derivative and the coupling reagent can be delivered directly to the reaction vessel, bypassing the preactivation vessel. Following the coupling reaction and wash of the polymeric support, a resin sampler can be programmed to automatically remove aliquots of the resin for ninhydrin or amino acid analysis. In this instance an alternative resin sampling reaction vessel must be used.

Two different types of activation are generally used on the Model 433A, namely HBTU or HOBt–DCC. Activation with HBTU is also referred to as *FastMoc* chemistry. In the activation step of the 0.25 mmol and 0.10 mmol *FastMoc* cycles, 1.0 mmol of dry protected amino acid in the cartridge is dissolved in solutions of HBTU–HOBt and *N,N*-diisopropylethylamine (DIPEA) in DMF with additional NMP added. The *FastMoc* 1.0 mmol cycles use three 1.0 mmol cartridges, or a total of 3.0 mmol of Fmoc amino acid. The activated Fmoc amino acid is formed rapidly and the solution transferred directly to the reaction vessel. All washing steps are performed with NMP and Fmoc deprotection is achieved by treatment with piperidine–DMF (1:4). Cycle times are 24, 45, and 70 min for the 0.1, 0.25, and 1.0 mmol scales, respectively, with times potentially increasing when feedback monitoring is enabled. With the 0.1 and 0.25 mmol Fmoc-amino acid–HOBt–DCC cycles, 1.0 mmol of dry protected amino acid in the cartridge is dissolved during the activation step with a solution of NMP and HOBt. This solution is transferred to the preactivation vessel and DCC in NMP is added. Following the activation step, the active species is transferred to the reaction vessel where coupling of the activated amino acid to the support-bound peptide occurs. Washing steps are performed with NMP and Fmoc deprotection is achieved by treatment with piperidine–DMF (1:4). Cycle times are 60 and 108 min for the 0.1 and 0.25 mmol scales, respectively, with times potentially increasing when feedback monitoring is enabled. With the 0.5 mmol *t*Boc protocols, 2.0 mmol of the dry protected amino acid is dissolved with a solution of NMP and HOBt in NMP. This solution is transferred to the preactivation vessel and DCC in NMP is added. After the activation occurs, the active species is transferred to the reaction vessel and dimethylsulphoxide (DMSO) is added to enhance and improve solvation of the peptide resin as well as increase coupling yields for difficult sequences. During the final coupling stage DIPEA is also added to further disrupt peptide–peptide hydrogen bonds, increase solvation, and neutralize any protonated amine groups to accelerate coupling reactions. Subsequent to coupling and washing, a capping reaction occurs using an acetic anhydride–DIPEA–NMP solution. With 0.1 mmol *t*Boc protocols, 1.0 mmol of the protected amino acid is used. Washing steps use either DCM or NMP, *t*Boc group removal is typically via treatment with a trifluoroacetic acid (TFA)–DCM (1:1) solution (with a subsequent neutralization step with DIPEA), and the total cycle time is 65 and 104 min for the 0.1 and 0.5 mmol scales, respectively.

The system incorporates nine solvent/reagent reservoirs to enable delivery of the NMP main wash, neat piperidine deblock (diluted with main wash on line), 0.45 M HBTU–HOBt–DMF activator solution, 2.0 M DIPEA–NMP, MeOH final wash, DCM wash, alternative activators (e.g. 1.0 M DCC–NMP), capping solution, and solvent exchange for Fmoc/*t*Bu chemistry. In the case of Boc/Bzl-based methods, DCM and NMP are used as the main wash solvents, TFA (diluted with DCM on-line) for deblock, 1.0 M DCC–NMP and 1.0 M HOBt–NMP for activation, DMSO–NMP (4:1) for improved solvation, acetic anhydride for capping, and MeOH for final wash. A gas positive-pressure chemical delivery system controls the flow of all reagents and solvents. Gas pressure is provided by an external tank of nitrogen.

In the Model 433A SynthAssist® Software, based on a Macintosh® platform, there are 152 functions, designated with a number and a name, which control all the processes required for synthesis. Some functions activate a switch or set of switches to deliver solutions or to perform a specific task.

For Fmoc chemistry, the Model 433A features automated conductimetric monitoring with feedback control (35). Conductivity monitoring is based on $N^{\alpha}$-Fmoc removal with piperidine–NMP (1:4) to generate dibenzofulvene and a conductive carbamate salt of piperidine. The extent of deprotection is determined by comparing the conductivity of two samples of deprotection solution. In one algorithm, if the user sets a deprotection loop value of 4% (acceptable range 3–10%), then successive $N^{\alpha}$-Fmoc removal treatments within the same cycle will continue until the last conductivity reading is within 4% of the value for the previous sample reading. In a second algorithm, the extent of deprotection is determined by comparing the conductivity of the last sample to the first sample in the current cycle. With this algorithm, a pre-determined value for the deprotection loop (i.e. 4%) requires $N^{\alpha}$-Fmoc removal to continue until the last conductivity value is less than or equal to 4% of the first conductivity value. Intelligent dynamic feedback control is activated on encountering difficult Fmoc deprotections. A number of cycle modifications such as extended deprotection, double or extended couplings, capping, and solvent exchange, which can be selected at initial set-up by the user, can be automatically incorporated into the synthesis if required. Two additional monitoring channels are available to enable UV monitoring and user-definable monitoring.

## 2.2 Protein Technologies, Inc., SONATA/Pilot™ peptide synthesizer

The SONATA/Pilot (*Figure 2*) design is based on a matrix valve component similar to that used in the Protein Technologies, Inc. SYMPHONY/ Multiplex™ peptide synthesizer (discussed in Section 4.2). SONATA/Pilot is a floor-standing system with seven primary reagent reservoirs and 20 amino acid reservoirs. The system uses a single reactor vessel capable of scales of 0.1–50 mmol (100 g of solid support). Transfer of reagents is via gas pressure except for the primary solvent that is delivered by a pumping system capable of operating at flow rates up to 400 ml/min for higher scales. Similar to the SYMPHONY/Multiplex system, the amino acid derivatives are incorporated as predissolved solutions and the system is compatible with both Fmoc and *t*Boc chemistries.

## 2.3 Protein Technologies Inc., Model PS3™ peptide synthesizer

The Model PS3 batch peptide synthesizer (*Figure 3*) is designed for use with prepackaged reagents. Activating agents, typically BOP or HBTU, are pre-packed in monomer vials in conjunction with protected amino acids for point-of-use dissolution. Following addition of a base solution, activation occurs in the vial and the activated amino acid derivative is transferred to the reaction vessel containing the solid support. Fluid transfer and agitation of the poly-meric support is achieved via gas pressure. Three glass reaction vessels perform in a sequential manner at a scale range of 0.1–0.25 mmol with either *t*Boc or Fmoc chemistries. The system has four solvent/reagent reservoirs and runs by a stand-alone control with programming via a liquid crystal display

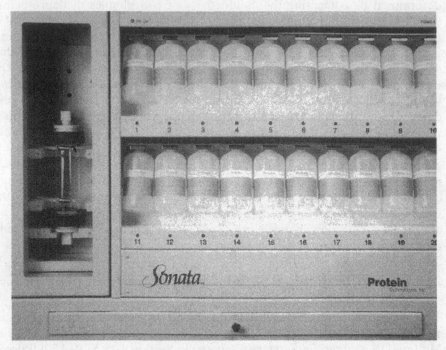

**Figure 2.** Protein Technologies, Inc., SONATA/Pilot™ peptide synthesizer.

(LCD) display on the instrument. Operation is menu-driven with six standard stored programs and nine user-modifiable coupling programs.

## 2.4 Advanced ChemTech Model 90 peptide synthesizer

The Advanced ChemTech (ACT) Model 90 (*Figure 4*) is a bench-top, dual-reactor, *t*Boc/Fmoc instrument suitable for assembling two different peptides concurrently at scales of 0.1–3 mmol in glass reaction vessels. Alternatively, scales of 3–10 mmol can be performed in a single reactor configuration. To operate across the complete scale range, 25, 50, 100, 200, 300, and 500 ml reactors are required. Fluid transfer is achieved by nitrogen pressure and operator-controlled wrist-action shaking and/or nitrogen bubbling accomplishes mixing in the reaction vessels.

There are four 90 ml reservoirs in which to place predissolved amino acid solutions or amino acids as dry powders for on-line dissolution. Depending on the selected coupling method, protected amino acids can be preactivated in the amino acid reservoirs followed by transfer to the reaction vessel for acylation. Alternatively, amino acids can be added directly to the reaction vessel for *in situ* activation. The system is supplied with seven solvent/reagent reservoirs (three 1 gallon and four 1 litre) to allow solutions for general

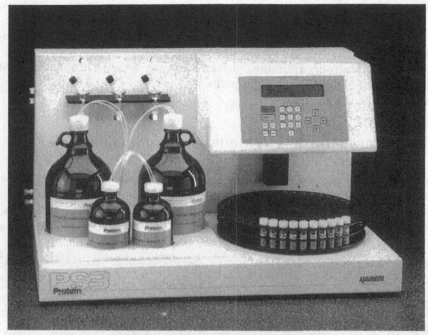

**Figure 3**. Protein Technologies Inc., Model PS3™ peptide synthesizer.

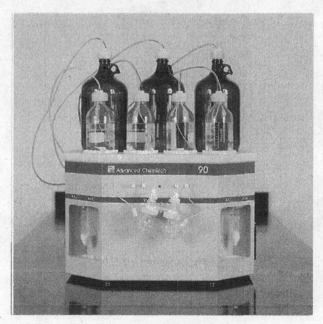

**Figure 4**. Advanced ChemTech Model 90 peptide synthesizer.

washing, $N^\alpha$-$t$Boc/Fmoc group removal, activation, and capping. Activation chemistries that can be incorporated include *in situ* activation via DCC, DIC, TBTU, PyBOP, or HBTU. Preactivated symmetrical anhydrides, HOBt esters, and pentafluorophenyl esters can also be used.

The system operates via mouse-driven software with a DOS-based IBM PS/2™ compatible computer equipped with standard protocols that can be further modified by the user.

## 2.5 Advanced ChemTech Model 400 production-scale synthesizer

Large-scale solid phase synthesis of peptides can be performed on the floor-standing ACT Model 400 batch system (*Figure 5*) that operates in a single reactor mode with 0.1–5 kg of starting solid support (36). All operations and

**Figure 5**. Advanced ChemTech Model 400 production-scale synthesizer.

liquid transfer occur under nitrogen pressure. Additional nitrogen bubbling and/or oscillating enables fluid/resin mixing in the reaction vessel. The system is compatible with Boc/Bzl and Fmoc/tBu methods and can be linked up to four 200-litre solvent drums. Amino acids can be added as solids to 5-litre/10-litre flasks and automatically dissolved just prior to addition to the solid support. Alternatively, the amino acid can be predissolved, stored in 5-litre/10-litre flasks and added automatically when required. The instrument is equipped with a filtration device that allows for the removal of solid pre-activation by-products prior to addition to the reaction vessel. This feature enables compatibility with the common activation chemistry used.

The system is usually customized to cover the specific needs of a laboratory (additional solvents, safety features, and integration into existing laboratory infrastructure) and is of stainless steel construction with only glass and Teflon contact areas to allow operation in a clean room environment. Software and hardware are designed specifically for compliant operation.

## 2.6 ABIMED EPS 221 synthesizer

The EPS 221 (*Figure 6*) is an automated Fmoc peptide synthesizer based on a robotics system with a motor-driven syringe for reagent transfer. Movement of the robot in an X–Y–Z direction allows the system to perform all solvent/reagent delivery steps. The system runs at synthesis scales up to 100 μmol on 1–3 columns with a total of 45 amino acid residue positions distributed over all column locations. Reagents are delivered to the column by a needle that is connected to a motor-driven syringe. The standard protocol for the instrument begins with a 2–3 ml DMF wash delivery to the solid support to swell the resin for 10 min. A piperidine–DMF (1:4) solution is then taken up from the derivatives rack and pushed slowly through the resin, followed by another DMF wash. Amino acids are activated using a 0.9 M solution of PyBOP and a 4 M solution of N-methylmorpholine (NMM). A fivefold excess of the activated amino acid solution is pushed slowly through the resin bed. Gas is blown through the resin to achieve mixing. After the coupling reaction the solid support is again washed with DMF. The synthesizer is operated from the keyboard controller of the sample preparation robot equipped with an eight-line display and floppy disk drive.

## 3. Continuous-flow peptide synthesis

In traditional batch peptide synthesis, all reactions are carried out essentially in a discontinuous fashion with sequential treatment of acylating and deprotecting agents in a shaken, stirred, vortexed, or bubbled reaction vessel. The excess of reagents is removed by filtration and repeated washing steps. In continuous-flow peptide synthesis (37), systems were developed originally with the aim of improving the efficiency and speed by removing excess

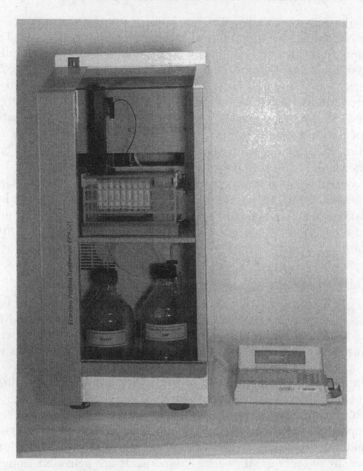

**Figure 6.** ABIMED EPS 221 synthesizer.

reagents with a continual solvent flow through a packed column containing a solid support. A glass column with filters at the top and bottom replaces the reaction vessel. The system incorporates a positive displacement pump to enable continuous fluid flow at a variety of flow rates. Valves provide for selection of solvent and reactant flow through the resin bed to waste as well as for the introduction and recirculation of the acylating species during the coupling reaction.

Under the pumped flow solvent conditions in continuous-flow instrumentation, polystyrene resins can create high pressures that can halt a synthesis. To use these polymeric supports in continuous-flow systems, a minimum of a 4:1 ratio of glass to polystyrene bead is required. The incorporation of polyethylene glycol spacers into polystyrene provides a support, namely PEG–PS, which is compatible with and recommended for continuous-flow systems.

PEG–PS supports have a defined and reproducible structure that comprises approximately 60 wt% of polyethylene glycol providing a support that is physically robust (38). PEG–PS resins swell appreciably in a range of solvents and are superior to polystyrene supports for the construction of more complex peptides, including those containing cyclic (lactam and disulphides), branched, and post-translational modified moieties.

Continuous-flow instrumentation was designed for Fmoc/*t*Bu-based methods as N$^\alpha$-protecting group removal proceeds under milder conditions (piperidine) than Boc/Bzl chemistry (TFA) and would, therefore, be more compatible with a reciprocating pump. Another advantage to Fmoc chemistry is the absorbance of the fluorene chromophore in an accessible region of the UV spectrum (39). Liquid flow effluent from the reaction column can be passed through a UV detector system to enable continuous spectrophotometric monitoring of both acylation and deprotection reactions. Additionally, the detection method provides confirmation that the various operations are proceeding normally and that the instrument function is correct.

In a typical UV trace, the recirculation of the soluble, activated amino acid derivative is indicated by a characteristic oscillating pattern with an initial damped harmonic motion-type pattern (40, 41). The maximum absorbance decreases as the concentration of the Fmoc amino acid in solution is reduced due to coupling to the solid support. Diffusion processes result in slight broadening of successive UV peaks. After several recirculations through the resin bed, a uniform reactant concentration and a consistent UV absorbance throughout the system are observed. The wash-out of excess reagents with DMF is very rapid as shown by a sharp fall in the optical density of the column effluent during the washing step. The facile release and elimination of the Fmoc species, generally observed as a sharp rise and fall in UV absorbance, can easily monitor the deprotection step. The height and shape of this peak provides valuable data with regard to the rate of Fmoc group removal, which can indicate the difficulty of assembling a particular sequence. Troublesome Fmoc deprotections can produce UV data peaks which are characteristically broadened and reduced in height.

## 3.1 PE Biosystems Pioneer™ peptide synthesis system

The Pioneer (*Figure 7*) peptide synthesis system (42) was introduced recently as a successor to the 9050Plus instrument, which was first commercialized in 1988. The Pioneer combines dual simultaneous column capability with a high throughput multiple peptide synthesis option (discussed in Section 4.1). It is an Fmoc/*t*Bu continuous-flow system that in its standard mode of operation functions with two columns operating simultaneously with independent control of solvent, reagent, and amino acid delivery. Fluid flow is via programmable positive displacement pumps that deliver with a flow rate of up to 50 ml/min and require no flow-rate calibration. Each column can function

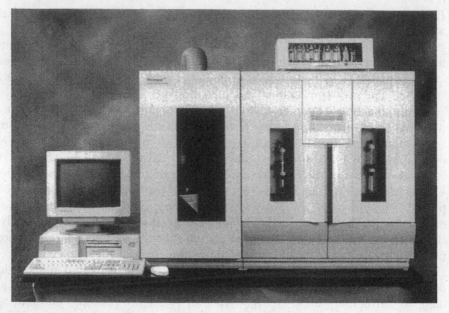

**Figure 7.** PE Biosystems Pioneer™ peptide synthesis system.

at 0.02–2.0 mmol scale range. The two separate pump systems enable simultaneous operation of each column at a different scale. An integral UV spectrophotometer monitors in real time all reactions that occur on both columns.

The solid support is placed in glass reaction columns that have filters at the top and bottom. Adjustable end pieces allow for different scales of synthesis to be performed within the same reaction column. The standard column size is 15 × 100 mm, capable of containing up to 2.5 g of PEG–PS resin. A large-scale column (25 × 150 mm) is also provided to enable synthesis scales of up to 2.0 mmol.

Fmoc-amino acid derivatives are stored as dry powders in vials positioned in order of the sequence in the racks of a transport system. Similar to a fraction collector, the transport system moves in the X,Y-direction, one position per cycle, and a probe is used for point-of-use dissolution of a fourfold excess of the Fmoc-amino acid residues. The probe delivers activator solution to the vial and then bubbles the derivative with nitrogen to form a homogeneous solution. The pump transfers the amino acid solution via the probe through the column for a recirculation/coupling step. To ensure high concentration, a low recirculation volume is used. The end of the probe is equipped with a filter to ensure that no particulates enter the system and to eliminate blockages. Both of the two transport systems possess a maximum of 101 amino acid positions for scales of synthesis not greater than 0.2 mmol. For scales up to a maximum of 0.5 and 2.0 mmol, alternative rack geometries are provided with 68 and 24 amino acid positions, respectively. A range of activation chemistries can be used, including carbodiimide (DIC), penta-fluorophenyl ester, aminium salts (TBTU, HBTU, and HATU) and phosphonium salts (PyBOP and PyAOP).

The system incorporates eight solvent/reagent reservoirs including bottles for three activators to perform mixed acylation strategies, a main wash, piperidine–DMF (1:4), capping, solvent exchange, and additional wash. Pre-programmed protocols are provided to allow for automated removal of allyl-based protecting groups to enable a fully automatic Fmoc-based synthesis of cyclic and branched chain peptides (43). Additionally, programmes are available for fast (22-min cycle time), standard (45-min cycle time), extended (75-min cycle time), capping, solvent exchange, and reduced diketopiperazine cycles.

The system can be operated from stand-alone control via an LCD and a membrane keypad screen on the instrument. Seven pre-programmed proto-cols are provided and a maximum of 36 sequences containing 101 amino acids in length in each of the eight user profiles is allowed. Alternatively, for more flexibility, full protocol editing, chemical editing, and database capability, the system can be controlled from a computer workstation with icon-driven easy-to-use Windows® 95-based software.

Real-time on-line UV monitoring is provided for both columns with software enabling full feedback control. If the UV data determine a difficult deprotection, the system can automatically extend the deprotection and the subsequent coupling of the next incoming amino acid (44).

A unique characteristic of the Pioneer peptide synthesis system is the option of a multiple peptide synthesis (MPS) accessory with 16-column capacity. A maximum of two MPS units can be added to each Pioneer, extending the capability for simultaneous synthesis from 2 to 32 different peptides. The novel design allows independent operation of both columns, MPS accessories, or a combination of the two. The MPS unit possesses scales at 0.025, 0.05, and 0.1 mmol. Twenty-four different predissolved amino acid derivatives can be incorporated to elongate a sequence. The appropriate amino acid solution is delivered in eightfold excess to the columns, together with the activator solutions.

# 4. Multiple peptide synthesis systems

A range of MPS systems is available commercially which generally synthesize at lower scales and with limited flexibility in terms of protocol and chemical editing capability (45–48).

## 4.1 PE Biosystems Pioneer MPS option

The Pioneer MPS option, described in Section 3.1, is a 16-column accessory that can be added to the PE Biosystems Pioneer peptide synthesis system. A total of two MPS options can be added on to each Pioneer, extending its column capacity from 2 to 32, running at scales of 0.025, 0.05, or 0.1 mmol.

## 4.2 Protein Technologies, Inc. SYMPHONY/Multiplex™ peptide synthesizer

The SYMPHONY/Multiplex (*Figure 8*), introduced in 1993, is a batch *t*Boc/Fmoc peptide synthesizer with 12 independent reaction vessels (49). The key component is the patented matrix valve system that enables different solutions to be transferred to each of the 12 reactors simultaneously. This feature allows the assembly of 12 different peptides under optimal conditions for each sequence with regard to scales and customized coupling reactions. Solution flow is controlled via chemically resistant membrane valves arranged to provide a matrix system with common solution ports but separate delivery lines for each reaction vessel. A combination of gas pressure and an external vacuum pump operates the valve system. Solution transfer is achieved by nitrogen pressure.

The reusable glass reaction vessels for the solid supports have an internal volume of 16 ml and enable synthesis scales up to 0.1 mmol. Alternatively, disposable plastic vessels with 24 ml volume can be used. Resin agitation is

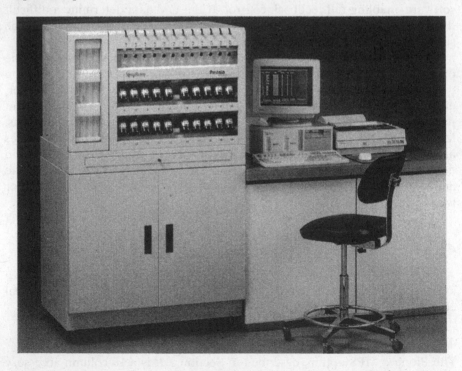

**Figure 8.** Protein Technologies, Inc. SYMPHONY/Multiplex peptide synthesizer.

accomplished by occasional nitrogen bubbling through a 40 μm filter at the bottom of the reaction vessel.

The system is a floor-standing instrument that includes a vented cabinet for storage of solvents and reagents. Six solvent/reagent reservoirs are available typically for DMF or NMP as a main wash solvent, piperidine–DMF (1:4) for deprotection, DCM as a secondary reagent, HBTU solution for activation, acetic anhydride as a capping solution, and a TFA-based cleavage cocktail.

The SYMPHONY/Multiplex system includes 20 × 100 ml reservoirs for predissolved amino acid solutions. The standard 100 mM amino acid concentration provides a fivefold excess. The software is designed primarily for *in situ* HOBt ester activation using HBTU, TBTU, PyBOP, or BOP activators.

In the SYMPHONY/Multiplex system, the activator solution consists of HBTU or TBTU at a concentration equimolar with the amino acid solutions in conjunction with two- to fourfold equivalents of a base such as NMM. For a specified coupling step, the system transfers a determined number of aliquots of amino acid solution to the solid support, followed by the same volume of activator solution. Subsequent to completion of addition, the resin is agitated by gas bubbling for the specified coupling time.

Control of the synthesis parameters is by a stand-alone IBM-compatible personal computer. The software is based on 'point and click' pull-down menus, allowing access to Peptide, Program, Reaction Vessel operations, Manual operations, Set-up, and Data sections.

After chain elongation, peptides are released from the solid support in the reaction vessels according to cleavage protocols selected in the software. Crude products are transferred to centrifuge tubes in a vented compartment for ether precipitation and centrifugation.

## 4.3 Advanced ChemTech Models 348, 396, and 357 bimolecular synthesizers

Advanced ChemTech provides a range of robotic bench-top MPS systems that vary in the number of peptides that can be constructed and the scale range that can be performed (50). All systems are batch, *t*Boc/Fmoc compatible and use syringe pumps for fluid delivery. Model 357 (*Figure 9*) has a maximum of 36 × 8.0 ml reaction-well reactor capacities (5–250 mmol scale); Model 348 (*Figure 10*) has a maximum of 96-well capacity (5–150 μmol scale); and Model 396 (*Figure 11*) has a maximum of 96-well capacity (5 μmol–1 mmol). In Models 348 and 396 it is also possible to use 8-, 16-, and 40-well Teflon reactor blocks. Additionally, Model 357 can serve as a large-scale single reactor system capable of handling up to 36 g of resin. The 600 ml chamber allows for pooling and dividing of the solid support for the 'mix and split' technique (51–53).

A variable speed orbital mixer and/or nitrogen bubbling (only on the Model 357) is incorporated to stir solvents/reagents with the solid supports. A

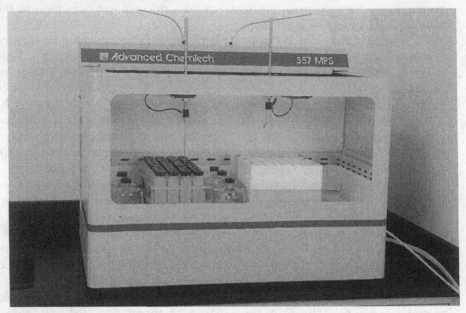

**Figure 9.** Advanced ChemTech Model 357.

maximum of 128 amino acid reservoirs can be used, containing predissolved amino acid residues. Solutions are delivered by a single-needle-based robotic arm system to individual wells in the reactor block for *in situ* activation. Five solvent/reagent reservoirs allow for washing, deprotection, and activation. An optional on-board cleavage unit is available for Model 348 and 396, allowing for automated release of the peptides from the resin. Operation of these systems is via a Pentium® Class PC. Both models 348 and 396 are equipped with Windows®-based software with built-in standard protocols which can be further modified by the user as required.

Models 348 and 396 have been introduced recently in an Omega version that incorporates an enclosed cabinet with a front window and hood together with safety features for CE and UL compliance.

## 4.4 ABIMED AMS 422 multiple peptide synthesizer

The AMS 422 (*Figure 12*) (54–56) is an automated Fmoc-based batch instrument for simultaneous construction of 48 different peptides at scales of 5–50 μmol. Solvents and reagents are delivered by nitrogen pressure and removed by an aspiration system composed of two stainless steel vessels separated by a solenoid valve and an all-Teflon membrane vacuum pump. The instrument is based on an X–Y–Z robotics system for liquid handling and contains a multiple column reaction module. A valve manifold controls each

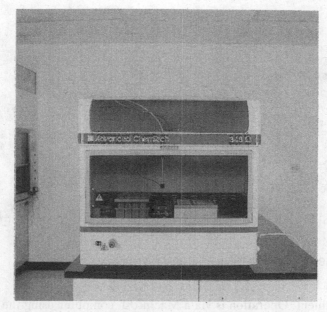

**Figure 10.** Advanced ChemTech Model 348.

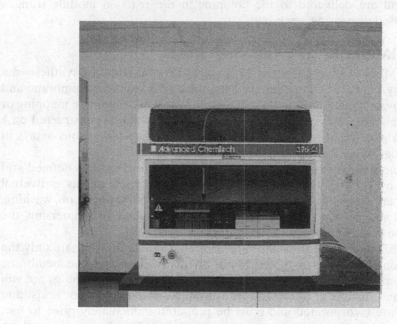

**Figure 11.** Advanced ChemTech Model 396.

**Figure 12.** ABIMED AMS 422 multiple peptide synthesizer.

column effluent. Operation is via a Macintosh computer using the Supercard graphics package. Resin is contained in individual open-top column reactors fitted with filters at the bottom. Fmoc-amino acids, activator, and base are stored as predissolved solutions in DMF in an array of septum-sealed containers and are delivered to the columns in the reaction module using a needle-based dispensing assembly.

## 4.5 ABIMED ASP 222 Auto-Spot Robot

The ABIMED ASP 222 (*Figure 13*) uses the Spot-method of synthesis developed by Frank (57). Peptides are assembled on a cellulose membrane and remain covalently attached to the polymer for subsequent epitope mapping or receptor binding assays. Several hundred sequences can be constructed on a microtitre plate-sized membrane and used repetitively for various assays in immunology, biochemistry, and drug research.

The membrane is derivatized with a linker amino acid within a defined grid followed by capping the surrounding area. These spots act as individual reactors performing the synthesis by repeated cycles of deprotection, washing, coupling, and washing. These steps are accomplished by immersing the membrane in the appropriate solvent bath.

ABIMED developed the Auto-Spot Robot ASP 222 to automate only the distribution of activated amino acids onto the derivatized cellulose membrane in a defined grid. Bulk operations such as deprotection and washings are still carried out manually. Predissolved solutions of Fmoc-amino acids and activator are incorporated and must be prepared immediately prior to use. The membranes contain 400 peptide spots and fit directly into the system.

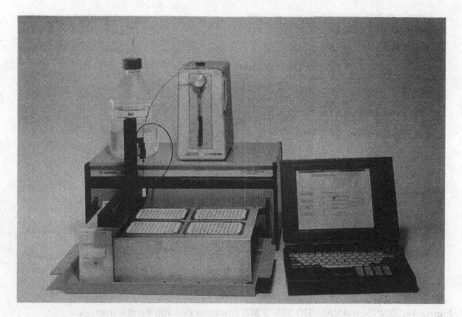

**Figure 13.** ABIMED ASP 222 Auto-Spot Robot.

Typical membrane loadings are 0.5–0.6 $\mu$mol/cm$^2$. Distribution of the 44 positions of the predissolved activated amino acids can be performed on four membranes simultaneously. Delivery of the amino acids to the membrane is via a robotic arm system with a needle-based assembly connected to a motor-driven syringe. The ASP 222 is operated using Windows®-based software on an IBM-compatible 383/484 computer to allow sequence entry manually or from external databases.

## 4.6 ZINSSER ANALYTIC SMPS 350 multiple peptide synthesizer

Developed in 1989 in collaboration with Boehringer Ingelheim, the SMPS 350 (*Figure 14*) can synthesize 3 × 48 peptides in a simultaneous manner using Fmoc chemistry. Two X–Y–Z robot arm systems with sample tips and two dispenser/dilutor systems perform the liquid handling and washing steps. Chain elongation occurs on polystyrene resin positioned in small disposable plastic reaction vessels with conical bottoms placed in supports similar to a microtitre plate. A 3 × 48 microreactor or 3 × 30 minireactor capable of containing a maximum of 15 and 50 mg, respectively, of solid support can be used. The system has a rack for 32 predissolved amino acid residues as well as three solvent/reagent reservoirs for washing, deprotecting, and activating reagents. Operation of the instrument is via DOS-based software on an IBM-compatible computer system.

297

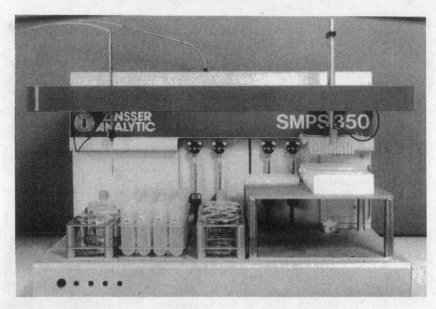

**Figure 14.** ZINSSER ANALYTIC SMPS 350 multiple peptide synthesizer.

## 4.7 ZINSSER ANALYTIC SOPHAS solid phase synthesizer

SOPHAS (*Figure 15*) is an automated instrument capable of assembling 864 sequences in a single run. Reactors can be selected from a 96-well plate format to 25 ml vessels. Mixing is conducted by mechanic agitation. The system uses four independent pipetting devices, each supported by two dilutors and a time/pressure delivery channel to transfer 4–5 solvents/reagents or nitrogen. The system is operated by Windows® NT software.

## 4.8 SHIMADZU PSSM-8 peptide synthesizer

The PSSM-8 is a batch Fmoc instrument with eight disposable reaction vessels comprised of 2.5 ml syringes with filters at the bottom. The synthesizer operates at a 5–50 μmol/reactor scale. The system includes an amino acid station equipped with $20 \times 2.0$ ml Eppendorf tubes/reactor positions for predissolved derivatives. Fmoc-amino acids, solvents, and reagents are delivered by syringes to the reactors via a needle-based assembly. Nitrogen bubbling agitates the solid support. Each reactor position is supplied with three 30-ml reactor bottles to allow for activation and/or capping. The main wash solvent and deprotecting reagent are located in reservoirs external to the instrument. Syntheses are programmed via a separate computer system followed by transfer of the data to a memory card.

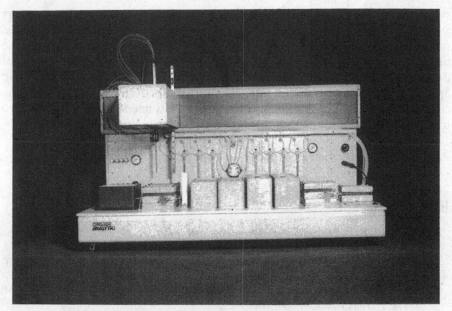

**Figure 15.** ZINSSER ANALYTIC SOPHAS solid phase synthesizer.

## 5. Conclusions

Since the original work of Merrifield, the field of solid phase peptide synthesis has evolved enormously to sophisticated automated equipment. Today's modern instrument has been engineered to a level at which a researcher only has to input a sequence into a computer and allow the machine to produce the desired target peptide. Contrary to the belief of researchers who have just entered the field of solid phase peptide synthesis and claim that the field is a mature science, there is unfortunately no guarantee that an individual instrument will prepare a desired sequence effectively, due to chemical and sequence constraints. Although synthesizers have removed the tediousness of repetitive synthetic operations, chemists still must decide the appropriate chemical pathway, select an instrument that will satisfy their demands, and interpret the data generated.

An interesting development in solid phase synthesis is the awareness of this technique in drug discovery. Standard organic reactions have been optimized on solid phase and novel handles have been developed to release compounds from a solid support by various chemical methods and with additional functionality (58). Once a synthetic pathway has been developed, diversity is added to generate libraries of potential lead candidates for high-throughput screening assays. To assist in the production of small-molecule libraries, automated machinery has been developed. The original instruments were based on minor modification to commercially available multiple peptide synthesizers. Currently, there are a variety of solid phase organic synthesizers available commercially (59–63). Similar to peptide instrumentation, these units each possess unique features with regard to the level of automation, the number of reactors, and the scale. While this field is still at an early stage of development, engineering designs that build versatility and ingenuity will be applied to peptide instrumentation. Finally, as improvements continue in the chemistry for solid phase synthesis, machinery will need to adapt with regard to valving and plumbing to be able to implement these methods.

All registered trademarks are the properties of their respective owners.

# References

1. Merrifield, R. B. (1963). *J. Am. Chem. Soc.*, **85**, 2149.
2. Bodanszky, M. and Bednarek, M. A. (1989). *J. Protein Chem.*, **8**, 461.
3. Carpino, L. A., Beyermann, M., Wenschuh, H., and Biernert, M. (1996). *Acc. Chem. Res.*, **29**, 268.
4. Wenschuh, H., Beyermann, M., Rothemund, S., Carpino, L., and Bienert, M. (1995). *Tetrahedron Lett.*, **36**, 1247.
5. Sheehan, J. C. and Hess, G. P. (1955). *J. Am. Chem. Soc.*, **77**, 1067.
6. Sarantakis, D., Teichman, J., Lien, E. L., and Fenichel, R. (1976). *Biochem. Biophys. Res. Commun.*, **73**, 336.
7. König, W. and Gieger, R. (1970). *Chem. Ber.*, **103**, 788.
8. Carpino, L. (1993). *J. Am. Chem. Soc.*, **115**, 4397.
9. Dormoy, J. R. and Castro, B. (1979). *Tetrahedron Lett.*, **20**, 3321.
10. Coste, J., Le-Nguyen, D., and Castro, B. (1990). *Tetrahedron Lett.*, **31**, 205.
11. Albericio, F., Cases, M., Alsina, J., Triolo, S. A., Carpino, L. A., and Kates, S. A. (1997). *Tetrahedron Lett.*, **38**, 4853.
12. Dourtoglou, V., Ziegler, J. C., and Gross, B. (1978). *Tetrahedron Lett.*, **19**, 1269.
13. Knorr, R., Trzeciak, A., Bannwarth, W., and Gillessen, D. (1989). *Tetrahedron Lett.*, **30**, 1927.
14. Carpino, L. A., El-Faham. A., Truran, G. A., Minor, C. A., Kates, S. A., Griffin, G. W., Shroff, H., Triolo, S. A., and Albericio, F. (1994). In *Innovation and perspectives in solid phase synthesis – Peptides, proteins and nucleic acids* (ed. R. Epton), p. 95. Mayflower Worldwide Ltd., Birmingham, UK.
15. Carpino, L. A., El-Faham, A., Minor, C. A., and Albericio, F. (1994). *J. Chem. Soc., Chem. Commun.*, 201.
16. Kaiser, E., Colescott, R. L., Bossinger, C. D., and Cook, P. I. (1970). *Anal. Biochem.*, **34**, 595.
17. Hodges, R. S. and Merrifield, R. B. (1975). *Anal. Biochem.*, **65**, 241.
18. Horn, M. and Novak, C. (1987). *Am. Biotech. Lab.*, **5**, 12.
19. Krchnák, V., Vágner, J., Safár, P., and Lebl, M. (1988). *Coll. Czech. Chem. Commun.*, **53**, 2542.
20. Young, S. C., White, P. D., Davies, J. W., Owen, D. E. I. A., Salisbury, S. A., and Tremeer, E. J., (1990). *Biochem. Soc. Trans.*, **18**, 1311.
21. Chang, C. D. and Meienhofer, J. (1978*). Int. J. Pept. Protein Res.*, **11**, 246.

22. Meinenhofer, J., Waki, M., Heimer, E. P., Lambros, T. J., Makofske, R. C. and Chang, C. D. (1979). *Int. J. Pept. Protein Res.*, **13**, 35.
23. Chang, C. D., Felix, A. M., Jimenez, M. H., and Meienhofer, J. (1980). *Int. J. Pept. Protein Res.*, **15**, 485.
24. Schafer-Nielson, C., Hansen, P. H., Lihme, A., and Heegard, P. M. H. (1989). *J. Biochem. Biophys. Methods*, **20**, 69.
25. Fox, J., Newton, R., Heegard, P., and Schafer-Nielson, C. (1990). In *Innovation and perspectives in solid phase synthesis – Peptides, proteins and nucleic acids* (ed. R. Epton), p. 141. SPCC Ltd., Birmingham, UK.
26. McFerran, N. V., Walker, B., McGurk, C. D., and Scott, F. C. (1991). *Int. J. Pept. Protein Res.*, **37**, 382.
27. Ozols, J. (1990). *Methods Enzymol.*, **182**, 587.
28. Edman, P. (1950). *Acta. Chem. Scand.*, **4**, 277.
29. Edman, P. (1950). *Acta. Chem. Scand.*, **4**, 283.
30. Laursen, R. A. (1971). *Eur. J. Biochem.*, **20**, 89.
31. Barany, G., Albericio, F., Kates, S. A., and Kempe, M. (1997). In *Chemistry and biological application of polyethylene glycol* (ed. J. M. Harris and S. Zalipsky), p. 239. American Chemical Society Books, Washington, DC.
32. Rapp, W. (1996). In *Combinatorial peptide and nonpeptide libraries* (ed. G. Jung), p. 425. VCH, Weinheim, Germany.
33. Geiser, T., Beilan, H., Bergot, B. J., and Otteson, K. M. (1988). In *Macromolecular sequencing and synthesis: selected methods and applications* (ed. D. H. Schlesinger), p. 199. Alan R. Liss, New York.
34. Fields, C. G., Lloyd, D. H., Macdonald, R. L., Otteson, K. M., and Noble, R. L. (1991). *Peptide Res.*, **4**, 95.
35. PE Applied Biosystems Peptide Synthesis Technical Bulletin. Feedback monitoring kit for ABI431A peptide synthesizer.
36. Birr, C. (1990.) In *Innovation and perspectives in solid phase synthesis* (ed. R. Epton), p. 155. SPCC Ltd., Birmingham, UK.
37. Atherton, E. and Sheppard, R. C. (1989). In *Solid phase peptide synthesis: a practical approach*, p. 87. IRL Press, Oxford.
38. Zalipsky, S., Chang, J. L., Albericio, F. and Barany, G. (1994). *Reactive Polymers*, **22**, 243.
39. Atherton, E. and Sheppard, R. C. (1987). In *The peptides, analysis, synthesis, biology* (ed. E. Gross and J. Meienhofer), Vol. 9, p. 1. Academic Press, New York.
40. Atherton, E., Brown, E., Sheppard, R. C., and Rosevear, A. (1981). *J. Chem. Soc., Chem. Commun.*, 1151.
41. Dryland, A. and Sheppard, R. C. (1986). *J. Chem. Soc., Perkin Trans. 1*, 125.
42. Daniels, S. B., Solé, N. A., Hantman, S. F., Gibney, B. R., Rabanal, F., and Kates, S. A. (1998). In *Peptides 1996: Proceedings of the Twenty-fourth European Peptide Symposium* (ed. R. Ramage and R. Epton), p. 323. Mayflower Worldwide Ltd., Birmingham, UK.
43. Kates, S. A., Daniels, S. B., and Albericio, F. (1993). *Anal. Biochem.*, **212**, 303.
44. PerSeptive Biosystems Technical Bulletin (1998). Pioneer™ peptide synthesis system: monitoring peptide synthesis.
45. Holm, A. and Meldal, M. (1989). In *Peptides 1988* (ed. G. Jung and E. Bayer), p. 208. Walter de Gruyter, Amsterdam.

46. Meldal, M., Holm, C. B., Bojesenm G., Jakobsen, M. H., and Holm, A. (1993). *Int. J. Pept. Protein Res.*, **41**, 250.
47. Wang, Z. and Laursen R. A. (1992). *Pept. Res.*, **5**, 275.
48. Jung, G. and Beck-Singer, A. (1992). *Angew. Chem., Int. Ed. Engl.*, **31**, 367.
49. Crawford, J. K. (1997). In *The 1997 American peptide symposium program and abstracts, Nashville, Tennessee*, pp. 2-103.
50. Saneii, H. H., Shannon, J. D., Miceli, R. M., Fischer, H. D., and Smith, C. W. (1994). In *Proceedings of the Thirteenth American Peptide Symposium: Peptides, chemistry, structure and biology* (ed. R. S. Hodges and J. A. Smith), p. 1018. Escom, The Netherlands.
51. Furka, A., Sebestyén, F., Asgedom, M., and Dibó, G. (1991). *Int. J. Pept. Protein Res.*, **37**, 487.
52. Houghten, R. A., Pinilla, C., Blondelle, S. E., Appel, J. R., Dooley, C. T., and Cuervo, J. H. (1991). *Nature*, **354**, 84.
53. Lam, K. S., Salmon, S. E., Hersh, E. M., Hruby, V. J., Kazmierski, W. M., and Knapp, R. J. (1991). *Nature*, **354**, 82.
54. Gausepohl, H., Boulin, C., Kraft, M., and Frank, R. W. (1992). *Pept. Res.*, **5**, 315.
55. Gausepohl, H., Behn, C., Kreuzer, O., and Buettner, J. A. (1993). In *Proceedings of the 2nd Japanese Symposium on Peptide chemistry; Peptide chemistry 1992* (ed. N. Yanaihara), p. 107. Escom, The Netherlands.
56. Gausepohl, H., Behn, C., Schöpfer, R. and Frank, R. W. (1997). *In Innovation and perspectives in solid phase synthesis and combinatorial libraries* (ed. R. Epton), p. 87. Mayflower Scientific, Birmingham, UK.
57. Frank, R. W. (1992). *Tetrahedron*, **48**, 9217.
58. Blackburn, C., Albericio, F., and Kates, S. A. (1997). *Drugs of the Future*, **22**, 1007.
59. Gerhardt, J. and Rapp, W. (1998). In *Proceedings of the Twenty-fourth European Peptide Symposium* (ed. R. Ramage and R. Epton), p. 150. Mayflower Scientific, Birmingham, UK.
60. Bouitn, J. A. and Fauchere, J.-L. (1996). *In Proceedings of the 1995 International Symposium on laboratory automation and robotics*, p. 36.
61. Krchnak, V. and Lebl, M. (1997). In *Innovation and perspectives in solid phase synthesis and combinatorial libraries* (ed. R. Epton), p. 93. Mayflower Scientific, Birmingham, UK.
62. Hoperich, P. D. (1996). *Nature Biotechnology*, **14**, 1311.
63. Rivero, R. A., Greco, M. N., and Maryanoff, B. E. (1997). In *A practical guide to combinatorial chemistry* (ed. A. W. Czarnik and S. H. DeWitt), p. 201. American Chemical Society, Washington.

# 14

# Manual multiple synthesis methods

B. DÖRNER, J. M. OSTRESH, R. A. HOUGHTEN,
RONALD FRANK, ANDREA TIEPOLD, JOHN E. FOX,
ANDREW M. BRAY, NICHOLAS, J. EDE, IAN W. JAMES, and
GEOFFREY WICKHAM

## 1. Simultaneous multiple peptide synthesis—the T-bag method

### 1.1 Introduction

Simultaneous multiple peptide synthesis enables the parallel synthesis of large numbers of peptides. The T-bag (tea-bag) method (1) was developed along with other methods, e.g. pin synthesis, synthesis on paper plates, synthesis on parallel columns, and synthesis on cellulose (for reviews see ref. 2), as technology to facilitate simultaneous multiple synthesis. Large numbers of peptides, peptidomimetics, and small organic molecules have been prepared using the T-bag method to address different research fields, such as conformational analysis, structure activity analysis, synthesis methodologies, and antibody–antigen interaction studies (for reviews see refs 2 and 3). Using the T-bag method, more than 150 peptides can be prepared in parallel in flexible amounts, with easily enough material for biological tests and analytical studies. The synthesis of peptides of length of up to 26 amino acid residues has been reported (4). Moreover, the T-bag technology is easy to apply in practice and requires very little special equipment.

### 1.2 The T-bag method™

T-bags (*Figure 1*) are prepared by containing solid phase resins within polypropylene mesh material. Polypropylene is rather chemically inert as well as fairly thermally stable (to 150°C), allowing a wide range of chemical reactions to be used for solid phase synthesis without affecting the bag material. Polystyrene cross-linked with 1% divinylbenzene, 100–200 mesh, is mainly used as the solid support, but other types of base resin can be used as well, e.g. TentaGel (5). The size of the resin beads must exceed the size of the pores of the polypropylene mesh material of the T-bags to avoid resin loss during synthesis. Syntheses are carried out manually (1), using semi-

## B. Dörner et al.

automation (6), or within a multiple peptide synthesizer (5). The preparation of T-bags for solid phase synthesis, starting with 100 mg resin per bag, is described in *Protocol 1*.

---

**Protocol 1. Preparation of polypropylene mesh bags**

*Equipment and reagents*

- Polypropylene mesh material (mesh with 74 μm openings, available from Polymer Group, Inc., PO Box 308, Benson, NC 27504, USA)
- Cardboard, about 30 cm × 30 cm, with a template outlined showing rectangles in the size of the T-bags to be prepared
- Scissors

- Ruler
- Black ink pen (ink should have a high carbon content; available through art supply stores
- Sealer (¼ inch seal, available from McMaster Carr, 9630 Norwalk Blvd., Los Angeles, CA 90054, USA)
- Solid phase synthesis resin

*T-bag preparation*

1. Cut two sheets of polypropylene mesh material of size large enough to fit all the T-bags needed for the multiple synthesis.

2. Mark a raster of 3 cm × 4 cm rectangles on one sheet of the polypropylene mesh, using the cardboard template.

3. Mark a number with the ink pen on the bottom of each rectangle.

4. Put the unmarked sheet on top of the sheet with the ink-marked labels and seal the lines between all the rectangles (the very top line remains open—three sides of the bags have to be sealed). The number is permanently sealed into the polypropylene to ensure identification of each bag.

5. Cut the sheet into strips to form pockets and fill each with 100 mg of resin.

6. Seal the open end of the strip and cut all the bags apart.

---

### 1.2.1 Simultaneous multiple peptide synthesis

Synthesis using the T-bag method can be performed using either Boc (2, 6) or Fmoc (7, 8) synthetic strategies. For all manipulations, enough solvent should be used to cover the T-bags (about 3–4 ml per bag containing 100 mg of resin). To enable efficient washings and reactions, the reaction vessels (polyethylene bottles) should be shaken vigorously, preferably through the use of a reciprocating shaker. Thus, during a T-bag synthesis of various sequences in parallel, the deprotection and washing steps can be performed with all bags combined in a single polyethylene bottle as outlined in *Figure 2*. For the amino acid couplings, the bags are separated depending on the different sequences to be prepared. Following the coupling reactions, two washing cycles are done separately before combining all the bags again for subsequent

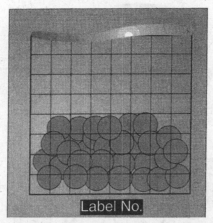

**Figure 1**. Labelled T-bag made out of polypropylene mesh material with a sealed edge, containing resin beads.

washing and deprotection steps. The final cleavage of the peptides from the resin can either be performed directly using the T-bags or the bags can be opened and the free resin portions treated with the necessary cleavage cocktails.

In addition to the parallel preparation of individual peptide sequences, simultaneous multiple peptide synthesis is used together with the divide, couple, recombine method (9) (also termed split-and-mix (10) or portioning-and-mixing (11)) to prepare peptide combinatorial libraries containing mixtures of thousands to millions of peptides (12).

### 1.2.2 Simultaneous multiple peptidomimetic synthesis

Simultaneous multiple peptidomimetic syntheses have been reported, including post-modification of resin-bound peptides (13, 14) and stepwise synthesis of peptidomimetic compounds (15, 16). Furthermore, small organic molecules, such as 2,4,5-trisubstituted thiomorpholin-3-ones (17), 4-amino-3,4-dihydro-2(1*H*)-quinolinones (18), and primary amines (19) have been prepared using the T-bag technology. In addition to amide formation, alkylation, reduction, reductive alkylation, thioether formation, nucleophilic additions using Grignard reagents, and [2 + 2]-cycloaddition are examples of reactions which can be performed on solid supports contained in T-bags.

## 2. Multiple peptide synthesis with the SPOT-technique

### 2.1 Introduction

SPOT-synthesis is a facile and very flexible technique for the simultaneous parallel solid phase chemical synthesis. Series of compounds or compound

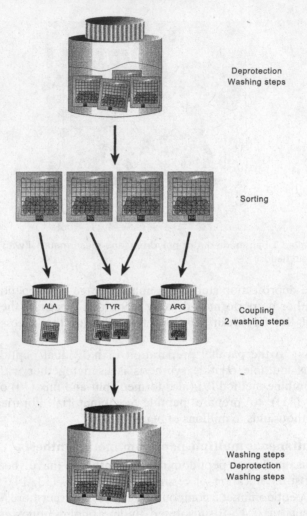

**Figure 2.** The principal steps in simultaneous multiple peptide synthesis using the T-bag technology. The resin-containing bags are treated together for washing and deprotection steps, whereas individual couplings are performed in parallel as required for the different peptide sequences.

libraries composed of series of sub-library pools can be assembled by manual or automated dispensing of small aliquots of solutions containing activated monomer derivatives onto predefined arrays of positions on a planar membrane support. The method provides rapid and low cost access to large numbers of compounds, peptides in particular, both as solid phase bound and solution phase products for systematic biological activity screening (20). Chemical and technical performance of this type of solid phase synthesis has so far been optimized for the assembly of arrays of peptide sequences up to a

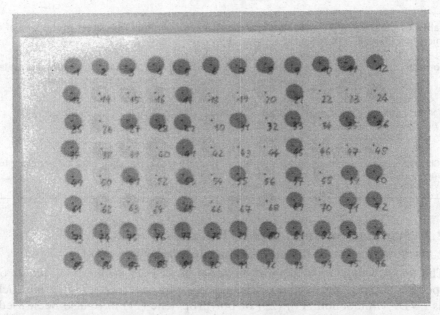

**Figure 3.** A microtitre plate-adapted format of a manually prepared 96-spot membrane. Dark spots indicate stained (blue) free amino functionalities, light spots have turned to yellow after coupling with an Fmoc-amino acid HOBt ester.

length of 20 residues using conventional Fmoc/tBu chemistry. Other syntheses based on amide bond formations, such as peptoids and polyamide nucleic acids (PNAs) (21), have been reported. In principle, chemistries that can be carried out at ambient temperature can be adapted to SPOT-synthesis.

SPOT-synthesis is particularly flexible with respect to numbers and scales that can be accomplished. The arrays are freely selectable to fit the individual needs of the experiment by the choice of membrane material, thickness, area specific anchor loading, and spot size (*Table 1*). The standard format used in manual SPOT-synthesis was adapted to the 8 × 12 array of a microtitre plate with 96 spots (*Figure 3*).

An automated SPOT-synthesizer, the ASP222, has been developed at ABIMED Analysen-Technik GmbH based on a Gilson pipetting workstation. This instrument can handle simultaneously up to six standard membrane sheets. Moreover, automated spotting can be exploited to reduce the size of spots and thus increase considerably the number per area (*Table 1*), which otherwise is extremely tedious to manufacture by hand. At present, up to 2200 spots can be generated by pipetting only 30 nl reaction volumes onto a microtitre plate-sized sheet (in total 13 000 spots per working area of the robot). The instrument so far performs only the pipetting work; all washing steps are carried out manually.

**Table 1.** Standard SPOT-array configurations (each one fitted to a sheet of the size of a microtitre plate, 8 × 12 cm)

| Format | Membrane type[a] | Spotted volume (μl)[b] | Spot size (mm) | Positional distance (mm) | Synthesis scale (nmol) |
|---|---|---|---|---|---|
| 8 × 12 = 96 | 540 | 0.5/0.7 | 7 | 9 | 25 |
| 7 × 10 = 70 | Chr1 | 1.0/1.5 | 8 | 10 | 50 |
| 17 × 25 = 425 | 540 | 0.1/0.15 | 3 | 4 | 6 |
| 40 × 50 = 2000 | 50 | 0.03/0.05 | 1 | 2 | 1 |

[a] Chromatography paper products from Whatman. Typical amino substitution is 0.2–0.4 μmol/cm$^2$ for type 50 and 540 paper, 0.4–0.6 μmol/cm$^2$ for type Chr1 paper.
[b] Volumes are given for array generation step/peptide assembly step.

## 2.2 Synthesis of peptide SPOT-arrays

### 2.2.1 Preparation of the membrane supports

The most widely used support material for SPOT-synthesis is made of specially selected, pure cellulose chromatography paper. Other materials such as PVDF (Immobilon AV) (22) or polypropylene (21) have also been used. All membranes require a chemical derivatization to carry free amino functions. Paper is easily derivatized by the esterification with an N-Fmoc protected amino acid (e.g. Fmoc-βAla) to available hydroxyl functions on the cellulose fibres of the whole sheet followed by Fmoc-cleavage (20). The arrays of spot reactors on such amino-membranes are then generated by coupling a second Fmoc-βAla or suitable linker/anchor acid, applying spotwise only small aliquots of activated solution as given in *Table 1* with either a micropipette or pipetting robot. After completion of this spot-reaction, all residual amino functions between spots are blocked by acetylation. This array formation step requires very accurate pipetting. During peptide assembly, slightly larger volumes are dispensed and the spots then formed exceed those initially generated in order to avoid incomplete couplings at the edges.

The choice of a linker/anchor compound determines the fate of the peptide products. Stable amide/ester bonds such as with the βAla–βAla spacer (*Protocol 3*) result in an array of immobilized peptides. Conventional linker chemistry can be incorporated to yield cleavable peptides in different formats. Esterification reactions are problematic in this open reactor system; in such cases, a preformed amino acid linker compound should be used. Such linkers should allow a solid phase bound deprotection/pre-purification followed by a mild final release reaction. Several options have been described (23–25). More sophisticated and superior amino membranes are now available from several sources (ABIMED, AIMS, Jerini, PerSeptive).

Free amino functions on the spots can be stained with bromophenol blue (26) prior to the coupling reactions. This allows visual monitoring of the performance of all the synthesis steps. Thus, a standard membrane for SPOT-synthesis displays an array of light blue spots on a white background. Each spot can be marked by writing a number in pencil next to it (*Figure 3*). These numbers refer to the corresponding peptide sequences that are assembled on them and are a guide for rapid manual distribution of the solutions of activated amino acid derivatives at each elongation cycle using a suitable pipetting protocol. For automated pipetting, no pencil marking is necessary. Exact fixing of the membranes is ensured by the perforation for the holder pins.

---

**Protocol 2.**   Generation of the SPOT-array

*Equipment and reagents*

- A kit that includes all necessary items for manual SPOT-synthesis is currently available from GENOSYS
- Flat reaction troughs with good closing lids made of chemically inert material (polyethylene, polypropylene) that have the dimensions slightly larger than the membranes used
- A micropipette adjustable from 0.5 to 10 μl (Eppendorf or Gilson) with corresponding plastic tips; alternatively, the ASP222 spotting robot from ABIMED
- Small (1.7 ml) plastic tubes (e.g. Eppendorf, safe-twist) as reservoirs for amino acid solutions
- A rocker table
- Two dispensers adjustable from 5 to 25 ml for DMF and alcohol stock containers
- Amino-cellulose membranes from ABIMED or AIMS

- Software: special DOS-PC computer programs for the generation of peptide lists and pipetting protocols are included in the synthesis kit and the operation software of the ASP222
- A stock solution of 10 mg/ml bromophenol blue indicator (Merck) in DMF
- N,N'-diisopropylcarbodiimide (DIC)
- 2% v/v acetic anhydride solution in DMF
- Amine-free N,N-dimethylformamide (DMF)[a]
- Alcohol (methanol or ethanol) of technical grade
- Amine-free N-methylpyrrolidinone (NMP)[b]
- 20% v/v solution of piperidine in DMF; piperidine is toxic and should be handled with gloves under a hood!
- N-hydroxybenzotriazole (HOBt)[c]

*Method*[d]

1. Generate the list of peptides to be prepared and correspondingly define the array(s) required for your particular experiment according to number, spot size, and scale (see *Table 1*).

2. Mark the spot positions on the membranes with pencil dots for manual synthesis and place the membranes in the reaction trough or fix them on the platform of the ASP222-robot.[e]

3. Prepare a solution containing activated anchor compound in NMP (e.g. 0.3 M Fmoc-βAla-OH, 0.45 M HOBt, and 0.35 M DIC; or 0.3 M Fmoc-βAla-OPfp, 0.075 M HOBt, and 0.35 M DIC).[f]

4. Leave this solution for 30 min and then spot aliquots of chosen volume to all positions according to the array configuration prepared.

5. Leave for 15 min.

6. Repeat spotting once and let react for 30 min.

---

## Protocol 2. *Continued*

7. Wash the membrane with 20 ml of acetylation mix for 30 s, once again for 2 min, and finally leave overnight in acetylation mix.

8. Wash the membrane three times with 20 ml DMF.

9. Incubate the membrane for 10 min with 20 ml piperidine/DMF.

10. Wash the membrane four times with 20 ml DMF.

11. Incubate the membrane with 20 ml of 1% bromophenol blue stock in DMF (must be a yellow solution!). Repeat if traces of remaining piperidine turn it into a dark blue solution. Spots should be stained only light blue!

12. Wash twice with 20 ml alcohol.

13. Dry with cold air from a hair-drier in between a folder of 3 mm paper.

14. Seal in a plastic bag and store at −20 °C.

[a] Amine contamination is checked by addition of 6 μl of bromophenol blue stock solution to 1 ml of DMF. If the resulting colour is yellow, then this batch can be used without further purification.
[b] Amine contamination is checked as under DMF. If the resulting colour is yellow, then the NMP can be used without further purification. Most commercial products, however, are not acceptable. Then add alumina oxide (for chromatography, Fluka) and agitate gently overnight. Pass through a 20 μm polyethylene sinter or cellulose filter. Divide the clear liquid into 100 ml portions and store tightly closed at −20°C.
[c] HOBt is only available as the monohydrate. Dehydrate in a desiccator over phosphorus pentoxide at 50°C and $10^{-3}$ bar for at least 3 days. Store in a tightly closed container at room temperature.
[d] All volumes given below hold for a standard paper sheet of 8 × 12 cm (*Table 1*) and have to be adjusted to more sheets, other paper qualities, or sizes. Solvents or solutions used in washing/incubation steps are agitated gently on a rocker plate and aspirated or decanted after the time indicated.
[e] Membranes can be cut easily into parts or pieces with scissors prior to or post synthesis as required for the particular experiment. Simply mark the cutting lines with a pencil. Spot positions should be also marked with a pencil dot if treated individually after the synthesis. The pencil marking is sufficiently stable during the synthesis procedure.
[f] Although more expensive, the pentafluorophenyl ester is preferred when working with the ASP222 robot (see Chapter 13, Section 4.5).

### 2.2.2 Peptide assembly

Any series of individual peptide sequences can be freely arranged as a two-dimensional array on a SPOT membrane for systematic analysis with, for example, overlapping fragments derived from a protein sequence (SPOTscan), stepwise N- or C-terminally truncated fragments (SPOTsize), substitution analogues (SPOTsalogue), etc. (20). Moreover, modern peptide library screening approaches allowing the a priori delineation of peptide epitopes (27) have been incorporated into SPOT-synthesis (28, 29). These approaches exploit the preparation of arrays of defined peptide mixtures (pools) by incorporation of randomized positions and the presentation of

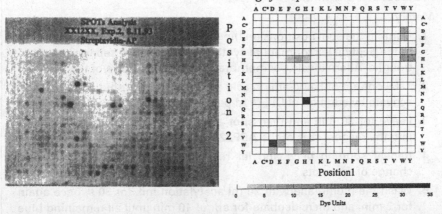

**Figure 4.** Example of a first screening round for the **a priori** delineation of peptides binding to streptavidin. Left: photograph of a membrane with 400 spots of a hexapeptide library in the format Ac-XX12XX- (plus 25 reference spots in the upper lane) after probing with a streptavidin–alkaline phosphatase conjugate. Bound protein was visualized with a phosphatase specific colour reaction (20). Right: spectral diagram display of Ac-XX12XX-showing the correlation of quantified signals to the defined amino acid residues at positions 1 and 2 (X positions include all 20 proteinogenic amino acids); C* = Cys(Acm). Clearly, the XXHPXX- pool gives the strongest signal.

entire peptide libraries (e.g. all 64 Mio-hexapeptides) as strategic sets of sub-libraries (*Figure 4*) . Randomized positions (X) within a peptide sequence assembled on a spot are introduced quite reliably by coupling with equimolar amino acid mixtures and applying these at a sub-molar ratio with respect to available amino functions on the spots. This is to allow all activated derivatives to react quantitatively during a first round of spotting, then completing all peptide elongations by two to three successive repeats of spotting. Using this coupling procedure, any position in a peptide sequence can be randomized easily without special considerations or increase in technical effort.

---

**Protocol 3.** Assembly of peptides on SPOTs

*Equipment and reagents*

- For general information see *Protocol 2*
- Monomer derivatives: [a] Fmoc-amino acids are available from several suppliers in sufficient quality. *In situ* prepared HOBt esters of these Fmoc-amino acids in NMP are used throughout for spotting reactions. Side-chain protection is Cys(Acm) or Cys(Trt), Asp(OtBu), Glu(OtBu), His(Boc) or

His(Trt), Lys(Boc), Asn(Trt), Gln(Trt), Arg(Pbf), Ser(tBu), Thr(tBu), Trp(Boc), and Tyr(tBu). Dissolve 1 mmol of each in each 5 ml NMP containing 0.3 M HOBt to give 0.2 M solutions. Divide into 100 µl aliquots in correctly labelled Eppentorf tubes. Close tightly, freeze in liquid nitrogen, and store at −70°C.

311

**Protocol 3.** *Continued*

*Method*

1. Number the spot positions on the membranes with a pencil according to the peptide list for manual synthesis and place in separate reaction troughs. Alternatively, fix membranes on the platform of the ASP222.

2. Take one set of Fmoc-amino acid stock solutions from the freezer and activate by addition of DIC (4 µl per 100 µl vial; *ca.* 0.25 M).[b]

3. Leave for 30 min and then spot aliquots of these solutions or place into the reagent rack of the ASP222 and start. Leave for 15 min.

4. Repeat spotting once and let react for 30 min. Check the colour change of the spots.[c]

5. Wash the membrane with 20 ml acetylation mix for 30 s, once again for 2 min, and then incubate for about 10 min until all remaining blue colour has disappeared.

6. Wash the membrane three times with 20 ml DMF.

7. Incubate the membrane for 5 min with 20 ml piperidine/DMF.

8. Wash the membrane four times with 20 ml DMF.

9. Incubate the membrane with 20 ml of 1% bromophenol blue stock in DMF (must be a yellow solution!). Repeat if traces of remaining piperidine turn it into a dark blue solution. Spots should be stained only light blue!

10. Wash twice with 20 ml alcohol.

11. Dry with cold air from a hair-drier in between a folder of 3 mm paper.

12. Go back to step 2 for the next elongation cycle.

13. After the final cycle, all peptides can be N-terminal acetylated by carrying out steps 3, 4, 10, and 11.

[a] Special chemical derivatives: free thiol functions of cysteine can be problematic because of post-synthetic uncontrolled oxidation. To avoid this, you can replace Cys by serine (Ser), alanine (Ala), or α-aminobutyric acid (Abu). Alternatively, choose the hydrophilic Cys(Acm) and leave protected. For the simultaneous preparation of peptides of different size with free amino terminus, couple the terminal amino acid as $N^\alpha$-Boc derivatives so that they will not become acetylated during the normal elongation cycle. Boc is removed during the final side chain deprotection procedure (*Protocol 4*). Special labels can be attached to the N-termini by spotting respective derivatives in an additional coupling cycle. We have successfully added biotin via the *in situ* formed HOBt-ester (normal activation procedure) or fluorescein via the isothiocyanate (FITC) dissolved in NMP.

[b] Assembly of peptide pools on SPOTs: this procedure follows essentially the steps of *Protocol 3* except that the amino acid stock solutions are diluted to 50 mM with NMP and activated with only one quarter of DIC. For X-couplings, an equimolar mixture of the amino acid derivatives to be presented at these positions is prepared by combining equal aliquots of the diluted stock solutions (e.g. all 20). This mixture is activated and applied in the same way as the other individual amino acid solutions. Spotting (step 2) per elongation cycle is repeated four times.

[c] Some peptide sequences and terminal amino acid residues such as Cys, Asn, or Asp give a particularly weak staining. The indication of free amino functions is quite sensitive and a colour change to green is sufficient for a good coupling.

---

**Protocol 4.** Side-chain deprotection of SPOT assembled peptides

*Equipment and reagents*

- For general information see *Protocol 2*
- Deprotection mix: 50% TFA, 3% triisobutyl-silane, 2% water, and 45% DCM (v/v).

Trifluoroacetic acid is very corrosive and should be handled with gloves under a hood!

*Method*

1. Prepare 40 ml of deprotection mix per standard membrane.

2. Place the dried membrane from step 10 of Protocol 3 in the reaction trough (only one membrane per trough; do not stack several membranes!).

3. Add 20 ml deprotection mix, remove bubbles from under the membrane, close the trough tightly and agitate for 1 h.

4. Replace the deprotection mix with the remaining 20 ml and agitate again for 1 h.

5. Wash the membrane four times for 2 min with 20 ml DCM.

6. Wash the membrane three times for 2 min with 20 ml DMF.

7. Wash the membrane three times for 2 min with 20 ml alcohol.

8. Wash the membrane three times for 10 min with 20 ml 1 M acetic acid in water.

9. Wash the membrane three times for 2 min with 20 ml alcohol.

10. Dry with cold air from a hair-drier in between a folder of 3 mm paper.

11. Seal in a plastic bag and store at –20°C.

---

## 2.3 Applications of peptide SPOT-arrays

Initially, SPOT-synthesis was developed for rapid analysis of linear peptide epitopes recognized by antibodies. After blocking unspecific binding sites, the whole membrane (or parts of it) is incubated with the anti-serum and detection of bound antibody is achieved by conventional immuno- or Western-blot procedures with colour dye development. If peptide spots have not become chemically, enzymatically, or otherwise irreversibly modified, peptide arrays on cellulose membranes are reusable many times (>20) when treated carefully. Signal patterns obtained from peptide arrays on SPOTs can be documented and quantitatively evaluated using modern image analysis systems as used with other two-dimensional analysis media such as electro-phoresis gels and blotting membranes.

The list of further applications, however, is rapidly expanding. These include peptide interactions with proteins in general (cytokines, enzymes, receptors, etc.), metal ions, and nucleic acids, as well as whole cells. Any labelling technique such as radioisotopes (e.g. introduced by *in vitro* transciption/translation of cloned DNA) or fluorescent dyes is fully compatible. A full report on these is beyond the scope of this chapter. The user can follow described procedures (20) or simply adapt his/her own assay.

# 3. Manual multipeptide synthesis in block arrays

Described here is a manual system to enable the synthesis of large numbers of peptides easily, cheaply, and in quantities of a few milligrams of each peptide. Approximately four to five coupling cycles can be completed in one working day with a typical set of 48 different peptides completed in a week.

## 3.1 Hardware

The apparatus and procedures described here are performed in a device built and supplied by Biotech Instruments Ltd. (30). The system is based on a set of 48 wells drilled in a two-part polythene block, each well being a reactor for a single peptide. The design of the system centres around trying to make a foolproof synthesis method, suited for either simple manual operation or some form of mechanization. The device allows up to 48 different peptides to be synthesized simultaneously within a few days, using virtually any type of resin or activation chemistry at scales of 5–10 μmol for each peptide. The organization of the synthesis is assisted by means of computer-generated printout, which guides the tracking of the synthesis and directs the operator during the coupling reactions.

The top block contains the 48 reaction wells with their centres spaced 19 mm apart; the layout is shown in *Figure 5*. This spacing of the wells is exactly double that of the wells in a microtitre plate, and thus enables alternate channels of a standard multichannel pipette to be used for reagent delivery if

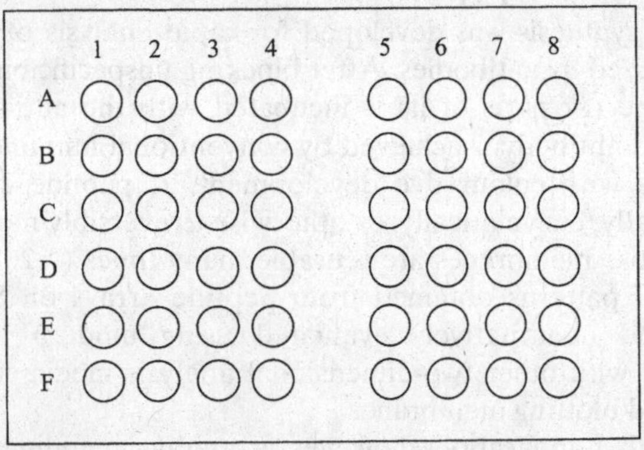

**Figure 5.** Layout of the reaction wells in the top block.

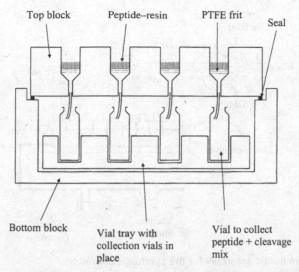

Top block      Peptide–resin      PTFE frit

Seal

Bottom block      Vial tray with collection vials in place      Vial to collect peptide + cleavage mix

**Figure 6.** Cross-section of the block system, showing the process of cleaving the peptides off the solid phase.

required. A short length of PTFE tubing fits into the outlet hole of each well to drain the wells and to collect the cleaved peptide in a vial during the final stage of synthesis. A PTFE frit, pore size 100 μm, is placed in the bottom of each 10-mm-diameter well to retain the peptide–resin.

*Figure 6* shows a cross-section of just four wells in the block, arranged as for the cleavage reaction. Reagents and solvents are added from the top and, after reaction, allowed to flow through into the chamber below with the assistance of a small, solvent-resistant vacuum pump.

The complete fluid system is shown in *Figure 7*, where the solvents are contained in pressurized bottles and are selected by a rotary valve. The flow to the four-way manifold is controlled by a two-way ON/OFF valve or pinch clamp in silicone tubing. The use of a manifold enables solvents to be delivered to four wells simultaneously. Alternatively, the solvents can be delivered by means of a multichannel pipette, operating with alternate channels delivering into four wells at a time.

After all the synthesis cycles are completed, the peptides are simultaneously cleaved off the resins and collected in glass vials placed under the reaction wells. The vials are held in a tray for stability and to maintain their exact positioning under the drain tube for each of the wells. These peptide solutions can then be dried and worked up in the usual way.

## 3.2 Chemistry

The method uses the standard Fmoc/*t*Bu strategy with any convenient activation procedure. For manual operation (*Protocol 5*), it can be somewhat

315

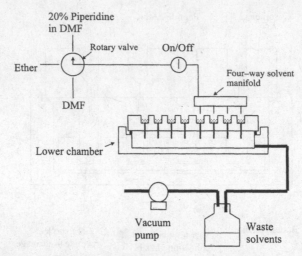

20% Piperidine
in DMF

Rotary valve    On/Off

Ether

Four–way solvent
manifold

DMF

Lower chamber

Vacuum
pump

Waste
solvents

**Figure 7.** Entire fluidic assembly for the synthesis operation.

easier to use the OPfp active esters, as only two reagents have to be added during the coupling cycle, the Fmoc-amino acid OPfp ester and 1-hydroxy-benzotriazole. Free amino acids are more stable in solution but require the mixing of three reagents. To ensure complete coupling reactions, a minimum of 5 × molar excess of the amino acids is used for all the synthesis steps, as the cost of the amino acids is not usually a major expense in the work. Any convenient solid phase can be used in this system, although the author has mainly worked with Wang resins for this scale of synthesis. Both peptide C-terminal acids and amides can be synthesized simultaneously and unusual amino acids can easily be incorporated. Typically, the peptides are cleaved off the resins using a cocktail of trifluoroacetic acid:1,2-ethanedithiol:water (95:2.5:2.5 v/v; *Protocol 6*), a mixture which is generally effective. As large numbers of peptides are being made simultaneously, the same cleavage mixture must be used for every peptide regardless of amino acid content. A philosophy of maximum protection has been used by the author, again because it is not usually viable to treat peptides individually.

The times and steps for a typical synthesis cycle is shown in *Table 2*. The times are not critical and if in doubt can be extended, as can the number of washes. After the deprotection step, it is advisable to include one wash step which almost fills the wells in order to completely wash the sides and remove all the piperidine. The ether wash removes all solvent from the resin and completely dries the PTFE frit. This prevents any solvent normally held within the resin from diluting the incoming activated Fmoc-amino acid and the high surface tension of the dry frit prevents the amino acid mixture from draining through the frit. The subsequent addition of the amino acid/reagent mixture must be done gently to prevent it from draining through the frit.

**Table 2.** The operations for one synthesis cycle

| Step no. | Solvent/reagent | No. of washes/dwell time [a] |
|----------|-----------------|------------------------------|
| 1 | Wash with DMF | 2 × |
| 2 | Deprotect with 20% piperidine | 4 min |
| 3 | Deprotect with 20% piperidine | 6 min |
| 4 | Wash with DMF | 3 × |
| 5 | 'Big' wash with DMF | 1 × |
| 6 | Wash with DMF | 1 × |
| 7 | Wash with ether | 2 × |
| 8 | Vacuum dry resin [b] | 2 min |
| 9 | Mix and add amino acids | |
| 10 | Acylation reaction [c] | 60 min |
| 11 | Wash with DMF | 3 × |
| 12 | | End of cycle |

[a] The exact times are not critical.
[b] Essential to prevent the reactants from draining through the frit.
[c] For short peptides, the acylation reaction time can be reduced to 30 min.

## 3.3 Software

A program written in Basic is used to edit and arrange all the 48 peptide sequences into an order. The problem of ensuring that the correct amino acid is added to any specific well is solved by the program printing a series of overlay sheets showing all the amino acids that are to be added during any one cycle of the synthesis. The position of the printing on these overlays corresponds exactly to the layout of the wells in the block and when pegged in place over the top of the block shows exactly where the amino acids are to be added into each well. When each amino acid is to be added to the deprotected resin, a pipette is used to pierce the paper and deliver the reagents into the well below. In this way, the overlay sheet not only serves to indicate which amino acid should go into any one well, but also serves to show which well has already been filled with reagents.

---

**Protocol 5.** The synthesis cycle

Safety notes: All the steps involving manipulation of solvents must be performed in an efficient fume hood.

*Reagents*

- Peptide synthesis grade DMF
- N-deblock solution: 20% v/v piperidine in DMF
- Amino acid-derivatized resins (5 μmol of each resin per reaction well)
- Solutions of the required Fmoc-amino acids. For each amino acid addition, dissolve 25 μmol of the protected amino acid in 150 μl of DMF

- A solution of TBTU in DMF at a concentration of 450 μmol/ml (145 mg of TBTU in 1.00 ml of DMF); use 50 μl of this solution per addition
- A solution of N-methyl morpholine (NMM) in DMF at a concentration of 500 μmol/ml (55 μl of NMM diluted to 1.00 ml with DMF); use 50 μl of this solution per addition
- Diethyl ether

---

317

**Protocol 5.** *Continued*

*Method (5 μmol synthesis scale)*

1. At the start of the synthesis, place caps over each well to avoid any spilt resin falling into the wrong reaction well.

2. Following the positions shown on the first overlay sheet, add 5 μmol of the correct amino acid resin to each of the wells in turn and replace the cap after each addition.

3. Remove the caps when all the resins have been added.

4. Use the software to calculate the amounts and volumes of amino acid solutions required. Weigh out the amino acids for the current day's work in labelled tubes and dissolve them in the required volume of DMF. Prepare the working volumes of TBTU and NMM solutions; these should be stable for the duration of the synthesis.

5. Following the procedure in *Table 2*, wash and N-deblock each of the amino acid resins.

6. After the ether wash and drying stage, the wells are ready for the addition of the incoming amino acid.

7. Place the appropriate coupling overlay sheet over the block and peg into position.

8. Working with one amino acid at a time, mix 150 μl of the amino acid solution with 50 μl of TBTU solution and 50 μl of NMM solution in an Eppendorf tube.

9. Pipette this reaction mix through the paper overlay into the correct well.

10. Repeat this procedure for all the additions for each amino acid in turn. When the process is complete, all the wells will have a puncture mark in the paper indicating that the additions are complete.

11. After the coupling time is over, the excess reagents can be pumped through to waste and the resins washed with solvent ready for the next cycle.

---

**Protocol 6.** Peptide cleavage

Safety notes: This process must be performed in an efficient fume hood. The cleavage solution is corrosive, volatile, and very mobile, and is therefore difficult to control in a pipette.

*Reagents*

- Prepare $N \times 2$ ml of cleavage mixture, where $N$ is the number of peptides to be cleaved, each containing 95% trifluoro- acetic acid + 2.5% 1,2-ethanedithiol + 2.5% v/v water

*Method*

1. Follow the procedure in *Table 2*, steps 1 to 8, to remove the final Fmoc-groups from the assembled peptides. Then wash and dry the peptide–resins.

2. Prepare the correct volume of the cleavage cocktail.

3. For each peptide to be cleaved, place a 4 ml glass collection vial scribed with the peptide code number in the collection rack in the lower chamber and carefully fit the block assembly together ensuring that the drain tubes enter all the vials. It is essential to scribe the tubes with a diamond pen as any writing with felt-tipped pen will be erased by the acid.

4. Add about 0.5 ml of the cleavage mix to each reaction well. A plastic Pasteur pipette is best used for this operation as the vigorous action of a stepper-type pipette can cause some of the resin to splash out of the wells. This solution will slowly drain through the frit into the vial below, simultaneously cleaving and extracting the peptide off the resin.

5. At approximately 20 min intervals, briefly apply the vacuum to drain any residual acid into the wells, then with the vacuum off.

6. Add a further 0.5 ml of the cleavage mixture, and repeat step 5.

7. When all the cleavage mix has been added and drained through, turn on the system vacuum pump. This will drain all the remaining acid into the vials and start to evaporate the acid by drawing a rapid stream of air over the surface of the cleavage mix in the vial.

8. Keep the vacuum pump operating for about 2-4 h until all the acid has been evaporated.

9. Add about 2 ml of cold diethyl ether to precipitate the peptides. If a light suspension is formed, the tubes can be centrifuged at about 3000 r.p.m. to ensure complete precipitation of the peptide.

10. Carefully decant the ether supernatant and repeat step 9.

11. Allow the peptides to air dry.

12. Dissolve the peptides in 1 ml of 10% acetic acid and allow to stand for about 1 h to complete the deprotection of any *N*-Boc-protected tryptophan.

13. Lyophilize the peptide solutions.

# 4. Synthesis of peptides by the Multipin™ method

## 4.1 Introduction

The Multipin™ method is an effective, low cost, simultaneous multiple peptide synthesis technology which gives researchers ready access to large numbers of peptides (31–38). Chiron Technologies currently markets a range of Multipin™ products for both peptide and non-peptide solid phase synthesis. This chapter gives an overview of a selection of the key procedures used for peptide synthesis with commercially available Multipin™ kits. Synthesis is performed on plastic pins arranged in the $8 \times 12$ microtitre plate format, thus allowing for the simultaneous preparation of sets of up to 96 peptides (or multiples thereof).

The 'pin' consists of a radiation-grafted polypropylene 'crown' fitted to an inert polypropylene 'stem'. Graft polymers used with the Multipin™ system include polystyrene, a methacrylamide copolymer (38), and poly(hydroxyethyl methacrylate) (34) (HEMA). We have found that the HEMA surface is best suited to peptide synthesis. Historically, peptides were prepared in a non-cleavable format on the crown surface for epitope mapping applications (39). Over the past decade, however, most peptides prepared by the Multipin™ method have been synthesized on cleavable linkers. The linkers used for peptide synthesis are outlined below.

Originally, Multipin™ peptide synthesis was performed using a Boc protocol (39); however, due to more streamlined handling (required for efficient multiple synthesis), Fmoc/tBu synthetic protocols are now used. The TFA-mediated cleavage method used with Fmoc/tBu synthesis is far more easily adapted for large numbers than the HF required with Boc synthesis.

Crowns with a loading of 20 μmol yield around 20 mg of a typical 10-mer peptide. Larger quantities can be accessed by preparing multiple copies of a target peptide, a simple process with simultaneous multiple synthesis. When small numbers of peptides are being prepared, synthesis is often performed on unmounted crowns in glass vials or polypropylene tubes. The synthesis cycle used with the Multipin™ method is: coupling, washing, Fmoc-deprotection, and washing. Coupling reactions can be performed in the wells of microtitre plates, in which case the washing and deprotection steps are performed in polypropylene baths. As with other forms of solid phase synthesis, the washing steps are critical. As noted in the protocols given below, the pins must be fully immersed and soaked for the given time. High quality solvents are always used for washing.

## 4.2 Linkers

*Figure 1* summarizes the linkers used for peptide synthesis by the Multipin™ method. Most carboxylic acids are prepared on the 4-(hydroxymethyl)-phenoxyacetic acid handle **1** (40). To avoid diketopiperazine formation during the synthesis of peptides containing a C-terminal Pro residue, the sterically hindered linker **2** is used (41). To minimize the risk of racemization during linker derivatization, Asp and Glu are linked via their side-chain functionality, as shown in **3**. Similarly, Asn and Gln are prepared by coupling Fmoc-Asp-OtBu and Fmoc-Glu-OtBu, respectively, to the Rink linker **4** (42).

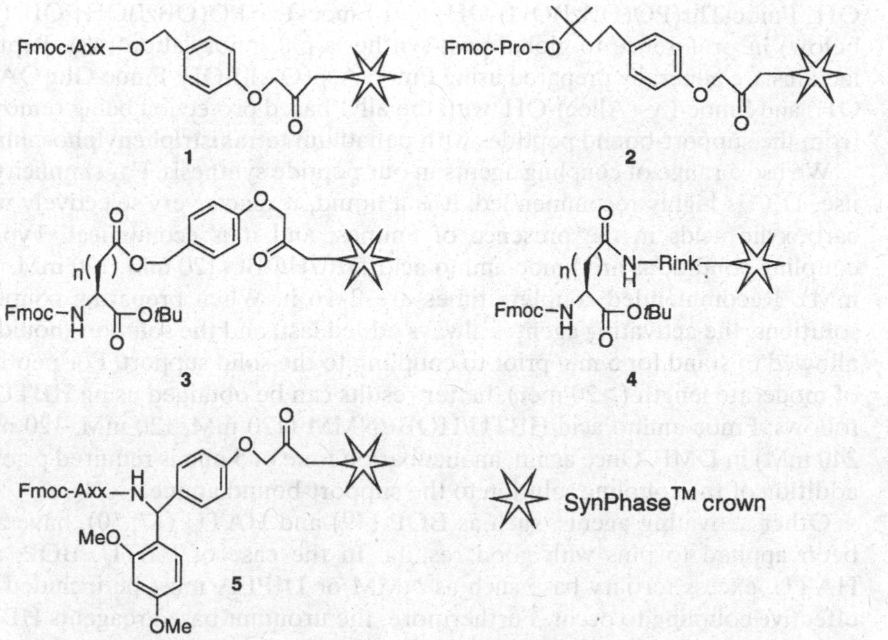

**Figure 8.** Linkers used on SynPhase™ crowns ($n$ = 1, Asp; $n$ = 2, Glu).

Primary peptide amides are prepared on the Rink linker **5**. Although not covered in this chapter, secondary amides have been prepared on the linker described by Barany and co-workers (43). Peptide aldehydes have been prepared on the oxazolidine linker (44). The trityl linker (45) has also been used for the preparation of protected peptide fragments and other specialized C-terminal endings such as hydroxamic acids (46).

## 4.3 Peptide synthesis

Although a variety of solvents have been reported for peptide synthesis, we use highly purified DMF for all coupling reactions (*Protocol 7*). Where Fmoc-protected amino acids require acid-labile side-chain protection, we recommend: Arg(Pbf), Asp(O*t*Bu), Cys(Trt), Glu(O*t*Bu), His(Boc), Lys(Boc), Ser(*t*Bu), Thr(*t*Bu), Trp(Boc), and Tyr(*t*Bu). Although Fmoc-His(Boc)-OH is thermally labile and must be stored at −20°C, we have found that it couples more efficiently than either Fmoc-His(Trt)-OH or Fmoc-His(Bum)-OH. Fmoc-Asn(Trt)-OH and Fmoc-Gln(Trt)-OH are generally used with HBTU, HATU, or BOP activation; the non-side-chain protected derivatives of Asn and Gln can be used with DIC/HOBt activation in short peptides or at the N-terminal coupling (47).

For phosphopeptide synthesis, we routinely use Fmoc-Ser(PO(OBzl)OH)-

OH, Fmoc-Thr(PO(OBzl)OH)-OH, and Fmoc-Tyr(PO(OBzl)OH)-OH (see below) in preference to global post-synthesis phosphorylation (48). Peptide lactams are generally prepared using Fmoc-Asp(OAll)-OH, Fmoc-Glu(OAll)-OH, and Fmoc-Lys(Alloc)-OH, with the allyl-based protection being removed from the support-bound peptides with palladium tetrakistriphenylphosphine.

We use a range of coupling agents in our peptide synthesis. For simplicity of use, DIC is highly recommended: it is a liquid, it reacts very selectively with carboxylic acids in the presence of amines, and it is economical. Typical coupling conditions are Fmoc-amino acid/DIC/HOBt (120 mM, 120 mM, 120 mM). Recommended coupling times are 2–16 h. When preparing coupling solutions, the activating agent is always added last, and the solution should be allowed to stand for 5 min prior to coupling to the solid support. For peptides of moderate length (>20-mer), better results can be obtained using HBTU as follows: Fmoc-amino acid/HBTU/HOBt/NMM (120 mM, 120 mM, 120 mM, 240 mM) in DMF. Once again, an incubation time of 5 min is required prior to addition of the coupling solution to the support-bound amine.

Other activating agents, such as BOP (49) and HATU (47, 50), have also been applied to pins with good results. In the case of HBTU, BOP, and HATU, excess tertiary base such as NMM or DIPEA must be included for effective coupling to occur. Furthermore, the uronium-based reagents HBTU and HATU should not be used in excess (with respect to amino acid) as they can react with support-bound primary amines (51). Consequently, BOP/DIPEA (100 mM, 200 mM) is the preferred reagent when preparing cyclic peptides via lactam formation on the pin surface. We have found HBTU/HOBt/DIPEA to be the activating agent of choice when coupling Fmoc-Ser(PO(OBzl)OH), Fmoc-Thr(PO(OBzl)OH)-OH, and Fmoc-Tyr(PO-(OBzl)OH)-OH (48). We recommend the use of TFFH/DIPEA (100 mM, 200 mM) for sterically demanding couplings such as those to N-alkylated amino acids and for couplings to unreactive amines such as anilines.

After each coupling reaction, the pins are subjected to a rigorous washing protocol. This is conveniently performed in a bath, so that simultaneous multiple washing is a straightforward process. Sufficient solvent is placed in the bath so that the crowns are fully covered when the block of 96 pins is placed in the bath. For consistency, each wash is timed. The standard washing protocol used after either coupling or Fmoc deprotection steps is: DMF, MeOH, followed by air drying. Note that the HEMA polymer is well solvated by MeOH. Fmoc deprotection is generally performed by soaking the pins in 20% piperidine in DMF (v/v) for 30 min. Like washing, this is conveniently performed in a bath. Following deprotection, excess reagent is shaken from the pin surface and the pins are then subjected to the standard washing protocol.

At the completion of peptide assembly, the peptides are side-chain deprotected and cleaved, usually concurrently. In the Multipin™ method, cleavage is performed using TFA solution containing scavengers. Cleavage can be performed either into either polypropylene tubes, which are racked into the 8 × 12 microtitre plate format, or into larger polypropylene tubes (*Protocol 8*). The latter method is convenient when multiple pins containing the one sequence are cleaved together, and when larger volumes of ether solution are required to extract thiol-based scavengers and by-products of side-chain deprotection from the cleaved peptides. The cleavage solutions are evaporated using either a solvent resistant vacuum centrifuge or a stream of dry nitrogen gas. The target peptides should be dissolved (or suspended) in aqueous acetonitrile and freeze dried to remove remaining volatile impurities prior to sampling for analysis and/or purification.

---

**Protocol 7.** Synthesis of peptides on pins

*Equipment and reagents*

- Multipin™ multiple peptide synthesis (MPS) kit (Chiron Technologies). The kit includes crowns and stems, plastic baths and reaction trays, software (for IBM or Macintosh), and a manual
- Solvent-resistant pipettors (e.g. Oxford, Gilson) and tips
- Fluorenylmethyloxycarbonyl (Fmoc)-protected amino acids with appropriate side-chain protecting groups
- N,N'-Diisopropylcarbodiimide (DIC) (Sigma)

- 1-Hydroxybenzotriazole hydrate (HOBt)
- BOP and HBTU
- Tetramethylfluoroformamidinium hexafluorophosphate (TFFH) (PerSeptive Biosystems)
- Miscellaneous reagents: N,N-diisopropyl-ethylamine (DIPEA), N-methylmorpholine (NMM)
- Solvents: N,N-dimethylformamide (peptide grade) (DMF), methanol (AR grade)
- Piperidine

*Method*

If multiple sets of peptides are being synthesized, it is advisable to refer to the manual provided with the Multipin™ Multiple Peptide Synthesis (MPS) kit. Use the software to generate the set of peptide sequences to be synthesized.

1. Mount crowns with attached stems into the block in the required format, as indicated by software.

2. Fmoc-deprotection procedure:[a] immerse crowns in 20% piperidine/DMF for 30 min. For multiple synthesis using the 8 × 12 format; the block containing up to 96 crowns on stems can be immersed in a bath containing the deprotection solution.

3. Post-Fmoc-deprotection washing procedure: wash the crowns in turn with DMF (2 × 5 min) and MeOH (2 × 5 min), and then air dry (15 min).

4. Coupling procedures

    *Procedure 1.* DIC/HOBt (standard coupling): prepare the coupling solution such that the final coupling concentrations are Fmoc-amino acid/DIC/HOBt (120 mM:120 mM:120 mM) in DMF. Allow at least 1.0–1.5 ml per I-series crown (loading = 20 μmol). Add coupling solution to crowns and couple for 2 h at room temperature.

**Protocol 7.** *Continued*

> *Procedure 2.* HBTU/HOBt/NMM (standard coupling): prepare the coupling solution such that the final coupling concentrations are Fmoc-amino acid/HBTU/HOBt/NMM (120 mM:120 mM:120 mM:240 mM) in DMF. Add coupling solution to crowns and couple for 2 h at room temperature.
>
> *Procedure 3.* HBTU/HOBt/DIPEA (phospho-amino acids): same as HBTU/HOBt/NMM except DIPEA replaces NMM.
>
> *Procedure 4.* TFFH/DIPEA (for sterically hindered unreactive amino acids): prepare the coupling solution such that the final coupling concentrations are Fmoc-amino acid/TFFH/DIPEA (120 mM:120 mM:240 mM) in DMF. Add coupling solution to crowns and couple for 2 h at room temperature.
>
> *Procedure 5.* BOP/DIPEA (intramolecular lactam formation): prepare the coupling solution such that the final coupling concentrations are BOP/DIPEA (120 mM:240 mM) in DMF. Add coupling solution to crowns and couple for 18 h at room temperature.

5. Post-coupling washing procedure: DMF (2 × 5 min), MeOH (2 × 5 min).

6. Air dry (15 min) the crowns.

[a] The procedures are illustrative of what is contained in the manual.

---

**Protocol 8.** Side-chain deprotection and cleavage of peptides

*Equipment and reagents*

- Additionally, depending on the type of cleavage, solvent for the extraction of peptide from the crowns or for washing of the dried peptide might be required. Appropriate solvents include acetonitrile (HPLC grade), diethyl ether, and hexane

- Trifluoroacetic acid (TFA)
- Scavengers and reducing agents: 1,2-ethanedithiol (EDT), anisole, and 2-mercaptoethanol

*Method*

1. Immerse of the crowns in TFA:EDT:anisole (38:1:1 v/v) for 2.5 h at room temperature. Use 1.5–2.0 ml of the mixture per crown for the 20 µmol scale.

2. Remove the crowns by filtration.

3. Evaporate the TFA filtrate containing the peptide, using a gentle stream of dry nitrogen. Alternatively, the solutions can be evaporated in a solvent-resistant vacuum centrifuge.

4. Triturate each peptide with 8 ml of a cold (4°C) mixture of ether:hexane:mercaptoethanol (1:2:0.003 v/v) for 30 min.

5. Place the suspensions in centrifuge tubes.

6. Centrifuge the tubes for 6 min at approx. 2000 r.p.m. in a spark-proof centrifuge. If there is a substantial pellet (the peptide), decant the supernatant and repeat the extraction with 4 ml cold ether:hexane (1:2 v/v) and centrifuge as above.

7. Decant the organic solvent wash and dry down the peptide pellet. If at any stage the peptide dissolves in the ether washes, do not discard the washes but dry them down to recover the peptide.

8. Reconstitute or suspend the peptides in 1% AcOH(aq) in 40% v/v MeCN/water.

9. Lyophilize. The procedure yields peptides which are relatively free of EDT, anisole, TFA, and deprotection by-products.

# References

1. Houghten, R. A. (1985). *Proc. Natl. Acad. Sci. USA*, **82**, 5131.
2. Jung, G. and Beck-Sickinger, A. G. (1992). *Angew. Chem., Int. Ed. Engl.*, **31**, 367.
3. Pinilla, C., Appel, J., Blondelle, S. E., Dooley, C. T., Dörner, B., Eichler, J., Ostresh, J. M., and Houghten, R. A. (1995). *Biopolymers*, **37**, 221.
4. Blondelle, S. E. and Houghten, R. A. (1991). *Biochemistry*, **30**, 4671.
5. Eichler, J., Lucka, A. W., Pinilla, C., and Houghten, R. A. (1995). *Mol. Div.*, **1**, 233.
6. Pinilla, C., Appel, J. R., and Houghten, R. A. (1994). In *Current protocols in immunology* (ed. J. E. Coligan, A. M. Kruisbeek, D. H. Margulies, E. M. Shevach, and W. Strober), p. 9.8.1. Wiley, New York.
7. Beck-Sickinger, A. G., Dürr, H., and Jung, G. (1991). *Pept. Res.*, **4**, 88.
8. Eichler, J., Lucka, A. W., and Houghten, R. A. (1994). *Pept. Res.*, **7**, 300.
9. Houghten, R. A., Pinilla, C., Blondelle, S. E., Appel, J., Dooley, C. T., and Cuervo, J. H. (1991). *Nature*, **354**, 84.
10. Lam, K., Salmon, S. E., Hersh, E. M., Hruby, V. J., Kazmierski, W. M., and Knapp, R. J. (1991). *Nature*, **354**, 82.
11. Furka, Á, Sebestyén, F., Asgedom, M., and Dibó, G. (1991). *Int. J. Pept. Protein Res.*, **37**, 487.
12. Ostresh, J. M., Blondelle, S. E., Dörner, B., and Houghten, R. A. (1996). In *Methods in enzymology* (ed. J. N. Abelson), Vol. 267, p. 220. Academic Press, San Diego.
13. Ostresh, J. M., Husar, G., Blondelle, S. E., Dörner, B., Weber, P. A., and Houghten, R. A. (1994). *Proc. Natl. Acad. Sci. USA*, **91**, 11138.
14. Ostresh, J. M., Schoner, C. C., Hamashin, V. T., Meyer, J.-P., and Houghten R. A. (1998). *J. Org. Chem.*, **63**, 8622.
15. Dörner, B., Husar, G. M., Ostresh, J. M., and Houghten, R. A. (1996). *Bioorg. Med. Chem.*, **4**, 709.
16. Nefzi, A., Dooley, C., Ostresh, J. M., and Houghten, R. A. (1998). *Bioorg. Med. Chem. Lett.*, **8**, 2273.

17. Nefzi, A., Giulianotti, M., and Houghten, R. A. (1998). *Tetrahedron Lett.*, **39**, 3671.
18. Pei, Y., Houghten, R. A., and Keily, J. S. (1997). *Tetrahedron Lett.*, **38**, 3349.
19. Katritzky, A. R., Xie, L., Zhang, G., Griffith, M., Watson, K., and Keily, J. S. (1997). *Tetrahedron Lett.*, **38**, 7011.
20. Frank, R. (1992). *Tetrahedron*, **48**, 9217.
21. Weiler, J., Gausepohl, H., Hauser, N., Jensen, O. N., and Hoheisel, J. D. (1997). *Nucleic Acids Res.*, **25**, 1792.
22. Gao, B. and Esnouf, M. P. (1996). *J. Immunol.*, **157**, 183.
23. Adler, S., Frank, R., Lanzavecchia, A., and Weiss, S. (1994). *FEBS Lett.*, **352**, 167.
24. Hoffmann, S. and Frank, R. (1994). *Tetrahedron Lett.*, **35**, 7763.
25. Volkmer-Engert, R., Hoffmann, B., and Schneider-Mergener, J. (1997). *Tetrahedron Lett.*, **38**, 1029.
26. Krchnák, V., Vágner, J., Safár, P., and Lebl, M. (1988). *Collect. Czech. Chem. Commun.*, **53**, 2542.
27. Geysen, H. M., Rodda, S. J., and Mason, T. J. (1986). *Mol. Immunol.* **23**, 709.
28. Frank, R. (1994). In *Innovations and perspectives in solid phase synthesis* (ed. R. Epton), p. 509. Mayflower Worldwide, Birmingham.
29. Kramer, A., Volkmer-Engert, R., Malin, R., Reineke, U., and Schneider-Mergener, J. (1993). *Peptide Res.*, **6**, 314.
30. Fox J. E. and Newton R. (1992). In *Solid-phase synthesis: 2nd International Symposium* (ed. R. Epton R), p. 267. Mayflower Worldwide, Birmingham.
31. Geysen, H. M., Meloen, R. H., and Barteling, S. J. (1984). *Proc. Natl. Acad. Sci. USA*, **81**, 3998.
32. Maeji, N. J., Bray, A. M., and Geysen, H. M. (1990). *J. Immunol. Methods*, **134**, 23.
33. Bray, A. M., Maeji, N. J., Valerio, R. M., Campbell, R. A., and Geysen, H. M. (1991). *J. Org. Chem.*, **56**, 6659.
34. Valerio, R. M., Bray, A. M., Campbell, R. A., DiPasquale, A., Margellis, C., Rodda, S. J., Geysen, H. M., and Maeji, N. J. (1993). *Int. J. Pept. Protein Res.*, **42**, 1.
35. Bray, A. M., Jhingran, A. G., Valerio, R. M., and Maeji, N. J. (1994). *J. Org. Chem.*, **59**, 2197.
36. Bray, A. M., Lagniton, L. M., Valerio, R. M., and Maeji, N. J. (1994). *Tetrahedron Lett.*, **35**, 9079.
37. Valerio, R. M., Bray, A. M., and Maeji, N. J. (1994). *Int. J. Peptide Protein Res.*, **44**, 158.
38. Maeji, N. J., Valerio, R. M., Bray, A. M., Campbell, R. A., and Geysen, H. M. (1994). *Reactive Polymers*, **22**, 203.
39. Rodda, S. J., Geysen, H. M., Mason, T. J., and Schoofs, P. G. (1986). *Mol. Immunol.*, **23**, 603.
40. Atherton, E., Logan, C. J., and Sheppard, R. C. (1981). *J. Chem. Soc., Perkin Trans. 1*, 538.
41. Akaji, K., Kiso, Y., and Carpino, L. A. (1990). *J. Chem. Soc., Chem Commun.*, 584.
42. Rink, H. (1987). *Tetrahedron Lett.*, **28**, 3787.
43. Jense, K. J., Alsina, J., Songster, M. F., Vagner, J., Albericio, F., and Barany, G. (1988). *J. Am. Chem. Soc.*, **120**, 5441.
44. Ede, N. J. and Bray, A. M. (1997). *Tetrahedron Lett.*, **38**, 7119.

45. Barlos, K. (1989). *Tetrahedron Lett.*, **30**, 3943.
46. Ede, N. J., James, I. W., Krywult, B. M., Griffiths, R. M., Eagle, S. N., Gubbins, B., Leitch, J. A., Sampson, W. R., and Bray, A. M. (1999). *Lett. Pept. Sci.*, **6**, 157.
47. Bray, A. M., Valerio, R. M., DiPasquale, A., Greig, J., and Maeji, N. J. (1995). *J. Pept. Res.*, **1**, 80.
48. Perich, J. W., Ede, N. J., Eagle, S. N., and Bray, A. M. (1999). *Lett. Pept. Sci.*, **6**, 91.
49. Fournier, A., Wang, C,-T., and Felix, A. M. (1988). *Int. J. Peptide Protein Res.*, **31**, 86.
50. Carpino, L. A. and El-Faham, A. (1994). *J. Org. Chem.*, **59**, 695.
51. Story, S. C. and Aldrich, J. V. (1994). *Int. J. Peptide Protein Res.*, **43**, 292.

## A1

# Equipment and reagents for peptide synthesis

## 1. Amino acid derivatives

Advanced Chemtech
Bachem
Neosystem Laboratoire
Novabiochem
PE Biosystems
Peninsular Laboratories
Peptides International
Pierce
Senn Chemicals

## 2. Resins

Advanced Chemtech (polystyrene, Tentagel)
Novabiochem (polystyrene, PEGA, Tentagel)
PE Biosystems (PEG–PS)
Peninsular Laboratories (polystyrene)
Polymer Laboratories (polystyrene, polyamide, and PEGA)
Peptides International (CLEAR)
Rapp Polymere (Tentagel and polystyrene)
Senn (polystyrene)

## 3. Special derivatives

Fmoc-Asp/Glu(ODmab)-OH, Fmoc-Asp/Glu-ODmab
*Bachem, Novabiochem*
Fmoc-Asn (Ac$_3$AcNH-β-Glc)-OH
*Bachem, Novabiochem*
Fmoc-Dab/Dpr/Lys(Ddiv)-OH (Ddiv is known commercially as ivDde)
*Novabiochem*

Fmoc-Ser/Thr/Tyr(PO(OBzl)OH)-OH
*Novabiochem*
Fmoc-Ser (Ac₃AcNH-α-Gal)-OH
Fmoc-Thr (Ac₃AcNH-α-Gal)-OH
*Bachem*
SAMA-OPfp
*Novabiochem*
4-Dodecylaminocarbonyl-fluoren-9-ylmethyl succinimidyl carbonate
4-[5-(biotinylamino)pentylaminocarbonyl]fluoren-9-ylmethyl succinimidyl
carbonate
*Italformaco SpA, Via dei Lavoratori 54, Cinisello Balsamo, 20092 Milan, Italy*

# 4. Solvents

Aldrich/Fluka
Merck
Rathburns

# 5. Equipment

## 5.1 For multiple manual synthesis

ABIMED (membranes for SPOT synthesis)
AIMS Scientific Products (membranes for SPOT synthesis)
Biotech Instruments (manual synthesis blocks)
Chiron Technologies (PINS and CROWNS)
Genosys Biotechnologies (SPOT kits)
Jerini Bio Tools GmbH, Rudower Chaussee 5, D-12489 Berlin, Germany
(membranes for SPOT synthesis)
PE Biosystems (membranes for SPOT synthesis)
Polymer Group, Inc., PO Box 308, Benson, NC 27504, USA (mesh for T-bags)

## 5.2 Automated instruments

ABIMED (multiple, SPOT)
Advanced Chemtech (batchwise, pilot scale, and multiple)
CS Bio (batchwise synthesis)
Multi Syn Tech (multiple)
PE Applied Biosystems (batchwise synthesis)
PE Biosystems (continuous-flow and multiple)
Protein Technologies (batchwise synthesis, pilot scale, and multiple)
Shimadzu (multiple)
Zinsser Analytic (multiple)

# A2

# List of suppliers

**ABIMED**, Analysen-Technik GmbH, Raiffeisenstrasse 3, 40764 Langenfeld, Germany.

**Advanced Chemtech**, 5609 Fern Valley Road, Louisville, Kentucky 40228, USA.

**Advanced Chemtech Europe**, Nieuwbrugstraat 73, B-1830 Machelen, Belgium.

**AIMS Scientific Products GbR**, Mascheroder Weg 1, D-38124 Braunschweig, Germany.

**Aldrich Chemical Company Inc.**, 1001 West Saint Paul Avenue, Milwaukee, WI 53233, USA.

**Amersham**

*Amersham International plc.*, Lincoln Place, Green End, Aylesbury, Buckinghamshire HP20 2TP, UK.

*Amersham Corporation*, 2636 South Clearbrook Drive, Arlington Heights, IL 60005, USA.

**Anderman**

*Anderman and Co. Ltd.*, 145 London Road, Kingston-Upon-Thames, Surrey KT17 7NH, UK.

**Bachem AG**, Hauptstrasse 144, CH-4416, Bubendorf, Switzerland.

**Beckman Instruments**

*Beckman Instruments UK Ltd.*, Oakley Court, Kingsmead Business Park, London Road, High Wycombe, Bucks HP11 1J4, UK.

*Beckman Instruments Inc.*, PO Box 3100, 2500 Harbor Boulevard, Fullerton, CA 92634, USA.

**Becton Dickinson**

*Becton Dickinson and Co.*, Between Towns Road, Cowley, Oxford OX4 3LY, UK.

*Becton Dickinson and Co.*, 2 Bridgewater Lane, Lincoln Park, NJ 07035, USA.

*Bio 101 Inc.*, c/o Statech Scientific Ltd, 61–63 Dudley Street, Luton, Bedfordshire LU2 0HP, UK.

*Bio 101 Inc.*, PO Box 2284, La Jolla, CA 92038–2284, USA

**Bio-Rad Laboratories**

*Bio-Rad Laboratories Ltd.*, Bio-Rad House, Maylands Avenue, Hemel Hempstead HP2 7TD, UK.

*Bio-Rad Laboratories, Division Headquarters*, 3300 Regatta Boulevard, Richmond, CA 94804, USA.

**Biotech Instruments Ltd.**, Biotech House, 75A High Street, Kimpton, Herts. SG4 8PU, UK.

**Boehringer Mannheim**

*Boehringer Mannheim UK* (Diagnostics and Biochemicals) Ltd, Bell Lane, Lewes, East Sussex BN17 1LG, UK.

*Boehringer Mannheim Corporation*, Biochemical Products, 9115 Hague Road, P.O. Box 504 Indianapolis, IN 46250–0414, USA.

*Boehringer Mannheim Biochemica*, GmbH, Sandhofer Str. 116, Postfach 310120 D-6800 Ma 31, Germany.

**British Drug Houses (BDH) Ltd**, Poole, Dorset, UK.

**Chiron Technologies Pty Ltd.**, 11 Duerdin Street, Clayton. 3168, Victoria. Australia.

**C. S. Bio Co.**, 1300 Industrial Road, San Carlos, CA 94070, USA.

**Difco Laboratories**

*Difco Laboratories Ltd.*, P.O. Box 14B, Central Avenue, West Molesey, Surrey KT8 2SE, UK.

*Difco Laboratories*, P.O. Box 331058, Detroit, MI 48232–7058, USA.

**Du Pont**

*Dupont (UK) Ltd.*, Industrial Products Division, Wedgwood Way, Stevenage, Herts, SG1 4Q, UK.

*Du Pont Co.* (Biotechnology Systems Division), P.O. Box 80024, Wilmington, DE 19880–002, USA.

**European Collection of Animal Cell Culture**, Division of Biologics, PHLS Centre for Applied Microbiology and Research, Porton Down, Salisbury, Wilts SP4 0JG, UK.

**Falcon** (Falcon is a registered trademark of Becton Dickinson and Co.).

**Fisher Scientific Co.**, 711 Forbest Avenue, Pittsburgh, PA 15219–4785, USA.

**Flow Laboratories**, Woodcock Hill, Harefield Road, Rickmansworth, Herts. WD3 1PQ, UK.

**Fluka**

*Fluka-Chemie AG*, CH-9470, Buchs, Switzerland.

*Fluka Chemicals Ltd.*, The Old Brickyard, New Road, Gillingham, Dorset SP8 4JL, UK.

**Gibco BRL**

*Gibco BRL (Life Technologies Ltd.)*, Trident House, Renfrew Road, Paisley PA3 4EF, UK.

*Gibco BRL (Life Technologies Inc.)*, 3175 Staler Road, Grand Island, NY 14072–0068, USA.

**Arnold R. Horwell**, 73 Maygrove Road, West Hampstead, London NW6 2BP, UK.

**Hybaid**

*Hybaid Ltd.*, 111–113 Waldegrave Road, Teddington, Middlesex TW11 8LL, UK.

*Hybaid, National Labnet Corporation*, P.O. Box 841, Woodbridge, NJ. 07095, USA.

**HyClone Laboratories** 1725 South HyClone Road, Logan, UT 84321, USA.

**International Biotechnologies Inc.**, 25 Science Park, New Haven, Connecticut 06535, USA.

**Invitrogen Corporation**

*Invitrogen Corporation* 3985 B Sorrenton Valley Building, San Diego, CA. 92121, USA.

*Invitrogen Corporation* c/o British Biotechnology Products Ltd., 4–10 The Quadrant, Barton Lane, Abingdon, OX14 3YS, UK.

**Jerini Bio Tools GmbH**, Rudower Chaussee 5, D-12489 Berlin, Germany.

**Sigma-Genosys Inc.**, 1422 Lake Front Circle, Suite 185, The Woodlands, TX 77380-3600, USA.

**Kodak: Eastman Fine Chemicals** 343 State Street, Rochester, NY, USA.

**Life Technologies**

*Life Technologies Inc.*, 8451 Helgerman Court, Gaithersburg, MN 20877, USA.

**Merck**

*Merck Industries Inc.*, 5 Skyline Drive, Nawthorne, NY 10532, USA.

*Merck, Frankfurter Strasse*, 250, Postfach 4119, D-64293, Germany.

**Millipore**

*Millipore (UK) Ltd.*, The Boulevard, Blackmoor Lane, Watford, Herts WD1 8YW, UK.

*Millipore Corp./Biosearch*, P.O. Box 255, 80 Ashby Road, Bedford, MA 01730, USA.

**MultiSynTech GmbH**, Wullener Feld 4, Witten, Germany.

**Neosystem Laboratoire**, 7 rue de Boulogne, 67100, **MultiSynTech GmbH**, Wullener Feld 4, Witten, Germany. Strasbourg, France.

**New England Biolabs** (NBL)

*New England Biolabs* (NBL), 32 Tozer Road, Beverley, MA 01915–5510, USA.

*New England Biolabs* (NBL), c/o CP Labs Ltd., P.O. Box 22, Bishops Stortford, Herts CM23 3DH, UK.

*Nikon Corporation*, Fuji Building, 2–3 Marunouchi 3-chome, Chiyoda-ku, Tokyo, Japan.

**Novabiochem**, Calbiochem-Novabiochem AG, Weidenmattweg 4, CH-4448, Läufelfingen, Switzerland.

**Nunc–Gibco**, Life Technologies Ltd., 3 Fountain Drive, Inchinnan Business Park, Paisley PA4 9RF, UK.

**Oncogene Research Products**, 84 Rodgers Street, Cambridge, MA 02142, USA.

**PE Applied Biosystems**, 850 Lincoln Centre Drive, Foster City, CA 94404, USA; Kelvin Close, Birchwood Science Park North, Warrington, Cheshire, WA3 7PB, UK.

**PE Biosystems**, 500 Old Connecticut Path, Framingham, MA 01701, USA; Kelvin Close, Birchwood Science Park North, Warrington, Cheshire, WA3 7PB, UK.

**Peninsular Laboratories Inc.**, 601 Taylor Way, San Carlos, CA 94070, USA.

**Peptides International, Inc.**, 11621 Electron Drive, Louisville, Kentucky 40299, USA.

**Pierce**, 3747 North Meridian Road, P.O. Box 117, Rockford, IL 61105, USA.

**Polymer Laboratories Ltd.**, Essex Road, Church Stretton, Shropshire SY6 6AX, UK.

**Perkin-Elmer**

*Perkin-Elmer Ltd.*, Maxwell Road, Beaconsfield, Bucks. HP9 1QA, UK.

*Perkin-Elmer Ltd.*, Post Office Lane, Beaconsfield, Bucks, HP9 1QA, UK.

*Perkin-Elmer-Cetus* (The Perkin-Elmer Corporation), 761 Main Avenue, Norwalk, CT 0689, USA.

**Pharmacia Biotech Europe** Procordia EuroCentre, Rue de la Fuse-e 62, B-1130 Brussels, Belgium.

**Pharmacia Biosystems**

*Pharmacia Biosystems Ltd.* (Biotechnology Division), Davy Avenue, Knowlhill, Milton Keynes MK5 8PH, UK.

*Pharmacia LKB* Biotechnology AB, Björngatan 30, S-75182 Uppsala, Sweden.

**Promega**

*Promega Ltd.*, Delta House, Enterprise Road, Chilworth Research Centre, Southampton, UK.

*Promega Corporation*, 2800 Woods Hollow Road, Madison, WI 53711–5399, USA.

**Protein Technologies, Inc.**, Mack Road, Box 4026, Woburn, MA 01888-4026, USA.

**Qiagen**

*Qiagen Inc.*, c/o Hybaid, 111–113 Waldegrave Road, Teddington, Middlesex, TW11 8LL, UK.

*Qiagen Inc.*, 9259 Eton Avenue, Chatsworth, CA 91311, USA.

**Rapp Polymere GmbH**, Ernst-Simon-Str. 9, D 72072 Tübingen, Germany.

**Schleicher and Schuell**

*Schleicher and Schuell Inc.*, Keene, NH 03431A, USA.

*Schleicher and Schuell Inc.*, D-3354 Dassel, Germany.

*Schleicher and Schuell Inc.*, c/o Andermann and Company Ltd.

**Senn Chemicals AG**, Industriestrasse 12, PO Box 267, CH-8157 Diesdorf, Switzerland.

**Shandon Scientific Ltd.**, Chadwick Road, Astmoor, Runcorn, Cheshire WA7 1PR, UK.

**Shimadzu Corporation**, Nishinokyo-Kuwabaracho 1, Nakagyo-ku, Kyoto 604, Japan.

**Sigma Chemical Company**

*Sigma Chemical Company (UK)*, Fancy Road, Poole, Dorset BH17 7NH, UK.

*Sigma Chemical Company*, 3050 Spruce Street, P.O. Box 14508, St. Louis, MO 63178–9916.

**Sigma-Genosys Inc.**, 1442 Lake Front Circle, Suite 185, The Woodlands, TX 77380-3600, USA.

**Sorvall DuPont Company**, Biotechnology Division, P.O. Box 80022, Wilmington, DE 19880–0022, USA.

**Stratagene**

*Stratagene Ltd.*, Unit 140, Cambridge Innovation Centre, Milton Road, Cambridge CB4 4FG, UK.

*Strategene Inc.*, 11011 North Torrey Pines Road, La Jolla, CA 92037, USA.

**United States Biochemical**, P.O. Box 22400, Cleveland, OH 44122, USA.

**Wellcome Reagents**, Langley Court, Beckenham, Kent BR3 3BS, UK.

**Zinsser Analytic GmbH**, Eschborner, Landstrasse 135, D-60489, Frankfurt, Germany.

# A3

# Useful tables

**Table 1.** Amino acid derivatives

| Amino acid | Molecular weight | Amino acid | Molecular weight |
|---|---|---|---|
| Fmoc-Ala-OH | 311.3 | Fmoc-His(Boc)-OH | 477.5 |
| Fmoc-βAla-OH | 311.3 | Fmoc-Hyp-OH | 353.4 |
| Fmoc-Arg(Mtr)-OH | 608.7 | Fmoc-Hyp(tBu)-OH | 457.2 |
| Fmoc-Arg(Pbf)-OH | 648.8 | Fmoc-Ile-OH | 353.4 |
| Fmoc-Arg(Pmc)-OH | 662.8 | Fmoc-Leu-OH | 353.4 |
| Fmoc-Asn-OH | 354.4 | Fmoc-Lys(Boc)-OH | 468.5 |
| Fmoc-Asn(Trt)-OH | 596.7 | Fmoc-Lys(Ddiv)-OH | 574.6 |
| Fmoc-Asp-OtBu | 411.5 | Fmoc-Met-OH | 371.5 |
| Fmoc-Asp(OtBu)-OH | 411.5 | Fmoc-Nle-OH | 353.4 |
| Fmoc-Cha-OH | 393.5 | Fmoc-Nva-OH | 339.4 |
| Fmoc-Cys(Acm)-OH | 414.5 | Fmoc-Orn(Boc)-OH | 454.5 |
| Fmoc-Cys(Tacm)-OH | 456.5 | Fmoc-Phe-OH | 387.4 |
| Fmoc-Cys(tBu)-OH | 399.5 | Fmoc-Pro-OH | 337.4 |
| Fmoc-Cys(tButhio)-OH | 431.6 | Fmoc-Ser(tBu)-OH | 383.4 |
| Fmoc-Cys(Trt)-OH | 585.7 | Fmoc-Ser(Trt)-OH | 569.7 |
| Fmoc-Gln-OH | 368.4 | Fmoc-Sta-OH | 397.5 |
| Fmoc-Gln(Trt)-OH | 610.7 | Fmoc-Thi-OH | 393.4 |
| Fmoc-Glu-OtBu | 425.5 | Fmoc-Thr(tBu)-OH | 397.5 |
| Fmoc-Glu(OtBu)-OH | 425.5 | Fmoc-Trp-OH | 426.5 |
| Fmoc-Gly-OH | 297.3 | Fmoc-Trp(Boc)-OH | 526.6 |
| Fmoc-His(Bum)-OH | 463.5 | Fmoc-Tyr(tBu)-OH | 459.6 |
| Fmoc-His(Trt)-OH | 619.7 | Fmoc-Val-OH | 339.4 |

**Table 2.** Protecting groups

| Name | Abbrev. | Formula | Mol.wt. |
|---|---|---|---|
| Acetamidomethyl | Acm | $C_3H_6NO$ | 72.087 |
| Acetyl | Ac | $C_2H_3O$ | 43.046 |
| Benzoyl | Bz | $C_7H_5O$ | 105.117 |
| Benzyl | Bzl | $C_7H_7$ | 91.134 |
| Benzyloxy | BzlO | $C_7H_7O$ | 107.126 |
| Benzyloxycarbonyl | Z | $C_8H_7O_2$ | 135.144 |
| t-Butoxy | tBuO | $C_4H_9O$ | 73.116 |
| t-Butoxycarbonyl | Boc | $C_5H_9O_2$ | 101.126 |

**Table 2.** *Continued*

| Name | Abbrev. | Formula | Mol.wt. |
|---|---|---|---|
| *t*-Butoxymethyl | Bum | $C_5H_{11}O$ | 87.143 |
| *t*-Butyl | *t*Bu | $C_4H_9$ | 57.117 |
| *t*-Butylthio | *t*Buthio | $C_4H_9S$ | 89.181 |
| 2,6-Dichlorobenzyl | 2,6-Di-Cl-Bzl | $C_7H_5Cl_2$ | 160.024 |
| 4,4'-Dimethoxybenzhydryl | Mbh | $C_{15}H_{15}O_2$ | 227.286 |
| 1-(4,4-Dimethyl-2,6-dioxocyclohexylidene)3-methylbutyl | Ddiv | $C_{10}H_{13}O_2$ | 207.295 |
| Fluorenylmethoxycarbonyl | Fmoc | $C_{15}H_{11}O_2$ | 223.254 |
| Formyl | For | CHO | 29.018 |
| 4-Methoxybenzyl | MeOBzl | $C_8H_9O$ | 121.160 |
| 4-Methoxy-2,3,6-trimethylbenzenesulphonyl | Mtr | $C_{10}H_{13}OS$ | 213.278 |
| 4-Methoxytrityl | Mmt | $C_{20}H_{17}O$ | 273.357 |
| 4-Methyltrityl | Mtt | $C_{20}H_{17}$ | 257.358 |
| 2,2,5,7,8-Pentamethylchroman-6-sulphonyl | Pmc | $C_{14}H_{19}O_3S$ | 267.369 |
| Tosyl | Tos | $C_7H_7O_2S$ | 155.197 |
| Trifluoroacetyl | Tfa | $C_2F_3O$ | 97.017 |
| Trityl | Trt | $C_{19}H_{15}$ | 243.331 |

**Table 3.** Amino acid residues

| Amino acid | 3-Letter code | Formula | Residue weight |
|---|---|---|---|
| β-Alanine | βAla | $C_3H_5NO$ | 71.079 |
| Alanine | Ala | $C_3H_5NO$ | 71.079 |
| 2-Aminobutyric acid | Abu | $C_4H_7NO$ | 85.106 |
| 4-Aminobutyric acid | γAbu | $C_4H_7NO$ | 85.106 |
| 6-Aminocaproic acid | εAhx | $C_6H_{11}NO$ | 113.161 |
| α-Aminoisobutyric acid | Aib | $C_4H_7NO$ | 85.106 |
| α-Aminosuberic acid | Asu | $C_8H_{13}NO_3$ | 261.314 |
| Arginine | Arg | $C_6H_{12}N_4O$ | 156.189 |
| Asparagine | Asn | $C_4H_6N_2O_2$ | 114.105 |
| Aspartic acid | Asp | $C_4H_5NO_3$ | 115.089 |
| 4-Chlorophenylalanine | Phe(pCl) | $C_9H_8ClNO$ | 181.624 |
| Citrulline | Cit | $C_6H_{11}N_3O_2$ | 157.173 |
| β-Cyclohexylalanine | Cha | $C_9H_{15}NO$ | 153.226 |
| Cysteine | Cys | $C_3H_5NOS$ | 103.145 |
| Cystine | Cys | $C_6H_8N_2O_2S_2$ | 204.271 |
| 3,4-Dehydroproline | Δ-Pro | $C_5H_5NO$ | 95.103 |
| 3,5-Diiodotyrosine | Tyr(3,5-di-l) | $C_9H_7I_2NO_2$ | 414.971 |
| 2-Fluorophenylalanine | Phe(2-F) | $C_9H_8FNO$ | 165.169 |
| 3-Fluorophenylalanine | Phe(3-F) | $C_9H_8FNO$ | 165.169 |
| 4-Fluorophenylalanine | Phe(4-F) | $C_9H_8FNO$ | 165.169 |
| Glutamic acid | Glu | $C_5H_7NO_3$ | 129.116 |
| Glutamine | Gln | $C_5H_8N_2O_2$ | 128.132 |
| Glycine | Gly | $C_2H_3NO$ | 57.052 |
| Histidine | His | $C_6H_7N_3O$ | 137.142 |
| Homocitrulline | Hci | $C_7H_{13}N_3O_2$ | 171.201 |
| Homoserine | hSer | $C_4H_7NO_2$ | 101.107 |

**Table 3.** *Continued*

| Amino acid | 3-Letter code | Formula | Residue weight |
|---|---|---|---|
| Hydroxyproline | Hyp | $C_5H_7NO_2$ | 113.117 |
| β-Hydroxyvaline | Val(βOH) | $C_5H_9NO_2$ | 115.133 |
| Isoleucine | Ile | $C_6H_{11}NO$ | 113.161 |
| Leucine | Leu | $C_6H_{11}NO$ | 113.161 |
| Lysine | Lys | $C_6H_{12}N_2O$ | 128.175 |
| Methionine | Met | $C_5H_9NOS$ | 131.198 |
| 4-Nitrophenylalanine | Phe(4-NO₂) | $C_9H_8N_2O_3$ | 192.176 |
| Norleucine | Nle | $C_6H_{11}NO$ | 113.161 |
| Norvaline | Nva | $C_5H_9NO$ | 99.134 |
| Ornithine | Orn | $C_5H_{10}N_2O$ | 114.148 |
| Penicillamine | Pen | $C_5H_9NOS$ | 131.198 |
| Phenylalanine | Phe | $C_9H_9NO$ | 147.178 |
| Phenylglycine | Phg | $C_8H_7NO$ | 133.152 |
| Proline | Pro | $C_5H_7NO$ | 97.118 |
| Pyroglutamine | Pyr,Glp | $C_5H_5NO_2$ | 111.101 |
| Sarcosine | Sar | $C_3H_5NO$ | 71.079 |
| Serine | Ser | $C_3H_5NO_2$ | 87.079 |
| Statine | Sta | $C_8H_{15}NO_2$ | 157.214 |
| β-(2-Thienyl)alanine | Thi | $C_7H_7NOS$ | 153.205 |
| Threonine | Thr | $C_4H_7NO_2$ | 101.106 |
| Tryptophan | Trp | $C_{11}H_{10}N_2O$ | 186.215 |
| Tyrosine | Tyr | $C_9H_9NO_2$ | 163.178 |
| Valine | Val | $C_5H_9NO$ | 99.134 |

339

# Index

trityl group,
    selective removal on solid phase 177
trityl resins,
    2-chlorotrityl chloride 17, 45, 50, 216
    4-(chloro(diphenyl)methyl)benzoyl 18, 45,
        50
tryptophan,
    $N^{in}$-Boc protection 25, 65, 67, 102
    modification by sulphenyl chlorides
        106
    preparation of peptides containing
        C-terminal 50
    sulphonation 68–9, 72

tyrosine
    $O$-2-Clt protection 24, 177, 225
    $O$-$t$Bu protection 24

UV monitoring of,
    Fmoc deprotection reaction 29, 117
    acylation reaction 289

Wang resin 17, 45
Weinreb amide-based linker 153–5

Printed in the United States
By Bookmasters